Motor Activity
and Movement Disorders

Contemporary Neuroscience

Sponsored by the International Behavioral Neuroscience Society

Motor Activity and Movement Disorders

Research Issues and Applications

Edited by

Paul R. Sanberg

*Division of Neurological Surgery,
Departments of Surgery, Psychiatry, Pharmacology, and Neurology,
University of South Florida, Tampa, FL*

Klaus-Peter Ossenkopp

*Neuroscience Program
and Departments of Psychology and Pharmacology and Toxicology,
University of Western Ontario, London, Ontario, Canada*

and Martin Kavaliers

*Neuroscience Program
and Departments of Oral Biology, Psychology, and Pharmacology and Toxicology,
University of Western Ontario, London, Ontario, Canada*

© 1996 Humana Press Inc.
999 Riverview Drive, Suite 208
Totowa, New Jersey 07512

Printed in the United States of America. 10 9 8 7 6 5 4 3 2 1

Library of Congress Cataloging in Publication Data
ISBN 0-89603-327-9
Motor activity and movement disorders; research issues and applications/edited by Paul
    Sanberg, Klaus-Peter Ossenkopp, Martin Kavaliers.
        p. cm.—(Contemporary neuroscience)
    Includes bibliographical references and index.
    ISBN 0-89603-327-9 (alk. paper)
    1. Movement disorders—Animal models.   2. Motor ability   3. Animal
locomotion.   I. Sanberg, Paul R.   II. Ossenkopp, Klaus-Peter.   III. Kavaliers,
Martin.   IV.   Series.
    [DNLM: 1. Motor Activity. 2. Movement Disorders.   WE 103 M9166 1996]
RC376.5.M645   1996
616.7—dc20
DNLM/DLC
for Library of Congress   95-40459
                                                                                                          CIP

Preface

Motor Activity and Movement Disorders:
Overview and Perspective

The measurement of motor activity is of major importance in many areas of biomedical research investigating brain function in both health and disease. In trying to understand brain processes it is important to remember that the ultimate function of the nervous system is to initiate behavior. Thus, behavioral approaches to brain research are essential (Robinson, 1983). The measurement of spontaneous motor activity is central to obtaining information about an animal's general behavior. Changes in motor activity have consequences for the measurement of all aspects of behavior, including spontaneous locomotor activity and other class-common and species-specific behaviors, as well as learned or conditioned behavior (cf. Kolb and Whishaw, 1983). Measurement of spontaneous motor activity can range from simple direct observation (e.g., Hutt and Hutt, 1970, 1983) to sophisticated automated procedures that provide simultaneous measurements of a number of behavioral variables (cf. Ossenkopp et al., 1995; Sanberg, 1985, 1986). In addition to picking a *reliable* and *valid* procedure for measuring motor activity, the researcher also needs to appreciate the influence of a variety of organismic and environmental factors that can affect the expression and interpretation of these behavioral measures.

Behavioral measurement can serve two different purposes in investigations of brain–behavior relationships. Behavioral variables can be used as indices of activity in the neural systems of interest, especially following exposure to toxic substances, therapeutic medications, damage to specifc brain structures, or other manipulations. On the other hand, analysis of behavior itself might be of primary interest. This approach tries to understand the functions of specific neural systems and the relationships among these systems in the initiation or modulation of behavior (Robinson, 1983). In many studies, both types of approaches may be required.

It is generally recognized that any single measure of motor activity will be confounded with other aspects of behavior. This has led to the suggestion that multiple, simultaneous measures (multivariate) of different aspects of spontaneous behavior be obtained in the same situation (cf. Robbins, 1977). This multivariate approach presents additional complexities in terms of statistical inference and interpretation of behavioral complexity (cf. Ossenkopp and Mazmanian, 1985). There are problems associated with studying several variables independently (cf. Frey and Pimental, 1978) since their interrelationships are not taken into account. However, a variety of multivariate statistical procedures, such as principal component and factor analysis, are available to deal with multiple measures. These analytical methods tend to promote "phenomenon realism" and provide more adequate characterization of general behavior processes.

Motor Activity and Movement Disorders addresses all of the preceding factors and is organized into two main sections. The first section deals with several basic research issues pertaining to the measurement and interpretation of motor activity and other aspects of movement. The second deals with the application of this methodology to preclinical research involving motor disorders, with an emphasis on motor asymmetries and turning behavior, as well as on dyskinetic movements. The primary focus is on methodology. Various chapters present a variety of quantitative approaches designed to assess specific aspects of motor activity, from oral movements to locomotor behavior, circling movements, tonic immobility, and catelepsy. Another feature emphasized throughout this volume is the use of automated measurement techniques that utilize computer technology and allow for detailed and objective approaches to the quantification of motor behavior.

The section on basic research issues starts with an examination of long-term habituation of a number of behaviors to a novel environment. This habituation model is examined in terms of behavioral plasticity for a nonassociative type of learning with the neural substrates, as well as electrophysiology, and neuropharmacology also being explored. Tonic immobility, a type of response inhibition found in a very broad range of species, from insects to birds and mammals, is examined in Chapter 2. The neural, physiological, and behavioral aspects of tonic immobility are examined and related to several forms of psychopathology, such as catatonic stupors. Endogenous and daily rhythms are intrinsic to the expression and organization of motor activity. Chapter 3 summarizes research conducted in this area, explaining the methodology and theoretical concepts used, and providing an overview of the phenomenology of rhythmic activity. Sexually dimorphic behaviors have received increasing attention over the last several decades. An overview of male–female differences in locomotor activity levels in rodents is provided in Chapter 4. Hormonal influences, both organizational and activational, as well as the reproductive ecology of the species are linked to many of these observed differences and their counterparts in humans. The study of these behavioral differences can help us to understand some of the selective forces that have helped produce them.

Chapter 5 presents a novel type of automated video-image analysis that can be used to study many aspects of motor activity, especially behavioral asymmetries and turning or circling behavior. A detailed description of the methodology is provided and data on these asymmetries, such as for thigmotactic scanning and locomotion, are illustrated. Effects of unilateral brain lesions and drug adminstration in intact animals are presented and discussed. Stereotyped behaviors are well-organized programs for normal species-typical behaviors that can be activated by a variety of conditions, including drug administration. Chapter 6 examines rodent gnawing behavior from this perspective. A variety of measurement procedures are reviewed and an automated procedure for quantifying gnawing in rodents is presented.

Part II contains chapters that have more direct clinical relevance, providing an overview of motor activity measurement approaches that either model certain aspects of human psychopathologies or provide experimental approaches to understanding brain function relevant to these clinical conditions. Catalepsy induced under laboratory conditions is one of the behavioral models used by neuroscientists to study the neurochemical systems involved in "active" immobility. Owing to the overlap of this behavioral condition with the symptoms in Parkinsonianism, catatonic schizophrenia, and other forms of psychopathology, it is of clear clinical importance. Chapter 7 reviews various methods used for measuring catalepsy and deals with factors that influence this phenomenon. An automated procedure that objectively quantifies this state and allows for better control and assessment in experimental investigation, is described. Tardive dyskinesia is a severe side effect of the therapeutic use of neuroleptics in humans. Chapter 8 presents an automated video assessment procedure that allows measurement of oral motor activity and possible dyskinesia. This chapter also presents data on neuroleptic-induced changes in these oral movements. A computer "Pattern Recognition System" for rodent spontaneous activity, as a tool to study central nervous system damage, is presented in Chapter 9. This system measures behavioral initiations, behavioral total time, and behavioral time structures. Data are presented that suggest the existence of brain damage-specific behavioral patterns. It is suggested that this methodology may be an ideal tool to study the effects of neurotoxic exposures.

The last four chapters focus on behavioral asymmetries and stereotypies as models for understanding brain function and various aspects of psychopathologies. Chapter 10 focuses on circling behavior in rodents and the asymmetry of this behavior, as a way of gaining insight into functional lateralization in the brain. This chapter reviews information on asymmetries in circling and turning behavior and their relationships to anatomical and biochemical asymmetries. Effects of unilateral brain lesions and drugs are also discussed. A new, nonpharmcological approach to assessing motor asymmetry is provided in Chapter 11. The "Elevated Body Swing Test" is described as a drug-free approach to examining the effects of unilateral brain lesions in animals. Data from studies on animal models of Huntington's and Parkinson's Disease are presented and evaluated. Spontaneous turning behavior in humans is the topic of Chapter 12. Rotational asymmetries are measured by means of a human rotameter and this methodology is then used to examine rotation asymmetries in autistic children. Suggestions for the application of this procedure to other patient groups is also given. The final chapter examines the clinical methodology used to study the behavioral aspects of Tourette's Syndrome. Clinical assessment procedures are reviewed and evaluated. These include self-report methods, videotape monitoring, and clinical rating scales. It is suggested that a broader examination of motor disturbances in these patients would facilitate research into the genetic and pharmacological aspects of this disorder.

Motor Activity and Movement Disorders does not attempt to provide an exhaustive review of these topics, but instead focuses on methodology and its applications to some current animal models and human disorders. To the extent that we stimulate further interest in these questions and procedures and contribute to a greater understanding of brain–behavior relationships, we consider this book to be a success.

We would like to acknowledge partial support by grants from the Smokeless Tobacco Research Council and the National Institute of Neurological Disease and Stroke (PRS), and the Natural Sciences and Engineering Research Council of Canada (KPO and MK) in the preparation of this volume. We also thank Duncan Innes for his much appreciated help in the preparation of the index and the many graduate students, postdoctoral fellows, and colleagues who have participated in research and discussions from which this volume has benefited.

Paul R. Sanberg
Klaus-Peter Ossenkopp
Martin Kavaliers

References

Frey, D. F. and Pimental, R. A. (1978) Principle component analysis and factor analysis, in *Quantitative Ethology* (Colgan, P. W., ed.), Wiley, New York, pp. 219–245.

Hutt, S. J. and Hutt, C. (1970) *Direct Observation and Measurement of Behavior*. Thomas, Springfield, IL.

Hutt, S. J. and Hutt, C. (1983) Why measure behavior? in *Behavioral Approaches to Brain Research* (Robinson, T. E., ed.), Oxford University Press, New York, pp. 14–26.

Kolb, B. and Whishaw, I. Q. (1983) Problems and principles underlying interspecies comparisons, in *Behavioral Approaches to Brain Research*

(Robinson, T. E., ed.), Oxford University Press, New York, pp. 237–263.

Ossenkopp, K.-P., Kavaliers, M., and Sanberg, P. R., eds. *Measuring Locomotion and Movement: From Invertebreates to Humans*. Landes, Austin, TX, in press.

Ossenkopp, K.-P. and Mazmanian, D. S. (1985) The measurement and integration of behavioral variables: aggregation and complexity as important issues. *Neurobehav. Toxicol. Teratol.* **7,** 95–100.

Robbins, T. W. (1977) A critique of the methods available for the measurement of spontaneous motor activity, in *Handbook of Psychopharmacology, Vol. 7, Principles of Behavioral Pharmacology* (Iversen, L., Iversen, S., and Snyder, S., eds.), Plenum, New York, pp. 37–82.

Robinson, T. E., ed. (1983) *Behavioral Approaches to Brain Research*. Oxford University Press, New York.

Sanberg, P. R., ed. (1985) Locomotor Behavior: New Approaches in Animal Research. Proceedings of a Satellite Symposium to the 14th Annual Meeting of the Society for Neuroscience. *Neurobehav. Toxicol. Teratol.* **7,** 70–100.

Sanberg, P. R., ed. (1986) Locomotor Behavior: Neuropharmacological Substrates of Motor Activation. Proceedings of a Satellite Symposium to the 15th Annual Meeting of the Society for Neuroscience. *Pharmacol. Biochem. Behav.* **25,** 229–300.

Contents

3. Circadian Organization of Locomotor Activity in Mammals
Ralph E. Mistlberger

4. Sex Differences in Rodent Spontaneous Activity Levels
Larissa A. Mead, Eric L. Hargreaves, and Liisa A. M. Galea

5. Automated Video-Image Analysis of Behavioral Asymmetries
R. K. W. Schwarting, J. Fornaguera, and J. P. Huston

6. Automated Methods of Measuring Gnawing Behaviors
Donald E. Moss and Edward Castañeda

10. Circling Behavior in Rodents: *Methodology, Biology, and Functional Implications*
Jeffrey N. Carlson and Stanley D. Glick

11. Asymmetrical Motor Behavior in Animal Models of Human Diseases: *The Elevated Body Swing Test*
Cesario V. Borlongan and Paul R. Sanberg

Contributors

CESARIO V. BORLONGAN • *Division of Neurological Surgery, Department of Surgery, The University of South Florida College of Medicine, Tampa, FL*

H. STEFAN BRACHA • *Departments of Psychiatry and Behavioral Sciences and Neurology, University of Arkansas for Medical Sciences; VA Psychiatry Service, North Little Rock, AK*

DAVID W. CAHILL • *Division of Neurological Surgery, Department of Surgery, The University of South Florida College of Medicine, Tampa, FL*

JEFFREY N. CARLSON • *Department of Pharmacology and Toxicology, Albany Medical College, Albany, NY*

EDWARD CASTAÑEDA • *Department of Psychology, Arizona State University, Tempe, AZ*

GAYLORD ELLISON • *Department of Psychology, University of California, Los Angeles, CA*

J. FORNAGUERA • *Institute of Physiological Psychology I, Heinrich-Heine University of Düsseldorf, Germany*

LIISA A. M. GALEA • *Laboratory of Neuroendocrinology, The Rockefeller University, New York, NY*

GORDON G. GALLUP, JR. • *Department of Psychology, State University of New York at Albany, NY*

JEFFREY W. GILGER • *Departments of Psychiatry and Behavioral Sciences and Neurology, University of Arkansas for Medical Sciences, North Little Rock, AK*

STANLEY D. GLICK • *Department of Pharmacology and Toxicology, Albany Medical College, Albany, NY*

ERIC L. HARGREAVES • *Department of Psychology, McGill University, Montreal, Quebec, Canada*

J. P. HUSTON • *Institute of Physiological Psychology I, Heinrich-Heine University of Düsseldorf, Germany*

xvii

ALAN KEYS • *Department of Psychology, University of California, Los Angeles, CA*

RODRIGO MARTINEZ • *Division of Neurological Surgery, Department of Surgery, The University of South Florida College of Medicine, Tampa, FL*

LARISSA A. MEAD • *Department of Psychology, University of Western Ontario, London, Ontario, Canada*

RALPH E. MISTLBERGER • *Department of Psychology, Simon Fraser University, Burnaby, British Columbia, Canada*

DONALD E. MOSS • *Laboratory of Psychobiochemistry, Department of Psychology, University of Texas at El Paso, TX*

PHYLLIS J. MULLENIX • *Department of Radiation Oncology, Harvard Medical School, Boston, MA*

DAWN R. RAGER • *Department of Psychology, University of Georgia, Athens, GA*

ADOLFO GUSTAVO SADILE • *Dipt. Fisiologia Umana and Funzioni Biologische Integrate, Seconda Università di Napoli, Italy*

PAUL R. SANBERG • *Division of Neurological Surgery, Departments of Surgery, Psychiatry, Pharmacology, and Neurology, The University of South Florida College of Medicine, Tampa, FL*

R. K. W. SCHWARTING • *Institute of Physiological Psychology I, Heinrich-Heine University of Düsseldorf, Germany*

R. DOUGLAS SHYTLE • *Division of Neurological Surgery, Departments of Surgery and Psychiatry, The University of South Florida College of Medicine, Tampa, FL*

A. A. SILVER • *Division of Child and Adolescent Psychiatry, Department of Psychiatry, The University of South Florida College of Medicine, Tampa, FL*

I. Basic Research Issues

CHAPTER 1

Long-Term Habituation of θ-Related Activity Components of Albino Rats in the Làt-Maze

Adolfo Gustavo Sadile

Introduction

Habituation, a response decrement with repeated or continuous presentation of indifferent stimuli, independent of muscular fatigue or receptor adaptation, is a relatively simple form of long-lasting behavioral plasticity that is present at all levels in the animal kingdom (*see*, e.g., Harris, 1943; Thompson and Spencer, 1966). A wide variety of behavioral responses of the intact organism habituate (Harris, 1943), as does spontaneous exploratory activity in a novel environment. When a rat is exposed to an unfamiliar environment, the isolation, the low level of pheromones and ultrasounds, the change in the illumination level and in temperature, and the environment itself elicit the expression of various motivational programs and sub-programs (Làt, 1973). There is an initial "freezing behavior" with locomotor inhibition that fades out within several seconds or few minutes with a high interindividual variability (*see* Fig. 1). It follows the "exploratory action program" with sniffing, walking about, and rearings on hindlimbs shown to be associated with hippocampal rhythmical slow-wave activity (RSA or θ;

From: *Motor Activity and Movement Disorders*
P. R. Sanberg, K. P. Ossenkopp, M. Kavaliers, Eds. Humana Press Inc., Totowa, NJ

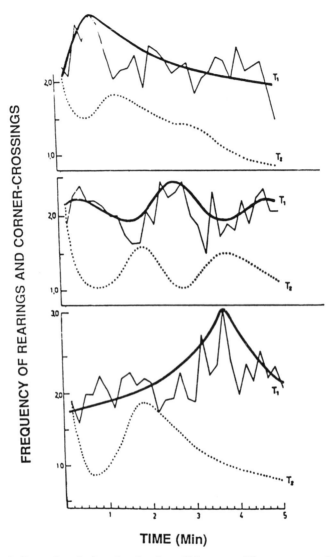

Fig. 1. Freezing behavior in the albino rat. The average frequency of corner-crossings + rearings during the first 5 min of two consecutive exposures, at 24 h intervals, to a Làt-maze is reported per 10-s blocks for random-bred rats reared under normal conditions (middle panel), and for rats with decelerated (top panel) or accelerated (bottom panel) body and brain growth.

Irmis et al., 1971; Vanderwolf, 1971; Whishaw and Vanderwolf, 1973) and to be atropine-insensitive (Vanderwolf, 1971). The behavioral arousal or activation elicited by the novel environment tends to wane as the environment itself loses its novelty properties (Stein, 1966). Thus, the exploratory program is followed by the "cleaning program," composed of various subprograms (face-washing, scratching, and so forth). This series of automatic behaviors is not associated with θ, but with hippocampal desynchronization, and it is known to be atropine-sensitive (Vanderwolf, 1971). Then comes the "sleeping program," provided external stimulation is held constant. The overall dynamics of behavioral arousal has been described as a nonlinear, complex exponential function (Làt, 1973, 1976), with an activation, a linear, and a deactivation phase, superimposed by oscillations occurring at exponentially decreasing, constant, and increasing intervals, respectively (*see* Fig. 2).

Animals show activity decrement during a single exposure, operationally defined as short-term habituation (STH), and on subsequent exposures, operationally defined as long-term habituation (LTH; Làt, 1973).

Using LTH as a relatively simple model of negative learning, a series of experiments was designed in the young and adult albino rat to investigate the consolidation process, its time course, and neural substrates by different approaches. Negative learning is demonstrated as such because it requires learning not to respond to biologically insignificant environmental stimuli.

In the young rat, mainly noninterference approaches can be followed that include maturation studies to follow the appearance of LTH during postnatal life. They are useful in tracing the temporal window for LTH appearance, indicative of the organizational level and of the "ergic" systems required for LTH. This approach has been carried out in so called control conditions, or after a number of established experimental procedures known to alter the rat's body and brain growth, because of its relative postnatal vulnerability (Dobbing, 1968).

In the adult rat, noninterference approaches include development studies that allow one to trace the formation of LTH using intertrial intervals of different length. They were carried out in the dark and in the light phase of the circadian cycle with consequent posttrial wakefulness (PEW) or sleep (PES), respectively. The correlative studies are also very informative and are aimed at extracting the informational content

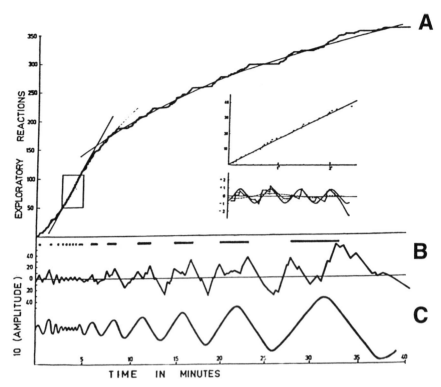

Fig. 2. Habituation dynamics. The behavioral arousal derived from the cumulative frequency of activity in a Làt-maze **(A)** is a complex, exponential curve with an activation, a linear, and a deactivation phase. Oscillations in activity can be detected at exponentially decreasing, constant, and increasing intervals in the three phases, respectively, and are shown as real **(B)** and interpolated data **(C)**. Taken with permission from Làt (1973).

of the within-group variability. The rationale underlying it is to break down the hard and soft constitutive elements of a structure thought to be involved in information processing and to reveal by multivariate analyses, implemented with correlation analyses, the components that are related to behavioral traits from those that are not (*see,* e.g., Cerbone et al., 1993a).

On the other hand, the interference approach was extensively used with treatments that have been extensively shown to influence more complex forms of learning and memory, by interfering with the hypothesized "consolidation process(es)"

(McGaugh, 1966) in associative learning paradigms. They indicate that the intertrial (ranging from 24 h to several weeks) activity decrement in a novel environment or LTH requires activation of immediate early genes (IEG) like *c-fos* and *c-jun*, DNA remodeling, polysome aggregation, protein synthesis, an intact hippocampus and neocortex, and both the slow-wave and paradoxical sleep phases (Cerbone and Sadile, 1994). Further, it is modulated by NMDA receptors (Gironi Carnevale et al., 1990; Pellicano et al., 1993), endogenous or exogenous vasopressin, but not by endogenous opioids (Cerbone and Sadile, 1994). In addition, lesion studies indicate that LTH is modulated by the dorsal noradrenergic bundle (Fallon and Loughlin, 1982; Loughlin et al., 1982), and by the cholinergic septohippocampal system (*see*, e.g., Jaffard et al., 1984).

This multiple approach is very useful and informative concerning LTH and may help in the understanding of more complex forms of learning and memory as well.

Methodology

Subjects

The exploration in the Làt-maze can be carried out on any animal species and even in humans, provided an appropriate setup is adjusted to the species and to the size of the subject. In the studies reported here, albino rats of a random-bred Sprague-Dawley stock (NRB) or of the Naples High (NHE) and Low-Excitability (NLE) lines (Sadile et al., 1983, 1984) have been used. They were grown in our own animal house under conditions that include litter size of nine pups per fostering lactating dam, weaning at 30 d, daily handling, and housing in groups of four per cage until the age of 60 d, followed by allocation of two per cage, with food and water *ad libitum*. Internal rhythms were synchronized by an artificial 12:12 light-dark cycle, with lights on from 6:00 AM to 6:00 PM that could be reverted to have animals with lights on from 6:00 PM to 6:00 AM. However, a reliable reversal of the LD cycle can be achieved only when pups are born and lactated by mothers with an inverted cycle. In studies during postnatal life, the biological age of the animals was chosen considering eyelid opening, which occurs between 13 and 14 postnatal days (PND) through the end of the fourth postna-

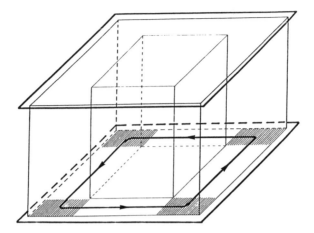

Fig. 3. Schematic diagram of the Làt-maze. The experimental box is described in detail in the Methodology section. The line with arrows on the floor indicates the path along the corridor between the two cubicles. Dashed squares point to the corners whose crossings monitored horizontal activity. The distance between two corners is 60 cm.

tal week, and maturation of activity inhibition or short-term habituation, which occurs at about 18 PND (*see* Maturation Studies). For studies in adult rats, the age was 60–80 PND, i.e., at least 2 wk after puberty. For experiments during aging, the timepoints to be chosen can be 12, 18, or 24 mo, depending on the phenomenon to be focused on. Before puberty both genders can be used, but in adulthood for experiments on LTH, only males can be used because of the short menstrual cycle of the female rat (3–4 d), and the great oscillations in CNS excitability caused by hormonal changes.

Apparatus

The experimental box can be an empty volume whose size is scaled down to the size of the animal. In our experiments we have been using the so called Làt-maze that the late Joseph Làt used to measure activity in order to index the construct of "nonspecific excitability level" (Làt, 1965) in rats. Since its structure has an elementary maze-like configuration, Sadile in his studies has named it "Làt-maze." As it is described in detail elsewhere (Làt, 1973) and shown in Fig. 3, for adult rats it consists, briefly, of an outer box (60 × 60 × 40 cm) containing a plastic transparent smaller one (30 × 30 × 40 cm). For pups the maze has to be scaled down to one third of its size. It can be, for example, 20 × 20 × 20 cm and can be inserted into the inner cubicle of the adult maze so that the illumination is provided by the same lamp from the roof of the adult maze, and it is better isolated from environmental noise. The material can be wood or plastic painted black or gray. The front wall of the outer box has to be transparent to allow behavioral observation through a one-way mirror, but it can be opaque if there is a detection system. However, it must be transparent in a double videocamera system to allow detection of horizontal activity from above and vertical activity from aside. Exploration can be measured under a diffuse, white light from the top in the illumination range of 0.1–0.2 μW cm^2, under a dim red light (not visible to rodents), or under complete darkness. In the latter two cases a special low threshold or infrared camera is necessary. It should be kept in mind that a strong white light inhibits activity through extraretinal inhibitory effects on suprachiasmatic nuclei (Moore, 1974). A white noise produced by a ventilator is necessary to attenuate environmental disturbances.

Procedure

The behavioral repertoire of the animal is measured in a room different from the animal house, to minimize disturbance to other animals and to increase the novelty properties of the box with a lower level of pheromones and ultrasounds. Animals are to be picked up by the base of the tail for adults, and by the skin over the neck for pups. Then they have to be introduced one at a time in a corner of the box, usually the upper right one, from above. Then, the cover containing the light source at its center is lowered, and exploration is allowed in the corridor delimited by the two boxes. The exposure can take place under a mild white light, under a red light or in complete darkness, depending on the modality and sensory channel focused on. In fact, rodents can map the environment with proprioceptive sensors from muscle spindles and vestibular channels, from olfactory cues and gustatory cues in the darkness, and with visual cues in a lighted environment.

In the interference approach, the treatment can be administered either before (pretrial) or after exposure (posttrial). In the pretrial approach, the treatment can alter the somatosensory integration and, thus, it has to be given before the retention test as well. Conversely, in the posttrial approach, which we have largely used, the treatment has to be given immediately after and at specified times following completion of the first exposure, e.g., 0.5, 1.0, 3.0, 6.0, 12, and 24 h, or 7, 14, 21, and 28 d later. In fact, when the gradient of retrograde amnesia/hyperamnesia is the focus of interest, the posttrial approach minimizes interference with learning and/or state-dependent learning processes that are carried over by the pretrial approach. However, the design requires several control groups to assess that the posttrial treatment does not carry over positive, rewarding or negative, punishing properties. This can be accomplished, for instance, by avoiding the administration of the treatment (usually through a needle injection) in the experimental room, but rather in a close-by place, or by using appropriate control groups that receive the treatment immediately after a short tube-drinking test with a sweetened solution of saccharin (0.1%). In fact, an aversive agent is likely to induce aversion for saccharin when tested 24 h later in a two-bottle saccharin–water choice test (*see*, e.g., Garcia and Ervin, 1968). Moreover, controls should also be included to test for proactive effects of drugs. A single noncontingent administration of the drug can be followed 24 h later by two 10-min exposures at 24 h interval. A drug having proactive effects would alter activity on first exposure compared to vehicle-treated animals. However, when the kinetics of the substance or drug used as treatment does not exclude proactive effects, the retention test should be made after an appropriate period of time, accordingly.

Activity Measurements

Activity detection can be accomplished by different means in different light conditions. In fact, all methods can be used in the lighted environment, whereas a dark environment requires an infrared camera or a low intensity threshold CCD camera, provided an infrared light source illuminates the field. With a videocamera, the entire behavioral repertoire of the animal can be monitored, including walking about, rearing on hindlimbs, sniffing, immobility, scratching, face-washing, leaning, and so forth. In our studies, the measured behavioral variables were horizontal (corner-crossing) and vertical activity (rearing on hind limbs), because they express most of the hippocampal rhythmical slow-wave activity (θ activity; *see*, e.g., Irmis et al., 1971; Vanderwolf, 1971). These variables can be monitored in different ways. J. Làt started with push-bottoms linked to an event recorder on paper, but only the frequency of events could be traced. Alternatively, an electrocapacitance system can be used to store the frequency of corner-crossings, and frequency and duration of rearings on punched strips. Using rats chronically implanted with electrodes in the dorsal hippocampus and neocortex, and monitoring by a 35-mm camera activity in a novel environment (empty 60 × 60 × 40 cm box) together with frequency bands of hippocampal electrical activity displayed in digital form on the roof of the box, Làt (1973) could study off-line by means of frame-by-frame analysis the relationship between hippocampal θ and various behavioral components, as shown in Fig. 4. However, when only the frequency of θ-related activity components is the focus of interest, a direct observation of the exploring animal through a one-way mirror is acceptable. The events are recorded on paper by an individual. However, reliability indexes need to be tested by correlation analysis among different observers looking at the same animal. In our studies the between-observer reliability coefficients were 0.995 for horizontal activity and 0.991 for vertical activity, both highly significant. Finally, the most appropriate method now is certainly the videotaping of activity by a low threshold CCD camera. The activity can be monitored off-line either by the direct method shown above or by a PC-based video tracking system. Among others, the one developed by Schwarting et al. (1993) and initially made for circling behavior appears to be satisfactory.

The duration of the exposure can be set at 5, 10, or 15 min or longer depending on whether the posttrial treatment is expected to interfere with the formation of LTH in a positive or a negative manner. In fact, a short exposure is likely to induce a more labile trace for habituation than a longer one. Therefore, using a short exposure gives one the best conditions for detecting an improvement in LTH, thus maximizing the possibility of positive treatment effects. Conversely, a longer exposure leads to a stronger LTH, allowing one to detect an

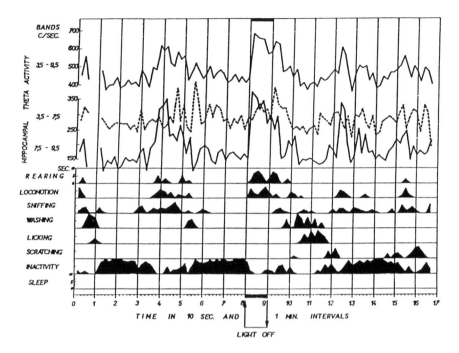

Fig. 4. Relationships between hippocampal electrical activity and behavioral components. Hippocampal θ activity in the 3.5–9.5, 3.5–7.5, and 7.5–9.5 bands Hz is plotted before and after lights-off stimulation along with frequency of different components of behavior in a Làt-maze. Taken with permission from Làt (1973).

impairment in LTH. During the interexposure period the rats have to be left undisturbed in the animal house. All members of a cage are to be tested simultaneously. Retention testing can be done at 24 h or later. The effect of different intervals is reported in the noninterference studies (*see* Development Studies). After each test, the floor has to be carefully cleaned with a wet sponge to eliminate any trace of urine, feces, and pheromones.

Emotionality Measurements

During exploration rats lay down fecal boli. A common interpretation attributes a mapping function to defecation. The studies of Broadhurst on defecation in a strongly illuminated open field (1960) also attribute defecation to emotionality. In the Làt-maze, the number of boli counted at the end of exposure can be taken to measure emotional habituation, as indexed by intertrial decrement.

Data Analysis

The number of corner-crossings, rearings, and fecal boli are considered separately for a number of reasons. First, these variables are attributed different meanings, have different neural substrates, and are thought to be controlled by different genes, because they have been dissociated in mice (Sanders, 1981) and rats (Van Abeelen, 1970). The total score for each variable during the entire testing period on at least two consecutive tests allows a rough estimate of LTH and of the overall amnesic or hypermnesic effect of the treatment in the interference approach. For corner-crossings and rearings, the integrals can be separated as relative to the first or the second part of exposure, with a prevailing noncognitive or cognitive meaning, respectively.

In order to give a quantitative evaluation of the treatment effect over baseline activity levels, the LTH index is computed as activity on retention test as delta percent of the activity on d 1. Different formulas can be used, such as [(d 1 – d 2/d 1) × 100]. However, a preliminary plot of the data is a necessary step before its processing for basic requirements, such as distribution and homogeneity of variance. Appropriate transformations $(x, x^{1/2}, (x_1)^{1/2} \cdot (x)^{1/2} + (x + 1)^{1/2}, \ln(x), \log x, x^{1/3}, 1/(x + 1), x^2)$ can help in fulfilling the requirements for parametric statistics. It is crucial to inspect the data for extremely low scores down to zero that may bias the intertrial decrement by yielding large percent numbers. In this case all numbers have to be transformed as $(x + 1)^{1/2}$.

Furthermore, in order to inspect the temporal pattern of habituation, plots can be made with 0.5 or 1-min epochs that allow a thorough analysis of the data. To better visualize the intertrial difference, an average difference can be computed at each timepoint, giving rise to a plot similar to the one reported in Fig. 5. Experimental groups can then be compared among themselves or vs controls.

Several indicators of spontaneous recovery (SR) and long-term habituation (LTH) can be extrapolated from the uninter-

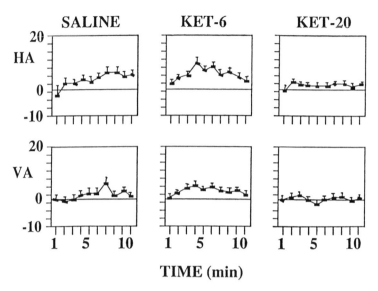

TIME (min)

Fig. 5. Effect of NMDA receptor blockade by ketamine and LTH. Testing–retesting absolute difference (d 1–d 2) in the number of corner-crossings (HA: upper row) and rearings (VA: lower row). Mean ± SEM per 1-min blocks for saline (SAL), ketamine 6 mg/kg (KET6), or 20 mg/kg (KET20). Taken with permission from Gironi Carnevale et al. (1990).

rupted habituation rate on successive exposures, according to Làt (1973). Sometimes the inspection of the activity curves suggests it is appropriate to compute LTH indexes based on the slope of the within-session activity decrement over the two sessions, as formerly pointed out by J. Làt (1973); *see also* ECS Treatment.

Statistical Methods

Behavioral arousal (BA) is defined as the sum of rearings and corner-crossings and may refer to the first part (BA11 or BA12), the second part (BA21 or BA22), or the entire period (BA31 and BA32) of the first or second exposure, respectively. STH and LTH are operationally defined as the within-exposure or between-exposure decrement in horizontal and vertical activity, respectively. STH can be measured as activity in the second part of the test as delta percent of that in the first part, whereas LTH of HA (LTH-H) and VA (LTH-V) can be measured as the score in the second test as delta percent of that in the first test, relative to the first part (LTH-H1 or LTH-V1) or to the second part (LTH-H2 or LTH-V2) of testing, respectively.

Factorial analyses of variance treatment × day (as a dependent variable) can be used to test effects on LTH or day × time blocks (as a dependent variable) to test time-dependent effects (Edwards, 1979). Alternatively, when the requirements for parametric statistics are not fulfilled, nonparametric one-way ANOVA by Kruskal-Wallis or two-way ANOVA by Friedman can be used (Siegel and Castellan, 1988). Between group comparisons can be made by posthoc two-tailed Student's *t*-test for independent parametric data, whereas Mann-Whitney U-test, Fisher's exact probability, or Wilcoxon test can be used for nonparametric data (Siegel and Castellan, 1988). When multiple comparisons are to be made, the α-level has to be adjusted according to Bonferroni procedure modified by Hölm (1979) to decrease the probability of type I error (false positive) in the multiple data handling.

Furthermore, whenever individual paired data are available, parametric or nonparametric correlative analyses by Pearson's product-moment (Edwards, 1979) or Spearman's rank-order (Siegel and Castellan, 1988), respectively, allow extraction of the informational content of within-group variability. Correlation coefficients can be compared by "z" transformation and Fisher-test (Edwards, 1979). A procedure for validating their significance can be followed, as described in detail elsewhere (Lipp et al., 1987; Patacchioli et al., 1989), to avoid spurious and chance findings resulting from the high number of expected correlations. Namely, this procedure is based on setting the α-level two-tailed; random elimination of single pairs by a Montecarlo-procedure, plotting the correlations, and preliminary testing of multiple correlations. To make scatterplots comparable across different scales, main correlations can be reported in standardized form with the zero coordinate representing the mean of either variable and the axes the deviations from the sample mean in units of standard deviation (Edwards, 1979).

Components of Habituation to Novelty

Exploration in the Làt-maze involves multimodally channelled information gathered from visual, olfactory, vestibular, and proprioceptive sensory systems. As a function of time, the novelty-arousing properties of the environment fade out, the neuronal firing in the locus ceruleus increases in response to the nonnoxious environmental stimulation (Aston-Jones and Bloom, 1981), and the exploratory drive decreases, accordingly. After retention testing, the environment is recognized as familiar and exploration decreases.

Moreover, walking and rearing in a novel environment is associated with hippocampal rhythmical slow activity (RSA or θ) of different frequency and amplitude (Vanderwolf, 1969; Whishaw and Vanderwolf, 1973). They have a quite different meaning, i.e., mainly spatial or cognitive for the walking about and emotional or noncognitive for the rearing activity (*see*, e.g., Gironi Carnevale et al., 1990; Montagnese et al., 1993; Sadile, 1994). Briefly, the existence of two main components has been hypothesized on the basis of multiple evidence, namely:

1. Rearing activity displays interindividual variability, reflecting differences in the motivational state, referred to as "nonspecific excitability level" by Làt (1965);
2. Rearing activity and θ amplitude, considered indexes of arousal, are highly correlated (Làt, 1970);
3. Rearing frequency in various learning situations reflects the general arousal level (Làt, 1970);
4. Walking about in a Làt-maze has a mainly spatial meaning, since its frequency decreases as a function of time in a task that is a no-solution problem;
5. Both walking and rearing activity have a mainly noncognitive meaning, be that attentional, emotional, or motivational, in the first part of exposure, compared to the cognitive (spatial) meaning prevailing in the second part;
6. The defecation score, i.e., the emotionality index (Broadhurst, 1960), clusters with rearing frequency in the factor analysis (Sadile, unpublished observations); and
7. The two activity components are controlled by different genes, since they have been genetically dissociated in mice (Sanders, 1981) and in rats (Van Abeelen, 1970).

After re-exposure, retention of the two components is differentially affected across different studies, as for instance:

1. In the sleep studies, slow-wave sleep (SS) followed by paradoxical sleep (PS) (but not by wakefulness) was positively

associated with different components of habituation, confirming the importance of both phases of sleep for the formation of LTH (Sadile et al., 1978b);

2. The long-term retention of the two components has recently been shown to be differentially sensitive to NMDA receptor blockade by noncompetitive and competitive NMDA receptor antagonist (Gironi Carnevale et al., 1990; Pellicano et al., 1993); and

3. Inhibition of the nitric oxide synthase (NOS) activity by N^6-nitro-L-arginine (L-NOARG; 10 mg/kg, ip) affects only retention of the emotional component, i.e., rearing activity (Papa et al., 1994).

Thus, there are several components of activity that undergo habituation and several mechanisms by which the animal habituates.

Different Strategies

Different strategies can be followed to investigate the mechanisms involved in the formation of LTH and its neural substrates, and to validate this plasticity phenomenon as a model for more complex forms of learning and memory processes.

Noninterference Studies

Noninterference studies can be carried out both in the young and adult rat, provided a Làt-maze of appropriate size is used.

Maturation Studies

Maturation studies can be run exclusively in the young rat to monitor the appearance of LTH during postnatal life. They are useful in tracing the temporal window of LTH appearance, indicative of the organizational level and "ergic" systems underlying it. This approach has been carried out in so-called control conditions, or after a number of established experimental procedures known to alter the rat's body and brain growth, because of its relative postnatal vulnerability (Dobbing, 1968). They include the litter size technique (Seitz, 1954; Kennedy, 1957), differential stimulation (Denenberg, 1964; Sadile et al., 1978c), and perinatal hypothyroidism (Sadile et al., 1978a).

In maturation studies, since exploration in the Làt-maze also depends on visual modality, it is necessary to start from age of eyelid opening rather than from date of birth. Thus, it is necessary to refer to the biological age of the animal. Behavioral testing is made within the period of time between eyelid opening and the end of the fourth postnatal week, which coincides with the occurrence of natural weaning. These treatments have been used in random-bred rats and in rats of the NLE and NHE rat strains (see Sadile et al., 1986).

As shown in Fig. 6, in so-called control rats that have been reared under standard conditions, including adjustment of litter size to 9/dam, daily postnatal handling, and weaning at 28 d of postnatal age, STH appeared at about 18 PND and LTH at the end of the fourth week of postnatal life.

Allowing a lactating dam to foster only 3 pups/litter (small litter) or 18 pups/litter (large litter) brings about a profound acceleration or deceleration of body and brain growth, respectively, and is an anticipated appearance (not shown; see Sadile et al., 1978d; Cerbone and Sadile, 1994) or delayed appearance of LTH during postnatal development, respectively, as shown in Fig. 6.

Moreover, perinatal hypothyroidism (see also Fig. 6; Sadile et al., 1978a) induced by perinatal treatment of the lactating dams with propylthiouracil (PTU; 0.2 g/kg) induces deceleration of body and brain growth and dissociation of STH and LTH with a delayed appearance of STH (Sadile et al., 1978a) and absence of LTH throughout postnatal life. As shown in Fig. 7, differential postnatal handling (Sadile et al., 1978c) with rats reared under normal conditions and either left undisturbed in their home cages (no postnatal stimulation; NO-PNS) or handled daily for 1 min during the first weeks of postnatal life (postnatal stimulation; PNS), significantly affected the appearance of LTH in normal rats, in fact, PNS rats handled daily showed LTH by 21 d, whereas NO-PNS rats showed LTH with a 1-wk delay (Sadile et al., 1978c).

The profound effect on LTH is associated with profound changes in the architecture of the nervous system (see, e.g., Denenberg, 1964). The apparent discrepancy between short duration of treatment (1 min/d) and magnitude of changes can be reconciled and explained by a cascade of neuroendocrine

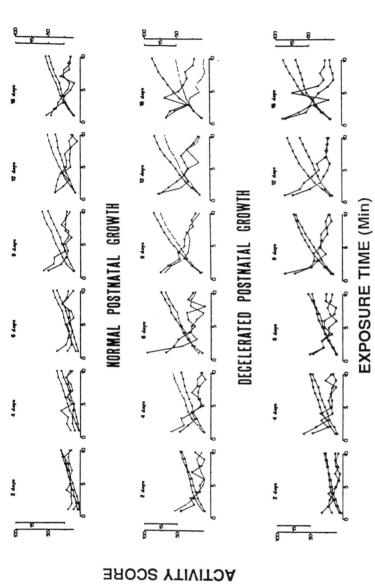

NORMAL POSTNATAL GROWTH

DECELERATED POSTNATAL GROWTH

ACTIVITY SCORE

EXPOSURE TIME (Min)

Fig. 6. Maturation of LTH during postnatal life: effect of altered body and brain growth. Average activity scores, i.e., corner-crossings + rearings per 1-min blocks, are reported as cumulative and differential form for rats reared under normal conditions (middle panel), or with decelerated postnatal body and brain growth (lower panel), or perinatal hypothyroidism (upper panel) on d 1 (closed symbols) and 2 (open symbols). Different groups were tested at the biological age of 2, 4, 6, 9, 12, or 15 d after eyelid opening with a transversal design.

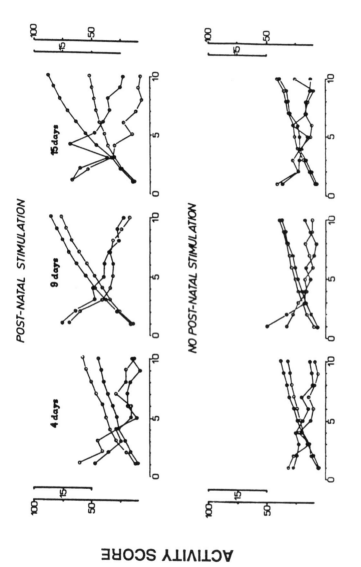

Fig. 7. Differential handling and maturation of LTH to novelty during postnatal life. Average activity scores, i.e., corner-crossings + rearings per 1-min blocks, are reported as cumulative and differential form for rats reared under normal conditions and handled daily or not handled during early postnatal life on d 1 (closed circles) and 2 (open circles). Rats were tested at the biological age of 4, 9, or 15 d after eyelid opening with a transversal design.

events that is triggered by the short handling period but continued even after its termination.

Development Studies

These noninterference approaches can be pursued both in the young and adult rat by studying the formation of LTH after intertrial intervals of different length. They have been run mainly in the adult rat and consist of testing followed by retention testing from 0.5 h–4 wk after the first test.

The development of LTH has been studied in male and female albino rats of NRB and of two strains therefrom derived, i.e., NHE and NLE, genetically selected for divergent behavioral arousal to novel environment (Sadile et al., 1983, 1984). These experiments have been carried out during the light or dark phase of a 12:12 LD cycle, with consequent PES or PEW, respectively. The interexposure intervals were 0.5, 1, 3, 6, and 12 h, and 1, 2, 7, 14, and 28 d with a transversal design and 10–12 rats/group.

The results are shown in Fig. 8A,B. Multivariate analysis of variance (MANOVA) showed significant main effects for strain, interexposure interval, and postexposure state of vigilance (sleep or wakefulness). Furthermore, analysis of the temporal pattern showed the formation of LTH to follow a nonlinear complex function.

Correlative Analyses

The within-group variability is very informative concerning individual differences that can be correlated to individual differences in some of the neuroanatomical components of the architecture of the information processing system under investigation. Although not indicative of a cause–effect relationship, strong correlations are informative of parallel processes taking place that are probably mediated by a third common factor. The rationale underlying this approach is that by decomposing the architectural structure of a system and analyzing a given behavioral trait, thought to depend on the integrity of that structure, we can then investigate which structural component is related to a specific behavioral variable. This allows one to cut informational redundancy by reducing a great number of variables to few ones that can be configured as dimensions. This morphogenetic approach has been followed in mouse strains

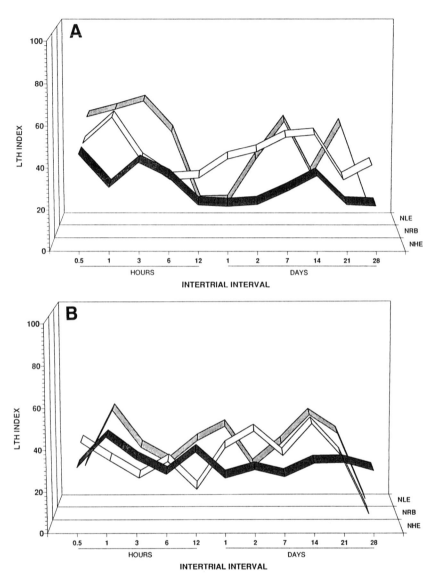

Fig. 8. Development of LTH with posttrial sleep **(A)** or wakefulness
(B). The average habituation index as activity on test trial 2 as percent
of test trial 1 is reported for NLE, NRB, and NHE rats with different
intertrial intervals using a transversal design (N = 10–12/group).

by Crusio et al. (1989) and van Daal et al. (1991a,b), in fish by
Gervai and Csanyi (1985), and in rats by Cerbone et al. (1993a).
Some examples have been presented in detail elsewhere, e.g.,

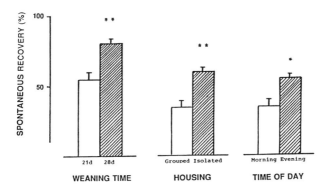

Fig. 9. Environmental conditions and LTH. LTH is reported as its reciprocal (spontaneous recovery), in rats weaned at 21 or 28 PND, differentially housed, or tested during the morning or the evening. $**p < 0.01$; $*p < 0.05$.

the number of intra + infrapyramidal mossy fibers (Cerbone et al., 1993a), the number of high-affinity hippocampal glucocorticoid receptors (Cerbone et al., 1993a), and the unscheduled DNA synthesis (UBDS; *see also* Papa et al., 1993a for a detailed description of the approach).

Interference Studies

The so-called interference approach has been extensively used for several decades to investigate the formation and the locus of learning and memory, mainly in rodents and fish (*see*, e.g., Rose, 1992). Namely, it is based on measuring the formation of LTH after posttrial administration of agents that are known to be effective in other types of learning and memory tasks. In fact, exploration in the Làt-maze is sensitive to a host of internal and external conditions that may affect performance and habituation, some of which have been reported previously (Sadile et al., 1978b). They include the genotype (Sadile et al., 1986), size of the litter (Sadile et al., 1986), degree of stimulation pups undergo during postnatal development (Sadile et al., 1986), time of weaning, differential housing, handling prior to experimental manipulation, time of day, and circannual seasonal variations (*see* Fig. 9).

The dependence of the behavioral test on multiple factors is very informative of its nature and underlying mechanisms.

A preliminary report on some of these treatments has been given previously (Sadile et al., 1978b). Some paradigmatic examples are given here to illustrate the aim, nature, and pitfalls of this approach. Different techniques can be used to tackle various aspects of the LTH phenomenon.

Scrambling of Patterned Neuronal Firing in Neural Networks and Formation of LTH

Patterned firing involving nerve cell assemblies is thought to be organized during acquisition and to continue as reverberatory activity in closed neuronal circuits (Glickman, 1961). Temporary reverberation is inferred primarily by the interference with consolidation exerted by several agents, such as convulsions and spreading depression, able to alter neural firing and surface potential, which have been proposed to be involved in the consolidation of learning (Albert, 1966a,b).

CORTICAL AND HIPPOCAMPAL SPREADING DEPRESSION

Spreading depression (SD) is a functional, reversible technique that has been demonstrated to disrupt retention of various learning tasks, such as active and passive avoidance, discrimination problems, and taste aversion when applied to the neocortex or hippocampal formation (Schneider, 1973; Bures et al., 1974). The SD-induced memory disturbance has been attributed to cortical potential disturbances, such as reversal of the normal surface-positive potential gradient of the cortex; disturbance of intracellular potentials; change in distribution of ions; and direct or indirect influence on neuronal metabolism. Although SD is not a pure decortication, a functionally depressed cortex brings about a temporary sensitization of subcortical structures, as well as the release of hypothalamic–neurohypophyseal peptide hormones (Bures et al., 1974) that are known to have peripheral and central influences on behavior (De Wied, 1966).

Few studies have focused on SD and behavioral habituation. First, LTH has been shown to be interfered with by unilateral and bilateral SD (Nadel, 1966). Second, transfer of habituation by SD has been shown in a reversible split-brain experiment (Squire, 1966). Third, habituation to intense acoustic stimulation has been disrupted when presented during CSD (Van der Staak and Fisher, 1976). Since habituation of acoustic startle response and decline of activity might be mediated by different processes (Williams, 1974), experiments are designed to investigate the role of the cortex and hippocampus in the formation of LTH.

In Experiment 1, a 30-min duration of SD was initiated at 0.0, 0.5, 1.0, and 3.0 h on completion of first exposure to Làt-maze. SD was induced by local application of 25% KCl over the cortex exposed through bilateral trephine openings (B-CSD group), whereas sham-CSD controls received 0.9% NaCl. SD induction was controlled by the disappearance of postural reflexes (Bures et al., 1974). The CSD group received 25% KCl immediately after the first and 0.9% NaCl after the second exposure to the box (internal control). The sham-CSD group received 0.9% NaCl immediately after completion of the first and the second testing period. As demonstrated in Fig. 10A, SD-induced functional decortication had a disruptive effect on LTH. The CSD group did not show response decrement on the second test compared to the first test, whereas on the third day it did (internal control). The time-course of the disruptive effect of B-CSD indicates that the steepest retrograde amnesia gradient takes place within the first 30 min after completion of the exposure, proceeding at a lower rate.

In Experiment 2, rats received an intrahippocampal microinjection of KCl either immediately on completion of the first exposure (posttrial experiment) or immediately before (pretrial experiment). As shown in Fig. 11, LTH was impaired in the HSD-groups compared with sham-HSD controls (upper panel), but the impairment was greater in the pre-trial group (middle panel) than in the posttrial group (lower panel). Analysis of the horizontal and vertical component showed that in both cases the effect pertained to the vertical component, indicating that the formation of memory for emotional information components starts during the experience and continues afterward.

These data suggest that LTH requires intact, functioning cortex and hippocampus. For the latter, the involvement lasts as long as the experience and immediately afterward, whereas for the former it lasts for several hours after presentation of indifferent stimuli. The disruptive effect of CSD is more marked when it is elicited immediately after the exposure, but it is present also later, indicating that it acts on a continuous pro-

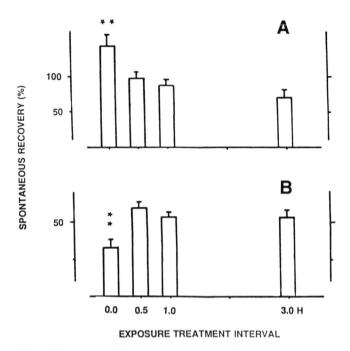

Fig. 10. **(A)** Functional decortication by bilateral spreading depression and LTH (upper panel). LTH of horizontal + vertical activity is shown in rats receiving bilateral CSD, 30 min duration, at 0.0, 0.5, 1.0, or 3.0 h following a 10-min test. LTH index is reported as its reciprocal (spontaneous recovery). **$p < 0.025$. **(B)** Exogenous vasopressin and LTH (lower panel). The LTH is shown for rats given a single posttrial sc injection of LVP (2 μg), immediately following first exposure. **$p < 0.025$ by two-tailed median test vs saline-injected controls. Taken with permission from Sadile et al. (1978b).

cess. Behavioral habituation, operationally defined, is the result of a balance between an excitatory and an inhibitory process. The SD treatment and the subsequent decortication would disinhibit subcortical centers, with resulting sensitization or dishabituation. The massive neuronal firing would then scramble patterned firing in subcortical circuits, with neither reverberation nor release to the cortex. Alternatively, SD has influence on protein synthesis (Bennett and Edelman, 1969; Krivanek, 1970), whose putative role in long-term memory storage is heavily supported (Barondes, 1970), as well as on many other biochemical aspects of metabolic neuronal machinery

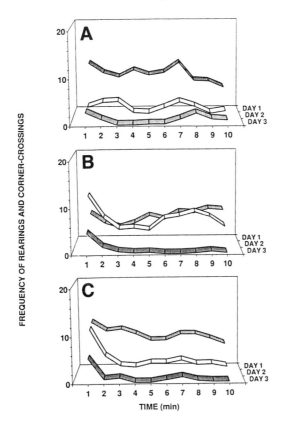

Fig. 11. Hippocampal spreading depression (HSD) and LTH. Average frequency of corner-crossings + rearings is reported per 1-min blocks for sham-injected rats **(A)** and for rats receiving an intra-hippocampal microinjection of 25% KCl in 0.5 µL immediately before **(B)** or after **(C)** a 10-min exposure. Retention testing was made 24 and 48 h later.

(water, electrolytes, energy metabolism, and so forth.) (*see* Bures et al., 1974).

SCRAMBLING OF PATTERNED FIRING BY ELECTRICAL
BRAIN STIMULATION

Brain stimulation at different sites has long been known to interfere with memory processes with different effects according to stimulation site, current parameters, and learning tasks, as indicated by a host of behavioral studies (Kesner and Wilburn, 1974). Moreover, a number of biochemical studies have shown that activation of the major ascending catecholamine systems by electrical stimulation produces a change in the content, release, or turnover of the transmitter at target sites (Dresse, 1966; Olds and Yuwiler, 1972; Squire, 1975).

Among others, electrical stimulation of the lateral hypothalamus at the level of the medial forebrain bundle (MFB), where several ascending pathways gather together in a small space at current intensities, subthreshold for rewarding intracranial self-stimulation (ICSS) have been reported to facilitate active and passive avoidance (Mondadori et al., 1976; Huston et al., 1977), and operant conditioning, producing time-dependent changes in enzymes involved in acetylcholine synthesis in the dorsal hippocampus (Destrade et al., 1978).

Experiments were designed to study the effect of scrambling neuronal patterned firing by electrical stimulation of the MFB (MFB-ES) on the formation of LTH. A group of rats received posttrial electrical stimulation (sine wave, 100 Hz, 0.2 s on and 0.8 s off, 30 s duration) through bipolar electrodes aimed at lateral hypothalamus (AP 5.4; H 2.6; L 1.6), immediately after a 10-min test in a Làt-maze. As shown in Fig. 12, there was no between-session activity decline for the MFB-ES group, in comparison with implanted, nonstimulated controls. Thus, the stimulation of all four catecholamine systems running in the MFB apparently disrupted the formation of LTH completely (0.0% vs 25% of controls).

The disruption of habituation by brain stimulation is in apparent contrast with the mentioned facilitatory effects on other learning tasks (Mondadori et al., 1976; Huston et al., 1977; Destrade et al., 1978). However, several hypotheses can be formulated to explain this effect, namely:

1. The stimulation of the MFB, although subliminal for ICSS, has positive, rewarding properties, and could act as conditioning stimulus (CS) with activity as unconditioned response. Thus, the absence of activity decline could be the consequence of searching for reward, although it can be excluded by inspection of the short-term habituation curves, showing decreasing instead of increasing activity level with time on retention test; and

2. Since habituation is processed at different CNS levels, the ES of the MFB artificially reproduces the physiological acti-

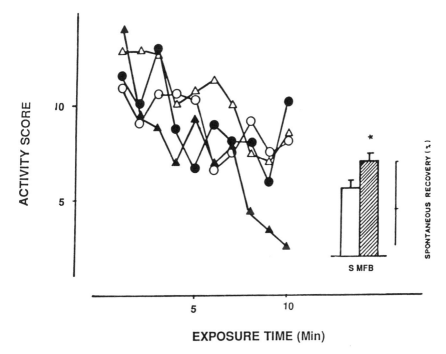

Fig. 12. Electrical stimulation of the medial forebrain bundle and LTH. Average activity scores (rearing + corner-crossings) per 1-min blocks are reported for sham-stimulated rats on d 1 (open triangles) and 2 (filled triangles), and for rats receiving a 30 s duration posttrial train of electrical stimulation (100 Hz, 0.2 s on and 0.8 s off, sine wave) on d 1 (open circles) and 2 (closed circles). In the insert the LTH index is depicted as its reciprocal (spontaneous recovery). *$p < 0.05$ (two-tailed t-test).

vation of the systems running in the MFB that takes place during REM sleep, when insignificant stimuli are thought to be erased (Hennevin and Leconte, 1971; Pearlman, 1971). Therefore, the disruptive effect of MFB-ES would be the consequence of accelerated processing of the trace of indifferent sensory stimuli.

According to the dual-process theory (Thompson and Spencer, 1966), habituation derives from the balance of two independent processes, one inhibitory and the other excitatory ("dishabituation"). The electrical stimulation of the latter would scramble reverberatory circuits (Glickman, 1961) at a time when the trace has not yet been transferred in a more stable form. This scrambling could take place at the stimulation site as well as in remotely activated neural systems. In addition, there would be an imbalance between the two systems that modulate arousal, with prevailing dishabituatory process and massive release of the subcortically confined trace to the cortex and inadequate storage of nonpatterned firing.

ELECTROCONVULSIVE SHOCK TREATMENT (ECS)

ECS has long been known to produce retrograde amnesia gradients (McGaugh, 1966) in a variety of learning tasks by mechanisms, such as polysome disaggregation, enkephalin release, and others. However, alternative explanations have also been proposed (Deutsch, 1973). Very little attention has been paid to the effects of ECS on habituation. Barrett and Ray (1969) showed attenuation of habituation by ECS treatment in mice, using step-through latencies in a two-compartment environment.

Thus, to investigate the effects of a single ECS application on the formation of LTH, two experiments were designed. As previously shown, ECS may have either a facilitatory or an inhibitory effect on LTH, depending on whether ECS is delivered immediately following a short or a longer exposure to novelty. This reveals that ECS has both amnesic and punishing effects that are difficult to discriminate, and require a number of control groups to exclude punishing from amnesic components.

In Experiment 1 rats were allowed to explore a Làt-maze during two 10-min exposures at 24 h intervals. A single ECS application (40 mA, 400 ms) at constant-current intensity was delivered immediately after the first test through saline-soaked padded alligators attached to the ears. Sham-ECS controls received a similar handling procedure without any stimulation. There was no significant difference between the two groups when the cumulative curves were considered. However, analysis of the differential curves per 2-min blocks, shown in Fig. 13A, revealed that the horizontal and vertical activity during the first 2-min of the second exposure was higher for ECS than for the control group ($p < 0.05$), indicating the disruption of the instrumental component of behavioral habituation, in accordance with Barrett's finding (Barrett and Ray, 1969). Furthermore, as shown in the insert of Fig. 13A, the median slope of activity decline (habituation rate) was indeed different between ECS and sham-ECS on d 1 and 2, with a steeper slope in the former group, indicating disruption of the classically conditioned component as well (see Sadile et al., 1978b).

In Experiment 2 rats received a single ECS application immediately on completion of a shorter exposure (5 min) and were given a 10-min retention test 24 h later. The controls received sham-ECS treatment. An additional group received ECS application 24 h before a 10-min exposure to investigate late effects on spontaneous exploratory activity. Finally, a group of rats received shock to the tail (40 mA, 400 ms) immediately after the first 10-min exposure and was retested 24 h later for an additional 10 min. All animals receiving ECS treatment went through tonic-clonic convulsions and recovered from post-ECS coma within 5 min, otherwise they were discarded. As shown in Fig. 13B, the decrement of exploratory reactions on the second exposure, as compared to the first one, was greater in the ECS group than in controls ($p < 0.01$). Thus, given after a short exposure (5 min; Exp. 2), the animals associate its negative effect with environmental cues. As a consequence, on retention testing, they show locomotor inhibition, which increases the more the animal explores the box (classically conditioned component). When given after a longer exposure (10 min; Exp. 1), the punishing effects tend to disappear, whereas the amnesic ones prevail. Thus, ECS treatment appears to have both amnesic and punishing effects on habituation, as in other learning tasks. If the ECS treatment had only punishing effects, rats would have shown strong locomotor inhibition on a retention test, as was the case of tail-shocked rats ($p < 0.01$). Moreover, ECS administered 24 h before testing showed no effect on spontaneous locomotor activity.

These data can be interpreted as direct influence of ECS on biochemical processes involved in the formation of LTH (Kumar et al., 1991). Alternatively, they could be explained by the hypothesis of control of behavior by reward and punishment (Huston et al., 1977). ECS administered after short exposures would punish high internal activation states, whereas at longer intervals would punish decline of activity and internal deactivation states. Thus, different central states would then be classically conditioned (Stein, 1966).

A

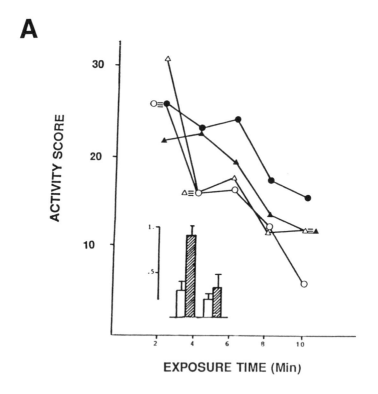

EXPOSURE TIME (Min)

B

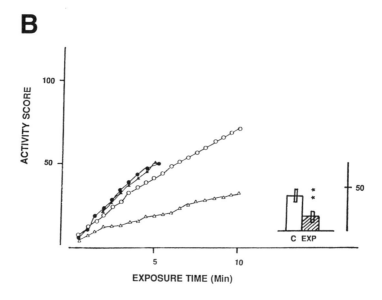

EXPOSURE TIME (Min)

Role of Posttrial Sleep in the Formation of LTH

The involvement of sleep in general, and of REM sleep in particular, in information processing and storage is suggested by several reports (Leconte and Bloch, 1970; Pearlman, 1971, 1974; Pearlman and Greenberg, 1973). However, the studies on REM sleep are based mainly on the deprivation method and a great controversy has questioned the specificity of REM sleep deprivation effects on retention testing. A host of studies were designed to explore the effect of sleep deprivation on the formation of LTH in rats. Basically, two strategies were followed. The first one reported here takes advantage of drugs that are known to interfere with PS, REM, or SS. The other strategy with a correlative approach investigates the involvement of sleep phases by monitoring EEG sleep before and after exposure to a spatial novelty. The EEG is then analyzed and frequency and duration of SS and PS episodes are correlated to parameters of LTH. (*see*, e.g., Montagnese et al., 1993).

Thus, following the first strategy, experiments were run in rats to explore the role of specific sleep stages (SS and/or PS) by selective pharmacological inhibition and/or enhancement (*see* Fig. 14), in particular, in Experiment 1A, four groups of rats received a strong overhead white light (50 μW/cm^2) or complete darkness in the home cages during the first 3 h following a 5- or 10-min exposure to a Làt-maze, exploiting the inhibition that a strong overhead illumination exerts on behavior by

---

Fig. 13. *(previous page)* Electroconvulsive shock and LTH. **(A)** The average sum of corner-crossings and rearings per 2 min-blocks is reported in differential form during testing for sham-ECS on d 1 (closed circles) and 2 (open circles) and ECS group on d 1 (closed triangles) and 2 (open triangles), respectively. The inserted graph shows the mean ± SEM slope by regression coefficient ("b") for sham (left pair) and ECS group (right pair) on d 1 (white column) and 2 (dashed column). **p < 0.01 (two-tailed *t*-test). **(B)** The average sum of corner-crossings and rearings per 30-s blocks is reported in cumulative form during testing for sham-ECS on d 1 (closed circles) and 2 (open circles) and ECS group on d 1 (closed triangles) and 2 (open triangles), respectively. The inserted graph shows the mean ± SEM of spontaneous recovery for sham- (C) and ECS group (EXP). **p < 0.025.

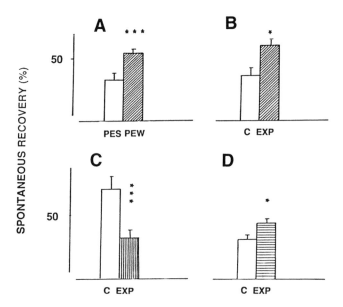

Fig. 14. Role of sleep in the formation of LTH. LTH as its reciprocal, i.e., spontaneous recovery (SR; Mean ± SEM) is reported in different groups. **(A)** PES or PEW; **(B)** vehicle (Con) or amphetamine-induced EEG-desynchronization (EXP); **(C)** vehicle (Con) or barbiturate-induced EEG synchronization (EXP); **(D)** vehicle (Con) or apomorphine-induced REM sleep deprivation (EXP). ***$p < 0.01$; *$p < 0.05$.

facilitating EEG-synchronization and, conversely, the facilitation on behavioral arousal and EEG desynchronization exerted by darkness (Fishman and Roffwarg, 1972; Borbèly and Huston, 1975). This effect is mediated by the suprachiasmatic nucleus with its inhibitory projections to the medial hypothalamus (Moore, 1974). Thus, manipulating the postexposure period by light and darkness, we attempted to investigate the role of sleep processes in the formation of LTH. Rats were all tested for retention 24 h later for 10 min in both cases. Two groups for each exposure length (5 or 10 min) were tested at the beginning and two at the end of the day-night cycle. Although rats had not been implanted with EEG and EMG electrodes, visual inspection of the rats in the light and dark (under a dim red light invisible to rats), allowed us to determine that 80 and 90% of animals spent most of the postexposure time asleep or awake, respectively.

The results indicated that postexposure light-induced sleep (PES) facilitated LTH independently of the exposure length, but with a stronger effect after longer exposures ($p < 0.01$, two-tailed median test). On the first exposure (d 1) there were no differences among the groups (Kruskal-Wallis H-test) suggesting that the novelty-induced, arousing properties of the environment equalized motivational levels across groups tested at different circadian times, thus ruling out state-dependent learning or altered stimulus control. Since a light-induced selective enhancement of slow-wave sleep (SWS) takes place (Borbèly and Huston, 1975) with relative reduction of REM phase (Fishman and Roffwarg, 1972), the data led us to infer that sleep as a whole is essential for transferring the habituation trace in a long-term memory store.

In Experiment 2, drugs were given posttrially to investigate drug-induced EEG changes on LTH. The posttrial administration assures no influence on performance, motivational state, or stimulus control. In addition, for every drug, a group received the treatment 24 h before the first test, to rule out proactive effects on somatosensory functions. Unless specified, no effect was detected.

Figure 14B reports the facilitatory effect ($p < 0.01$, two-tailed median test) on LTH of a single posttrial intraperitoneal injection of the fast acting barbiturate Nembutal at 25 or 40 mg/kg, which is known to induce within 300 s a SWS lasting for several hours or the flattening of EEG, interrupted by bursts of spindling activity, respectively. EEG synchronization appears to strengthen the subcortical trace by shutting off the neocortex that is essential in the processing of indifferent stimuli. Moreover, the flattening of EEG demonstrates that reverberatory electrical activity has no effect on consolidation of LTH, when induced in the postexposure period.

Figure 14C reports the impaired LTH ($p < 0.05$, two-tailed median test) by a single posttrial injection, immediately after a 10-min test, of a low dose (1 mg/kg, ip) of (+)-amphetamine (AMPH), which is known to increase EEG-desynchronization. Notwithstanding, only hypotheses can be formulated since the mechanism(s) responsible for the disruption of LTH is not known so far. First, AMPH induces a state of REM sleep deprivation (Bradley and Elkes, 1957). Second, low doses of AMP are known to cause the release of norepinephrine (NE) (Boakes and Bradley, 1972) with consequent prevalence of behavioral excitation that is likely to alter habituation dynamics (*see*, e.g., Groves and Thompson, 1970). Third, AMPH has a positive, rewarding effect on behavior, thus LTH impairment may be caused by a conditioning effect (*see*, e.g., Huston et al., 1977). Finally, AMPH inhibits unit firing in the locus ceruleus, which may have profound effects in forebrain sites through the dorsal noradrenergic bundle (Walters et al., 1974).

Figure 14D reports the impaired LTH ($p < 0.05$) by a single ip posttrial injection of low doses of the dopaminergic receptor stimulator apomorphine (1 mg/kg), which selectively reduces PS and increases SWS and total sleeping time (Gessa et al., unpublished observation). The disruptive effect of apomorphine was less pronounced than that of amphetamine.

Altogether, the results lend support to the hypothesis that normal sleep is essential for stabilizing the habituation trace or the transfer of the trace in long-lasting form, probably at brain sites different from those initially keeping the STH trace. Either process is facilitated by light-induced SWS enhancement (with no reduction of REM) and by barbiturate-induced EEG synchronization, whereas it is inhibited by wakefulness, apomorphine-induced REM sleep deprivation (with increased SWS), and AMPH-induced EEG desynchronization.

Therefore, it appears that habituation, which results from an excitatory and an inhibitory process (Groves and Thompson, 1970), requires a correct balance of the activation and deactivation phases during wakefulness as well as during sleep. It follows that the absolute or relative prevalence of either phase alters habituation dynamics with consequent incorrect formation of LTH.

Lesion Studies: Role of Noradrenergic and Cholinergic Modulation on LTH

Noradrenergic Bundles

Lesions and stimulation studies (Ungerstedt, 1971) point to a few hundreds neurons (about 1600) in the floor of the fourth ventricle configuring the locus ceruleus (LC) (Foote et al., 1983) as the origin of the widespread, fine network of NE-containing nerve terminals in the mammalian cerebral cortex (Fallon and

Loughlin, 1982; Loughlin et al., 1982). LC is undoubtedly involved in the initiation and maintenance of REM sleep (Jouvet, 1967), which may play a role in the transfer of newly acquired information in a long-term memory store (Leconte and Bloch, 1970; Pearlman, 1971, 1974; Pearlman and Greenberg, 1973). Moreover, amphetamine (AMPH), shown to disrupt LTH, reduces REM-S and it has been reported to produce 100% inhibition of LC unit firing (Walters et al., 1974). Thus, the AMPH-induced disruption of LTH may be caused by direct or indirect effect of temporary LC firing, and should be reproduced by bilateral lesion of LC. Finally, a pronounced activity of LC neurons has been demonstrated in nonanesthetized freely moving rats in response to nonnoxious environmental stimulation (Aston-Jones and Bloom, 1981).

In Experiment 1, rats were LC-lesioned bilaterally by passing 1 mA of DC current for 20 s through monopolar stainless-steel electrodes, using the Fifkova and Marsala (1967) coordinates (AP 1.8; L 1.0; V 4.0). After 3 wk of recovery, the animals were exposed to the Làt-maze for 10 min at 24-h intervals for 3 d. Out of 30 rats only 6 were considered LC-lesioned on the basis of NE content (70–90% reduction) and histological inspection. LC-lesioned rats showed normal activity decline within a single exposure or STH, indicating that the latter is not cortically mediated. Conversely, LC-lesioned rats showed no between-session response decrement or LTH, supporting the hypothesized role LC neurons play in the transfer of information from a short-term to a long-term storage.

In Experiment 2, rats were sham-lesioned (DB-S) or lesioned in the dorsal noradrenergic bundle (DB-L) bilaterally by intrabundle injection of the neurotoxin 6-OH-dopamine (6-OH-DA). Rats were tested in the morning (M) or in the evening (E) and retested 0.5 h later, or were tested in the evening and retested 24 h later. As shown in Fig. 15A, DB-S rats did not differ from DB-L in the morning. Conversely, DB-L were impaired in LTH when tested in the evening and retested 0.5 or 24 h later. Thus, the noradrenergic innervation of the forebrain plays a role with both its components, the one running in the dorsal as well as that running in the ventral bundle. In fact, knocking out mainly the dorsal bundle by *in situ* injection of the neurotoxin yielded only an impaired LTH, whereas com-

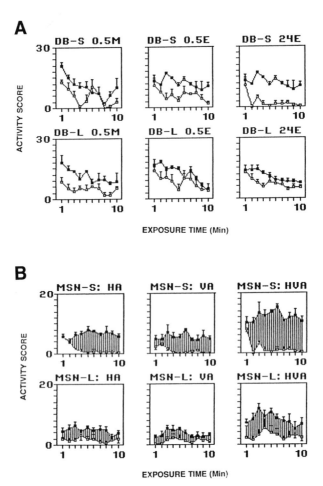

Fig. 15. **(A)** Dorsal noradrenergic bundle and LTH. Activity scores (mean ± SEM), i.e., corner-crossings + rearings per 1-min blocks are reported for sham-lesioned rats (DB-S) and for rats with 6-OH-dopamine lesion of the dorsal bundle (DB-L). Rats were tested in the morning (M) or in the evening (E) on test 1 (closed symbols) and test 2 (open symbols) with an interval of 0.5 h or 24 h. **(B)** Cholinergic septohippocampal pathway and LTH. Average scores (mean ± SEM) of horizontal (HA), vertical (VA), and total activity (HVA) are reported per 1-min blocks for sham-lesioned rats (MSN-S) and for rats with bilateral electrolytic lesion of the medial septal nuclei (MSN-L) on test 1 (closed symbols) and test 2 (open symbols) with a 24 h interval. The dashed area indicates the between-exposure decrement or LTH.

plete noradrenergic denervation by electrolytic destruction of the neurons of origin in the LC completely abolished LTH measured over three consecutive days. Thus, the two bundles modulate LTH acting at different levels and probably at different times on different stages of the multiple process of LTH.

Septohippocampal Cholinergic System

The cholinergic fiber system originating in the medial septal nuclei (MSN) relays electrical activity from the reticular activating system into the hippocampal formation and is a major modulatory system of the electrical activity of the hippocampus (*see*, e.g., Jaffard et al., 1984 for a review). To address the question of whether the cholinergic septohippocampal system is involved in the formation of LTH, albino rats were sham-lesioned or lesioned in the MSN by passing a direct current of 1 mA and 20 s duration. One week after electrolytic lesion or sham-lesion, rats were tested for two 10-min tests at 24 h intervals in a Làt-maze. As shown in Fig. 15B, LTH of both horizontal and vertical activity components was impaired compared to sham-lesioned controls, thus indicating a role in the formation of LTH.

Modulatory Role of Neuropeptides on LTH

A great deal of attention has been focused during the last decade on the role of neuropeptides acting both at peripheral and central sites on learning and memory processes (*see*, e.g., McGaugh et al., 1984).

Neurohypophyseal Peptides

The pioneering work of DeWied with extracts of neurohypophysis (De Wied, 1966, 1971) has led to the demonstration of a diffuse network of vasopressinergic and oxytocinergic nerve terminals of extrahypothalamic origin innervating the entire forebrain and spinal cord (Buijs, 1978; Buijs et al., 1978; Buijs and Swaab, 1979). Although vasopressin (VP) has been shown to delay extinction of instrumental conditioning (De Wied, 1966, 1971), given a great controversy spanning several years about the interpretation of VP effects, a series of experiments were designed to test the effect of vasopressin on LTH. These experiments involved the peripheral or intraventricular administration of exogenous vasopressin, or the blockade of endogenous peripheral or central VP on vasopressin V_1 receptor by the specific antagonist [d(CH_2)$_6$Ty$_2$ (ME)AVP], or the blockade of CSF vasopressin by an antivasopressin antibody. These experiments were described in details by Cerbone and Sadile (1994), and were inspired by the earlier findings (Fig. 10B) that 2 µg of LVP posttrially facilitated LTH only when given immediately on completion of exposure (time 0.0 h). Thus 2 µg of LVP facilitate LTH, supposedly acting on a process with its highest rate immediately following the experience, since at 0.5 h it was no longer susceptible to facilitation. Altogether, the findings described by Cerbone and Sadile (1994) clearly indicate that VP exerts a dose and genotype-dependent effect on LTH (Sadile et al., 1981, 1982). Since its action on LTH has been shown also to depend on the time of the year (Cerbone et al., 1983a), a modulatory role on LTH is inferred. Moreover, the central and peripheral administration studies reveal a dual, differential role` of vasopressin with the circulating hormone acting as signal to adjust the sensitivity of central receptors to the modulatory action of the neuropeptide form (Sadile, 1982; Cerbone and Sadile, 1984, 1985a).

Endogenous Opioid Peptides

Opioid peptides have been reported to act as neuromodulators at central and peripheral synapses and to affect learning and memory processes (*see,* e.g., Izquierdo et al. [1991], McGaugh et al. [1984]), in particular, Izquierdo (1982) has proposed a physiological erasing mechanism operating through endogenous opioids. Furthermore, using posttrial naloxone on habituation of activity in a novel environment a facilitatory effect has been inferred (Izquierdo, 1979) unraveling an amnestic action of opioids and supporting the hypothesized erasing mechanisms operating through endogenous opioids. To further address the question of their involvement in the formation of LTH, experiments were designed with rats exposed for 10 min to a Làt-maze twice at 24 h intervals and given posttrial injection of 0.1 or 1.0 mg/kg of the opioid receptor antagonist naloxone (NLX) or vehicle. NLX-treated rats were not different from vehicle controls (not shown). Thus, the acute blockade of endogenous opioids immediately after first exposure to a

Làt-maze did not affect the formation of LTH. These findings are in contrast with Izquierdo and can be explained in terms of habituation in tasks differing in complexity level; exposure-injection interval, since the role of opioids is perhaps mainly during the task than after it; and doses used, since the low range 0.1–1.0 mg was chosen to block high-affinity systems that have been hypothesized for hippocampus and neocortex (Sadile and Cerbone, 1984; Cerbone and Sadile, 1985b).

Molecular Events Associated with the Formation of LTH

The earlier findings of LTH impairment by inhibition of protein synthesis by posttrial icv injection of cycloheximide (Fifkova and Van Harreveld, 1977) pivoted a number of studies aimed at the neural substrates of LTH. They pertained to the expression of the immediate early genes (IEG) c-fos and c-jun, DNA remodeling, and the expression of effector proteins, such as the nitric oxide synthase activity (NOS).

IEG Expression

According to the molecular hypothesis, the neural substrates of learning and memory imply a variety of biochemical events that involve all messenger systems that transduce electrical signals into transcriptional regulation of gene expression, and eventually DNA remodeling (see Rose, 1991,1992). The IEGs are a numerous and heterogeneous class of third messengers coding for proteins that act on DNA as protein synthesis switches, and may be rapidly induced under a variety of conditions leading to the entry of Ca^{2+} ions into the cell (see, e.g., Morgan and Curran, 1991).

Immunocytochemistry for the IEGs c-fos and c-jun proteins in brain slices of random-bred rats under basal conditions, and on first (E) or repeated (R) exposure to a Làt-maze revealed in E-rats a time- and NMDA-dependent extensive c-fos and c-jun positivity. It involved the reticular formation, the hippocampus (granular and pyramidal neurons), amygdaloid nuclei, all layers of somatosensory cortex, and the granule cells of the cerebellum. In contrast, in R-rats the positivity was weaker than in E-rats but significantly higher than in basal (B) conditions (see Papa et al., 1993b for photomicrographs).

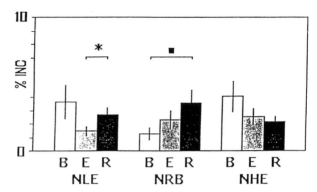

Fig. 16. Fast turnover DNA fraction in the hippocampus and exposure to novelty. The percent incorporation (% INC) of icv ^{3}H-thymidine into hippocampal DNA in NLE, NRB, and NHE rats is reported under basal conditions (B), and on exposure (E; d 1) or re-exposure (R; d 2) to a Làt-maze. (Taken with permission from Cerbone et al., 1993b.)

Genomic Remodeling

First proposed in eukaryotic cells by Baltimore (1985) as caused by amplification and transposition, the role of genomic remodeling in brain plasticity has been tested in several studies. As indexed by "unscheduled" brain DNA synthesis (UBDS), it is modulated by different tasks in mice (Reinis, 1972; Reinis and Lamble, 1972) and rats (Scaroni et al., 1983; Giuditta et al., 1986) (*see* Cerbone and Sadile, 1994 for a review). In rats, as shown, e.g., in Fig. 16, the incorporation into DNA of ^{3}H-methyl-thymidine given icv 15 min before test trial 1 or 2 after a pulse of 0.5 h in *ex vivo* homogenates of brain areas showed on test trial 1 but not 2, a significant increase in the hippocampus of E-rats but not R-rats, compared to B-rats. In addition, crosscorrelations among DNA synthesis across brain areas revealed a decreased synchronization on exposure (Sadile et al., in preparation).

Nitric Oxide Synthase (NOS)

A histochemical study (Papa et al., 1994) for NADPH-diaphorase (NADPH-d), known to be a nitric oxide synthase (NOS) enzyme (Dawson et al., 1991; Hope et al., 1991), was undertaken to investigate whether IEG expression and DNA remodeling led to the expression of the enzyme NOS that forms nitric oxide from L-arginine. Adult male Sprague-Dawley rats were tested for 10 min in a Làt-maze and were sacrificed at 0.0, 2.0, or 24 h thereafter. Rats exposed for the first time to the maze showed NADPH-d activity in the dorsal hippocampus (granule cells, few hilar neurons, and some CA1 pyramidal cells), the caudate-putamen complex, the cerebellum, and all layers of somatosensory cortex, compared to unexposed controls. The positivity was time-dependent since it was present by 2 h and decreased significantly by 24 h. Further, it was impaired by pretreatment with the NMDA receptor antagonist CPP (5 mg/kg) or the NOS inhibitor N^6-nitro-L-arginine (L-NOARG; 10 mg/kg).

Altogether, the data suggest the formation of LTH to involve a cascade of neurochemical events that include activation of IEGs, DNA remodeling, and activation of NMDA receptors in NOS neurons. Further, the widespread induction of IEGs and NADPH-d by exposure to novelty suggests that spatial and emotional information processing activates a neural network across different organizational levels of the CNS.

Summary and Conclusions

LTH to spatial novelty is a relatively simple and ubiquitous form of behavioral plasticity that allows the study of nonassociative learning in the awake freely-behaving organism by a number of approaches in the young as well as in the adult.

1. In adulthood, a series of strategies can be followed that comprise noninterference and interference studies. The former include development experiments where the formation of LTH is studied in random-bred albino rats and the NHE and NLE lines. They were carried out during the light or the dark phase of the circadian cycle, by using testing–retesting intervals of different length. Multivariate analysis of variance showed significant main effects for strain, intertrial interval, and state of vigilance in the PES or PEW. Moreover, analysis of the temporal pattern suggested that the formation of LTH follows a nonlinear complex function. In addition, retention relative to the first and second part of exposure showed a differential profile. Noninterference strategies also include correlative studies that reveal the relevance, for instance, of some components

of the architecture of the hippocampus to the expression of elementary behavioral traits, such as arousal and habituation. Further, common to different strategies is the dual nature of habituation components, the aspecific, emotional/attentional one prevailing in the first part of the test, accompanied by neurovegetative responses, and the specific, spatial one (cognitive) prevailing in the second part of the test. In addition, a classical approach has been used to investigate the making of habituation by manipulating the intertrial interval with agents of different nature that could interfere with the hypothesized "consolidation process(es)."

2. In young organisms, mainly noninterference maturation studies are extensively used. In particular, the appearance of LTH can be monitored during postnatal development in a scaled-down Làt-maze in normally reared rats and in rats with deranged rate of body and brain growth by the litter size technique, differential stimulation, or perinatal hypothyroidism.

LTH is a useful model to study the neural substrates of experience-induced nonassociative behavioral modifications, although its role in more complex forms of learning remains to be ascertained.

Habituation, broadly defined as response decrement, is widespread across species and organizational levels of the CNS. Across species it may take place independent of the nervous system, as in Protozoa (Thompson and Spencer, 1966). On the other hand, across organizational levels habituation has been demonstrated from the spinal cord to auditory system to behavioral habituation with an increasing complexity level. Thus, using the latter as a model system for learning and memory processes, the aim of the studies reviewed here was to show from the methodological point of view the relevance of LTH as a model, its neural substrates, and the "ergic" systems modulating it in the behaving animal. The experiments described here have all been carried out in the albino rat either random-bred or of the NLE/NHE strains. However, these kinds of experiments have been or can be carried out in cockroaches, fish, dogs, horses, and humans (Làt, 1965, 1973, 1976; Martinèk and Làt, 1968a,b).

The interference approach indicates that the formation of long-term habituation to a novel environment is a multistage process. In fact, there is an initial subcortical storage that can be sequestered and excluded from cortical processing, as indicated by the barbiturate treatment. Moreover, scrambling of neuronal patterned firing within neuronal networks by spreading depression, electrical stimulation, or ECS prevents the stabilization in a long-lasting form that takes place presumably and eventually in the neocortex. In fact, the elimination of the LC projections to forebrain sites and neocortex selectively inhibits LTH but not STH. The initial subcortical form of habituation is then transferred into a stage with direct and indirect cortical involvement. It is initially indirect in the modulation of arousal and in the release of the subcortical trace, whereas it is direct in the printing out of patterned firing in somatosensory cortex. Various treatments can interfere with printing out of habituation in a long-lasting form, at various levels in the multistage process leading to LTH. The time dependence of interference facilitatory or inhibitory effects points to storage processes (McGaugh, 1966). In fact, it requires polysome aggregation, as indicated by ECS treatment, protein synthesis, some neuropeptides involved in neurotubule assembly (Taylor et al., 1975), and posttrial sleep, as shown by the experiments described here and the different correlative profile between EEG parameters and LTH (Montagnese et al., 1993).

Moreover, the alteration in LTH induced by interference with SS, PS sleep phase, or both allows us to propose that habituation is a continuous multistage process, with continuous cortical-subcortical interactions that continue in a fixed temporal sequence during wakefulness as well as during sleep. The process is necessarily superimposed on circadian rhythms and modulated by the animal's own excitability level (Làt, 1973, 1976), which is the result of interaction between genetic and environmental determinants (Sadile et al., 1986).

The widespread activation of transcriptional regulators, such as c-fos and c-jun (Papa et al., 1993), and the enzyme NOS (Papa et al., 1994) at different organizational levels of the CNS following exposure to novelty (Papa et al., 1993) is very instructive about the nature of the behavioral task and the brain operations involved in its processing and storage. In fact, in a time-, NMDA (Papa et al., 1993), and nitric oxide-dependent manner (Papa et al., 1994) the first structures to be activated were the midbrain reticular formation, amygdaloid nuclei, vestibular nuclei, followed by the hippocampus, and later by the cerebral and cerebellar cortices. Thus, exposure to novelty, like more complex forms of behavioral plasticity, is multidimensional, because it is constituted of several aspecific (motivational, attentional, emotional) and specific (spatial mapping) components that activate different networks—the attentional, the emotional, and the spatial—processing nets in a way that is very difficult to disentangle, mainly because of their distributed and overlapping nature. However, there are strategies to dissociate these nets. One is the neurogenetic approach with genetic models defective in spatial processing, such as the NHE rat line. In fact, they show, for instance, no IEG expression following exposure to novelty (Papa et al., 1993a), and no inhibition of a fast turnover DNA fraction, in contrast to normal rats where LTH is associated to IEG expression and to changes in DNA remodeling, first demonstrated in eukaryotic cells (*see* Rose, 1991,1992) and then indirectly assessed in several learning paradigms by the so-called unscheduled DNA synthesis (UBDS; *see* Papa et al., 1995). In addition, lesion studies of the dorsal noradrenergic bundle suggest a key role in modulating UBDS in target areas, such as the hippocampus and the neocortex (Sadile et al., 1994).

Furthermore, experiments carried out with NMDA receptor antagonists (Gironi Carnevale et al., 1990; Pellicano et al., 1993) and with the nitric oxide synthase (NOS) inhibitor L-NOARG (Papa et al., 1994) have indicated that the emotional component is more labile and susceptible to disruption. In fact, the earlier study with the noncompetitive NMDA receptor antagonist at allosteric sites, ketamine, disrupted LTH of the emotional component of activity in the Làt-maze, i.e., rearing on the hindlimbs, whereas the spatial component, i.e., corner-crossings, was more stable (Gironi Carnevale et al., 1990). Therefore, it is tempting to speculate that the processing of these two elementary components of any behavioral task proceeds in parallel in different neuronal networks. Indirect evidence also comes from correlative studies among sleep parameters during SS and PS, and components of LTH (Montagnese et al., 1993).

The abovementioned effects of NMDA receptor antagonists and of the NOS inhibitor L-NOARG, affecting only emotional LTH, clearly indicate that LTH takes place only when the emotional and cognitive components are processed in a coupled manner.

Furthermore, exploration in a novel environment has been extensively shown to rely mostly on integrity of the hippocampal formation (Leaton, 1981; Chozick, 1983; Gray and McNaughton, 1983). The evidence comes mainly from correlative studies and from a neurogenetic approach that has been pursued by Sadile et al. in rats of the NHE and NLE (Sadile et al., 1983, 1984) lines. In fact, by selectively breeding animals with high and low scores in the Làt-maze, two lines were obtained, i.e., the NHE and NLE. The NHE rats are defective in spatial and emotional informational processing. They have been proposed as model to study the neural substrates of altered information processing and storage (*see* Cerbone et al., 1993b).

This multiple series of different approaches has already given sufficient information to suggest that LTH is a pervasive phenomenon involving and engaging different networks (attentional, motivational, emotional, and spatial) in computations oriented toward different operations across different organizational levels of the nervous system. Lesion studies of specific nodes within these diffuse, overlapping networks are likely to be less informative about how they work than about whether they are involved in specific operations. The maturation approach can bring relevant information to the understanding of the problem of LTH formation, since during postnatal maturation the temporal dissection of connections among different brain sites is a natural event during discrete temporal windows. Furthermore, the maturation approach indicates that the brain structures underlying STH and LTH are different and mature at different times, and that the two apparently independent processes can be dissociated, as shown by the acceleration and deceleration of brain growth, by perinatal hypothyroidism that knocks out the structures responsible for LTH while delaying the maturation of those underlying STH, and by the preservation of STH but not LTH in vasopressin-deficient Brattleboro rats (Cerbone et al., 1983b).

Although the relevance of LTH is intuitively of great importance, its role in the elaboration of more complex, associative forms of internal representations of the external world remains far from being elucidated.

Acknowledgments

This project has been supported across several years by grants from CNR, and MURST 40 and 60% grants, and by personal funds as well.

References

Albert, D. J. (1966a) The effects of depolarizing currents on the consolidation of learning. *Neuropsychology* **4**, 65–77.

Albert, D. J. (1966b) The effect of spreading depression on the consolidation of learning. *Neuropsychology* **4**, 49–64.

Aston-Jones, G. and Bloom, F. E. (1981) Norepinephrine-containing locus coeruleus neurons in behaving rats exibit pronounced responses to nonnoxious environmental stimuli. *J. Neurosci.* **8**, 887–900.

Baltimore, D. (1985) Retroviruses and retrotransposons: the role of reverse transcription in shaping the eukaryotic genoma. *Cell* **40**, 481,482.

Barondes, S. (1970) Cerebral protein synthesis inhibitors block long-term memory. *Int. Rev. Neurobiol.* **12**, 177–205.

Barrett, R. J. and Ray, O. S. (1969) Attenuation of habituation by electroconvulsive shock. *J. Comp. Physiol. Psychol.* **69**, 133–135.

Bennett, G. S. and Edelman, G. M. (1969) Amino acid incorporation into rat brain proteins during spreading depression. *Science* **163**, 393–395.

Boakes, R. J. and Bradley, P. B. (1972) A neuronal basis for the alerting action of (+)-amphetamine. *Br. J. Pharmacol.* **45**, 391–403.

Borbèly, A. A. and Huston, J. P. (1975) Selective enhancement of slow-wave sleep by light in the rat, in *Proceedings of the Second European Congress on Sleep* (Koella, W. P. and Levin, P., eds.), Karger, Basel, pp. 242–244.

Bradley, P. B. and Elkes, J. (1957) The effects of some drugs on the electrical activity of the brain. *Brain* **80**, 77–117.

Broadhurst, P. L. (1960) Experiments in psychogenetics: application of biometrical genetics to the inheritance of behavior, in *Experiments in Personality. Vol. 1. Psychogenetics and Psychopharmacology* (Eysenck, H. J., ed.), Routledge, Regan Paul, London, pp. 1–102.

Buijs, R. M. (1978) Intra- and extra-hypothalamic vasopressin and oxytocin pathways in the rat. Pathways to the limbic system, medulla oblongata and spinal cord. *Cell Tissue Res.* **192**, 423–435.

Buijs, R. M., Swaab, D., Dogterom, J., and Van Leeuwen, F. (1978) Intra- and extra-hypothalamic vasopressin and oxytocin pathways in the rat. *Cell Tissue Res.* **186**, 423–433.

Buijs, R. M. and Swaab, D. F. (1979) Immunoelectron microscopical demonstration of vasopressin and oxytocin synapses in the limbic system of the rat. *Cell Tissue Res.* **204**, 355–365.

Bures, J., Buresova, O., and Krivanek, J. (1974) *The Mechanism and Application of Leao's Spreading Depression of Electroencephalographic Activity*. Academic, New York.

Cerbone, A., Grimaldi, A., and Sadile, A. G. (1983a) Season-dependent differential effects of posttrial vasopressin upon behavioural habituation in the albino rat. *Neurosci. Lett.* **Suppl. 14,** S60.

Cerbone, A., Sacco, M., and Cioffi, L. A. (1983b) Behavioural habituation in vasopressin-deficient (Brattleboro) rats. *Neurosci. Lett.* **Suppl. 14,** S61.

Cerbone, A., Patacchioloi, F. R., and Sadile, A. G. (1993a) A neurogenetic and morphogenetic approach to hippocampal functions based on individual differences and neurobehavioral covariations. *Behav. Brain Res.* **55,** 1–16.

Cerbone, A., Pellicano, M. P., and Sadile, A. G. (1993b) Evidence for and against the Naples High- and Low-Excitability rats as genetic model to study hippocampal functions. *Neurosci. Biobehav. Rev.* **17,** 295–304.

Cerbone, A. and Sadile, A. G. (1984) The modulatory role of peripheral vasopressin on behavioral habituation in the rat is mediated by septal modulation of CA4–area dentata sensitivity to central vasopressin. *Soc. Neurosci. Abstr.* **10, Part II** , 1118.

Cerbone, A. and Sadile, A. G. (1985a) Peripheral vasopressin hormone modulates hippocampal sensitivity to central vasopressin peptide(s): an hypothesis to explain its effects on behaviour. *Behav. Brain Res.* **16,** 197.

Cerbone, A. and Sadile, A. G. (1985b) Differential effects of "delta" and "mu" opioid receptor activation upon hippocampal slow field depolarization in the albino rat. *Boll. Soc. It. Biol. Sper.* **61,** 39,40.

Cerbone, A. and Sadile, A. G. (1994) Behavioral habituation to spatial novelty: interference and noninterference studies. *Neurosci. Biobehav. Rev.* **18(4),** 497–518.

Chozick, B. S. (1983) The behavioral effects of lesions of the hippocampus: a review. *Int. J. Neurosci.* **22,** 63–80.

Crusio, W. E., Schwegler, H., and van Abeelen, J. H. (1989) Behavioral responses to novelty and structural variation of the hippocampus in mice. II. Multivariate genetic analysis. *Behav. Brain Res.* **32,** 81–88.

Dawson, T. M., Bredt, D. S., Fotuhi, M., Hwang, P. M., and Snyder, S. H. (1991) Nitric oxide synthase and neuronal NADPH diaphorase are identical in brain and peripheral tissues. *Proc. Natl. Acad. Sci. USA* **88,** 7797–7801.

De Wied, D. (1966) Inhibitory effects of ACTH and related peptides on extinction avoidance behavior in rats. *Proc. Soc. Exp. Biol.* **122,** 28–32.

De Wied, D. (1971) Long-term effects of vasopressin on the maintenance of a conditioned avoidance response in rats. *Nature* **232,** 58–60.

Denenberg, V. H. (1964) Critical periods, stimulus input, and emotional reactivity: a theory of infantile stimulation. *Psychol. Rev.* **77,** 335–351.

Destrade, C., Jaffard, L., and Cardo, B. (1978) Post-trial hippocampal and lateral hypothalamic electrical stimulation. Effects on long-term memory and on hippocampal cholinergic mechanisms. *Abh. Akad. Wiss. DDR* **5,** 189–201.

Deutsch, J. A. (1973) Electroconvulsive shock and memory, in *The Physiological Basis of Memory* (Deutsch, J. A., ed.), Academic, New York, pp. 115–124.

Dobbing, J. (1968) Vulnerable periods in developing brain, in *Applied Neurochemistry* (Davidson, A. N. and Dobbing, J., eds.), Davis, Philadelphia, pp. 287–316.

Dresse, A. (1966) Importance du système mesencephalo-telencephalique noradrenergique comme substratum anatomique du comportement d'autostimulation. *Life Sci.* **5**, 1003–1014.

Edwards, A. L. (1979) *Multiple Regression and the Analysis of Variance and Covariance.* Freeman, San Francisco.

Fallon, J. H. and Loughlin, S. E. (1982) Monoamine innervation of the forebrain: collateralization. *Brain Res. Bull.* **9**, 295–307.

Fifkova, E. and Marsala, J. (1967) Stereotaxic atlases for the cat, rabbit, and rat, in *Electrophysiological Methods in Biological Research* (Bures, J., Petran, M., and Zachar, J. eds.), Academic, New York, pp. 653–731.

Fifkova, E. and Van Harreveld, A. (1977) Long-lasting morphological changes in dendritic spines of dentate granule cells following stimulation of the entorhinal area. *J. Neurophysiol.* **6**, 211–230.

Fishman, R. and Roffwarg, H. P. (1972) REM sleep inhibition by light in the albino rat. *Exp. Neurol.* **36**, 166–178.

Foote, S. L., Bloom, F. E., and Aston-Jones, G. (1983) The nucleus locus coeruleus: new evidence of anatomical and physiological specificity. *Physiol. Rev.* **63**, 844–914.

Garcia, J. and Ervin, F. R. (1968) Gustatory-visceral and telereceptor-cutaneous conditioning: adaptation in internal and external milieus. *Commun. Behav. Biol.* **1**, 389–415.

Gervai, J. and Csanyi, V. (1985) Behavior-genetic analysis of the paradise fish, Macropodus opercularis. I. Characterization of the behavioral responses of inbred strains in novel environments: a factor analysis. *Behav. Genet.* **15**, 503–519.

Gironi Carnevale, U. A., Vitullo, E., and Sadile, A. G. (1990) Post-trial NMDA receptor allosteric blockade differentially influences habituation of behavioral responses to novelty in the rat. *Behav. Brain Res.* **39**, 187–195.

Giuditta, A., Perrone Capano, C., D'Onofrio, G., Toniatti, C., Menna, T., and Hydèn, H. (1986) Synthesis of rat brain DNA during acquisition of an appetitive task. *Pharmacol. Biochem. Behav.* **25**, 651–658.

Glickman, S. E. (1961) Perseverative neural processes and consolidation of the memory trace. *Psychol. Bull.* **58**, 218–233.

Gray, J. A. and McNaughton, N. (1983) Comparison between the behavioral effects of septal and hippocampal lesions: a review. *Neurosci. Biobehav. Rev.* **7**, 119–188.

Groves, P. M. and Thompson, R. I. (1970) Habituation: a dual-process theory. *Psychol. Rev.* **77**, 419–450.

Harris, J. D. (1943) Habituatory response decrement in the intact organism. *Psychol. Bull.* **40**, 385–422.

Hennevin, E. and Leconte, P. (1971) La function du sommeil paradoxal: faits et hypotheses. *Annèe Psychol.* **71**, 489–519.

Hope, B. T., Michael, G. J., Knigge, K. M., and Vincent, S. R. (1991) Neuronal NADPH diaphorase is a nitric oxide synthase. *Proc. Natl. Acad. Sci. USA* **88**, 2811–2814.

Hölm, S. (1979) A simple sequentially rejective multiple test procedure. *Scand. J. Stat.* **6,** 65–70.

Huston, J. P., Mueller, C., and Mondadori, C. (1977) Memory facilitation by posttrial hypothalamic stimulation and other reinforcers: a central theory of reinforcement. *Neurosci. Biobehav. Rev.* **1,** 143–150.

Irmis, F., Làt, J., and Radil-Weiss, T. (1971) Individual differences in hippocampal EEG during romboencephalic sleep and arousal. *Physiol. Behav.* **7,** 117–119.

Izquierdo, I. (1979) Effect of naloxone and morphine on various forms of memory in the rat: possible role of endogenous opiate mechanisms in memory consolidation. *Psychopharmacology* **66,** 199–203.

Izquierdo, I., Diaz, D. O., Perry, M. L., Souza, D. D., Elisabetsky, E., and Carrasco, M. A. (1982) A physiological amnesic mechanism mediated by endogenous opioid peptides and its possible role in learning, in *Neuronal Plasticity and Memory Formation* (Ajmone-Marsan, C. and Matthies, H., eds.), Raven, New York, pp. 89–111.

Izquierdo, I., Medina, J. H., Netto, C. A., and Pereira, M. E. (1991) Peripheral and central effects on memory of peripherally and centrally administered opioids and benzodiazepines, in *Peripheral Signaling of the Brain* (Frederickson, R. C. A., McGaugh, J. L., and Felten, D. L., eds.), Hogrefe & Huber, Toronto, pp. 303–314.

Jaffard, R., Galey, D., Micheau, J., and Durkin, T. (1984) The cholinergic septo-hippocampal pathway, learning and memory, in *Brain Plasticity, Learning and Memory, Advanced Behavioural Biololgy, vol. 28* (Will, B. E., Schmitt, P., and Dalrymple-Alford, J. C., eds.), Plenum, New York and London, pp. 167–181.

Jouvet, M. (1967) Neurophysiology of the states of sleep. *Physiol. Rev.* **47,** 117–177.

Kennedy, G. C. (1957) The development with age of hypothalamic restraint upon the appetite of the rat. *J. Endocrinol.* **16,** 9–17.

Kesner, R. P. and Wilburn, M. W. (1974) A review of electrical stimulation of the brain in the context of learning and memory. *Behav. Biol.* **10,** 259–293.

Krivanek, J. (1970) Effect of cortical spreading depression on the incorporation of ^{14}C-leucine into rat brain proteins. *J. Neurochem.* **17,** 531–538.

Kumar, K. N., Tilakaratne, N., Johnson, P. S., Allen, A. E., and Michaelis, E. K. (1991) Cloning of cDNA for the glutamate-binding subunit of an NMDA receptor complex (comments). *Nature* **354,** 70–73.

Làt, J. (1965) The spontaneous exploratory reaction as a tool in psychopharmacological studies. A contribution towards a theory of contradictory results in psychopharmacology, in *Pharmacology of Conditioning, Learning and Retention* (Mikhelson, M. Y. and Longo, V. C., eds.), Pergamon, London, pp. 47–66.

Làt, J. (1970) Report of the laboratory of physiology and pathophysiology of behavior, in *Annual Report of Psychiatric Research Institute,* Prague.

Làt, J. (1973) The analysis of habituation. *Acta Neurobiol. Exp.* **33,** 771–789.

Làt, J. (1976) The theoretical curve of learning and of arousal. *Activ. Nerv. Sup.* **18,** 36–43.

Leaton, R. N. (1981) Habituation of startle response, lick suppression, and exploratory behavior in rats with hippocampal lesions. *J. Comp. Physiol. Psychol.* **95**, 813–826.

Leconte, P. and Bloch, V. (1970) Deficiency in retention of conditioning after deprivation of paradoxical sleep in rats. *C. R. Acad. Sci. (D) (Paris)* **271**, 226–229.

Lipp, H. P., Schwegler, H., Heimrich, B., Cerbone, A., and Sadile, A. G. (1987) Strain-specific correlations between hippocampal structural traits and habituation in a spatial novelty situation. *Behav. Brain Res.* **24**, 111–123.

Loughlin, S. E., Foote, S. L., and Fallon, J. H. (1982) Locus ceruleus projections to cortex: topography, morphology, and collateralization. *Brain Res. Bull.* **9**, 287–294.

Martinek, Z. and Làt, J. (1968a) Interindividual differences in habituation of spontaneous reactions of dogs to a new environment. *Physiol. Bohemoslov.* **17**, 329–336.

Martinek, Z. and Làt, J. (1968b) Ontogenetic differences in spontaneous reactions of dogs to a new environment. *Physiol. Bohemoslov.* **17**, 545–552.

McGaugh, J. L. (1966) Time-dependent processes in memory storage. *Science* **153**, 1351–1358.

McGaugh, J. L., Liang, K. C., Bennett, M. C., and Sternberg, D. B. (1984) Hormonal influences on memory: interaction of central and peripheral systems, in *Brain Plasticity, Learning and Memory, Adv. Behav. Biol., vol. 28* (Will, B. E., Schmitt, P., and Dalrymple-Alford, J. C., eds.), Plenum, New York and London, pp. 253–259.

Mondadori, C., Huston, J. P., Ornstein, K., and Waser, P. G. (1976) Posttrial reinforcing hypothalamic stimulation can facilitate avoidance learning. *Neurosci. Lett.* **2**, 183–187.

Montagnese, P., Mandile, P., Vescia, S., Sadile, A. G., and Giuditta, A. (1993) Long-term habituation to spatial novelty modifies posttrial synchronized sleep in rats. *Brain Res. Bull.* **32**, 503–508.

Moore, R. Y. (1974) Visual pathways and the central neural control of diurnal rhythms, in *The Neurosciences, Third Study Program* (Schmitt, F. O. and Worden, F. G., eds.), MIT Press, Cambridge, MA, pp. 537–542.

Morgan, J. I. and Curran, T. (1991) Stimulus-transcription coupling in the nervous system: involvement of the inducible proto-oncogenes *fos* and *jun. Ann. Rev. Neurosci.* **14**, 421–451.

Nadel, L. (1966) Cortical spreading depression and habituation. *Psychon. Sci.* **5**, 119,120.

Olds, M. E. and Yuwailer, A. (1972) Effect of brain stimulation in positive and negative reinforcing regions in the rat on the content of catecholamines in hypothalamus and brain. *Brain Res.* **36**, 385–398.

Papa, M., Pellicano, M. P., Cerbone, A., Lamberti-D'Mello, C., Menna, T., Buono, C., Giuditta, A., Welzl, H., and Sadile, A. G. (1995) Immediate early genes and DNA remodeling in the Naples High and Low-Excitability rat brain following exposure to a spatial novelty. *Brain Res. Bull.* **37(2)**, 111–118.

Papa, M., Pellicano, M. P., Welzl, H., and Sadile, A. G. (1993) Distributed changes in *c-Fos* and *c-Jun* immunoreactivity in the rat brain associated with arousal and habituation to novelty. *Brain Res. Bull.* **32**, 509–515.

Papa, M., Pellicano, M. P., and Sadile, A. G. (1994) Nitric oxide and long-term habituation to novelty in the rat. *Ann. NY Acad. Sci. USA* **738,** 316–324.

Patacchioli, F. R., Taglialatela, G., Angelucci, L., Cerbone, A., and Sadile, A. G. (1989) Adrenocorticoid receptor binding in the rat hippocampus: strain-dependent covariations with arousal and habituation to novelty. *Behav. Brain Res.* **33,** 287–300.

Pearlman, C. A. (1971) Latent learning impaired by REM sleep deprivation. *Psychon. Sci.* **25,** 135,136.

Pearlman, C. A. (1974) REM sleep deprivation impairs bar-press acquisition in rats. *Physiol. Behav.* **13,** 813–817.

Pearlman, C. A. and Greenberg, R. (1973) Posttrial REM sleep, a critical period for consolidation of shuttle box avoidance. *Anim. Learn. Behav.* **1,** 49–51.

Pellicano, M. P., Siciliano, F., and Sadile, A. G. (1993) NMDA receptors modulate behavioral plasticity as assessed by dose and genotype-dependent differential effects of posttrial MK-801 and CPP on long-term habituation to spatial novelty in rats. *Physiol. Behav.* **54,** 563–568.

Reinis, S. (1972) Autoradiographic study of ^{3}H-thymidine incorporation into brain DNA during learning. *Physiol. Chem. Phys.* **4,** 391–397.

Reinis, S. and Lamble, R. W. (1972) Labelling of brain DNA by ^{3}H-thymidine during learning. *Physiol. Chem. Phys.* **4,** 335–338.

Rose, S. (1992) *The Making of Memory: From Molecules to Mind,* Bantam, London.

Rose, S. P. (1991) How chicks make memories, the cellular cascade from *c-fos* to dendritic remodelling. *Trends Neurosci.* **14,** 390–397.

Sadile, A., Cerbone, A., Lamberti, C., and Cioffi, L. A. (1984) The Naples High (NHE) and Low Excitable (NLE) rat strains: a progressive report. *Behav. Brain Res.* **12,** 228,229.

Sadile, A. G., De Luca, B., Cerciello, A., Messina, A., Trovato, C., and Cioffi, L. A. (1978a) Development of behavioral habituation in thyroidectomized rats. *Proc. Third Int. Congr. Pathophysiol., Varna* **S31,** 159.

Sadile, A. G., De Luca, B., and Cioffi, L. A. (1978b) Long-term habituation to novel environment: amnesic and hypermnesic effects of various postexposure treatments. *Abh. Akad. Wiss. DDR* **5,** 203–218.

Sadile, A. G., Trovato, C., Cerciello, A., Messina, A., Castaldo, G., and Cioffi, L. A. (1978c) Environmental stimulation and maturation of long-term habituation in the rat. *Proc. S. I. B. S.* S125.

Sadile, A. G., Trovato, C., Cerciello, A., Messina, A., Vivo, P., and Cioffi, L. A. (1978d) Altered ontogeny of behavioral habituation in rats with deranged brain growth during postnatal life, longitudinal and transversal analysis. *Proc. Third Int. Congr. Pathophysiol., Varna* **S04,** 146.

Sadile, A. G., Cerbone, A., and Cioffi, L. A. (1981) Vasopressinergic modulation of behavioural habituation. Dose-response curves in rats of the NSD-HE and NSD-LE strains. *Neurosci. Lett.* **Suppl. 7,** S51.

Sadile, A. G. (1982) Extrahypothalamic vasopressinergic projections to hippocampal formation are modulated by neurohypophyseal activity through short-term negative and long-term positive feed-backs. *Boll. Soc. It. Biol. Sper.* **55,** 29,30.

Sadile, A. G., Cerbone, A., and Cioffi, L. A. (1982) Vasopressinergic modulation of behavioural plasticity. Electrophysiological and behavioural study in the Naples High and Low Excitability rats. *Behav. Brain Res.* **5**, 117.

Sadile, A. G., Cerbone, A., Manzi, G., and Grimaldi, A. (1983) Bidirectional asymmetrical selection for behavioral arousal to novel environment: Naples High (NHE) and Low (NLE) Excitable rat strains. *Soc. Neurosci. Abstr.* **9**, 643.

Sadile, A. G., Cerbone, A., Grimaldi, A., Manzi, G., and Cioffi, L. A. (1986) Postnatal brain growth and behavior: evaluation of environmental factors. *Bibl. Nutr. Diet.* **38**, 194–205.

Sadile, A. G., Cerbone, A., Lamberti-D'Mello, C., Menna, T., Buono, C., Rafti, F., and Giuditta, A. (1995) The dorsal noradrenergic bundle modulates DNA remodeling in the rat brain following exposure to novelty. *Brain Res. Bull.* **37(1)**, 9–16.

Sadile, A. G. and Cerbone, A. (1984) High-affinity in vivo systems for Leu- and Met-enkephalin in CA4–area dentata subregion in the rat, differential maturation and plasticity with field slow depolarization as functional probe, in *Central and Peripheral Endorphins, Basic and Clinical Aspects, Frontiers in Neuroscience* (Muller, E. E. and Genazzani, A. R., eds.), Raven, New York, pp. 59–63.

Sanders, D. C. (1981) The Bethlem lines: genetic selection for high and low rearing activity in rats. *Behav. Genet.* **11**, 491–503.

Scaroni, R., Ambrosini, M. V., Principato, G. B., Federici, F., Ambrosi, G., and Giuditta, A. (1983) Synthesis of brain DNA during acquisition of an active avoidance task. *Physiol. Behav.* **30**, 577–582.

Schneider, A. M. (1973) Spreading depression: a behavioral analysis, in *The Physiological Basis of Memory* (Deutsch, J. A., ed.), Academic, New York, pp. 271–303.

Schwarting, R. K. W., Goldenberg, R., Steiner, H., Fornaguera, J., and Huston, J. P. (1993) A video image analyzing system for open-field behavior in the rat focussing on behavioral asymmetries. *J. Neurosci. Meth.* **49**, 199–210.

Seitz, P. F. D. (1954) The effects of infantile experiences on adult behavior in animal subjects, effects of litter size during infancy upon adult behavior in the rat. *Am. J. Psychiat.* **110**, 916–927.

Siegel, S. and Castellan, N. J. (1988) *Non Parametric Statistics for the Behavioral Sciences*, Second Edition, McGraw Hill, New York.

Squire, L. R. (1966) Transfer of habituation using spreading depression. *Psychon. Sci.* **5**, 261,262.

Squire, L. R. (1975) Inhibition of protein synthesis impairs long-term habituation. *Brain Res.* **97**, 367–372.

Stein, L. (1966) Habituation and stimulus novelty: a model based on classical conditioning. *Psychol. Rev.* **73**, 352–356.

Taylor, A., Moffly, R., Wilson, L., and Reaven, E. (1975) Evidence for involvement of microtubules in the action of vasopressin. *Ann. NY Acad. Sci. USA* **253**, 723–737.

Thompson, R. F. and Spencer, W. A. (1966) Habituation, a model phenomenon for the study of the neuronal substrates of behavior. *Psychol. Rev.* **73**, 16–43.

Ungerstedt, V. (1971) Stereotaxic mapping of monoamine pathways in the rat brain. *Acta Physiol. Scand.* **Suppl. 367,** 1–48.

Van Abeelen, J. H. F. (1970) Genetics of rearing behavior in mice. *Behav. Genet.* **1,** 71–76.

van Daal, J. H., Herbergs, P. J., Crusio, W. E., Schwegler, H., Jenks, B. G., Lemmens, W. A., and van Abeelen, J. H. (1991a) A genetic-correlational study of hippocampal structural variation and variation in exploratory activities of mice. *Behav. Brain Res.* **43,** 57–64.

van Daal, J. H., Jenks, B. G., Crusio, W. E., Lemmens, W. A., and van Abeelen, J. H. (1991b) A genetic-correlational study of hippocampal neurochemical variation and variation in exploratory activities of mice. *Behav. Brain Res.* **43,** 65–72.

Van der Staak, C. and Fisher, W. (1976) Habituation to intense acoustic stimulation during cortical spreading depression in rats. *Physiol. Behav.* **17,** 29–33.

Vanderwolf, C. H. (1969) Hippocampal electrical activity and voluntary movements in the rat. *Electroencephalogr. Clin. Neurophysiol.* **26,** 407–418.

Vanderwolf, C. H. (1971) Hippocampal electrical activity and voluntary movements. *Psychol. Rev.* **78,** 83–113.

Walters, J. R., Bunney, B. S., Aghajanian, C. D., and Roth, R. (1974) Locus coeruleus neurons: inhibition of firing by d- and l-amphetamine. *Fed. Proc.* **33,** 294.

Whishaw, I. Q. and Vanderwolf, C. H. (1973) Hippocampal EEG and behavior: changes in amplitude and frequency of RSA (theta rhythm) associated with spontaneous and learned movement patterns in rats and cats. *Behav. Biol.* **8,** 461–484.

Williams, J. M. (1974) Pharmacological and anatomical dissociation of two types of habituation. *J. Comp. Physiol. Psychol.* **87,** 724–732.

CHAPTER 2

Tonic Immobility
as a Model of Extreme States
of Behavioral Inhibition

Issues of Methodology and Measurement

Gordon G. Gallup, Jr. and Dawn R. Rager

Description

Tonic immobility (TI) (also commonly known as "animal hypnosis," "immobility reflex," and "contact defensive immobility") is a profound, but easily reversible condition of response inhibition found in a wide variety of different species, including various insects, crustacea, fish, amphibians, reptiles, birds, and mammals (for reviews, *see* Gallup, 1974a; Jones, 1986). Figures 1 and 2 depict chickens exhibiting the TI reaction.

TI can be produced in a number of different ways, but the common denominator to practically all effective techniques involves some form of physical restraint. Under laboratory conditions, the response is typically elicited by the application of manual restraint. In chickens, which have been widely used as subjects, manual restraint can be most effectively applied by holding the subject down on its side or its back (*see* Figs. 2 and 3). Initially in response to handling and restraint, most birds will struggle and try to escape. However, if one persists in holding the subject down, after a few seconds, these frantic reac-

From: *Motor Activity and Movement Disorders*
P. R. Sanberg, K. P. Ossenkopp, M. Kavaliers, Eds. Humana Press Inc., Totowa, NJ

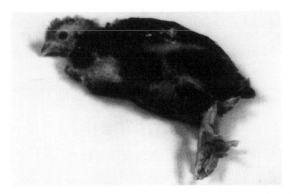

Fig. 1. A 3-wk-old domestic chicken displaying the TI response following the application of lateral restraint (*see* Fig. 3).

Fig. 2. TI in a chicken following ventral restraint (*see* Fig. 4).

tions subside, and the animal will go into a catatonic-like state, which after removing your hands will continue for some period of time in the absence of any further restraint. How long the response will last varies as a function of species and the experimental conditions that both precede and surround the testing situation. Any sudden change in stimulation, such as an abrupt movement by the experimenter or a loud sound, will often cause the response to terminate prematurely.

There can be large individual differences in the duration of tonic immobility, and in chickens the response may last from a few seconds to over an hour, with typical durations of 5–10 min in naive/untreated birds. Under some experimental conditions, however, average durations can last for an hour or more (Hicks et al., 1975).

Tonic Immobility

Fig. 3. Manual restraint applied laterally to a young chicken.

Table 1
Characteristics of TI in Chickens

Lack of movement
Unusual posture
Muscle hypertonicity
Waxy flexibility
Parkinsonian-like tremors
Diminished responsiveness to stimulation
Intermittent periods of eye closure
Mydriasis
Suppressed vocal behavior
Changes in heart and respiration rate
Altered EEG patterns
Changes in core temperature

As a profound, but reversible state of motor inhibition, TI is often accompanied by intermittent periods of eye closure, diminished vocal behavior, Parkinsonian-like tremors in the extremities, and waxy flexibility. In addition to these overt changes, other physiological concomitants include changes in respiration rate (hyperventilation), heart rate (bradycardia), core temperature (hypothermia), and altered EEG patterns (e.g., Klemm, 1971; Nash et al., 1976). Table 1 summarizes some of the changes that tend to accompany TI. Especially when animals close their eyes during TI, it may create the impression that the animal is dead or that it has simply gone to sleep, but despite the animal's outward appearance, it is now well established that the inhibition that occurs during TI is almost entirely efferent in nature, i.e., subjects in TI continue processing information and remain aware of events occurring in their immediate environment (Gallup et al., 1980).

This reaction has now been well documented in a large number of different species under a variety of different conditions. Indeed, in an attempt to develop a comprehensive bibliography of published papers on TI, Maser and Gallup (1977) were able to compile a list of almost 800 papers published over the past 300 years on the phenomenon, and since then, several hundred more papers have been added to the list. However, in spite of the interest in the reaction, serious scientific attention has only been focused on the response during the past 25 years.

Theories

There have been many attempts to explain TI. Until recently, one of the prevailing views was that TI was an animal analog to human hypnosis. However, it is generally conceded that the conditions associated with both the induction of TI and the characteristics of the response are so different from those that characterize human hypnosis that the similarity between the two states is purely superficial (Gallup, 1974a). One view that has received considerable recent support (*see* Gallup, 1977 for a review) is that the response may be a reaction to fear, and a large number of studies converge in showing that procedures designed to increase fear (e.g., electric shock, loud noise, conditioned aversive stimuli, adrenalin) serve reliably to prolong the duration of TI. Similarly, when steps are taken to alleviate fear (e.g., handling, familiarization, administration of tranquilizers), this has the effect of diminishing or abbreviating the response.

Even Charles Darwin and Ivan Pavlov speculated about the significance of TI. In the course of conducting some of his early research on salivary conditioning in dogs, Pavlov noted that on being restrained in preparation for conditioning, some of the dogs would go into TI, and he theorized that the reaction was the result of active motor inhibition descending from the cortex. Darwin was also aware of TI and thought that it might be an instance of death-feigning, i.e., not only the opossum plays opossum.

Methods of Induction

Different forms of immobility may be induced by a variety of stimuli, including experimenter/predator proximity or contact, restraint, rotation, pressure applied to the body with clips, flashing visual stimuli, hooding, and electrical brain stimulation (Carli, 1977; Crawford, 1977; Lefebvre and Sabourin, 1977b). Lefebvre and Sabourin (1977b) pointed out that immobility responses induced by these various methods may differ in terms of response characteristics, neural mechanisms, physiological correlates, and species specificity. A number of investigators have distinguished between "death-feigning" or "instinctive immobilization," which does not require physical contact, and "reflex immobilization," "immobility reflex," or "TI," which is typically induced by inversion and restraint (Klemm, 1977; Lefebvre and Sabourin, 1977b; Burghardt and Greene, 1988). In this chapter, we focus on immobility induced by inversion and restraint, i.e., TI.

Induction by Restraint

Ratner (1967) suggested that restraint is necessary for induction of the immobility response, although he used the term "restraint" in a very broad sense to include "entrapment" with or without physical contact with a predator or an experimenter. Most laboratory studies of TI have utilized manual restraint of the organism in either a dorsal/supine, lateral, or ventral/prone position (Gilman et al., 1950; Carli, 1977). Restraint following inversion to a lateral or dorsal/supine position is most commonly used (*see* Figs. 3 and 4), although evidence suggests that the vestibular and/or proprioceptive stimulation resulting from inversion is not essential for induction. For example, Gilman et al. (1949) reported that although immobilization without inversion (i.e., in a ventral/prone position) reduced the incidence and duration of TI, these reductions often failed to reach statistical significance. Moreover, the vestibular sense need not be intact for successful induction of TI (Klemm, 1977). Although TI can be induced by restraint without inversion, we and others (Ratner, 1967; O'Brien and Dunlap, 1975) have found that inversion makes it easier to identify response termination, which

Fig. 4. Manual restraint applied ventrally.

is typically defined by the execution of a righting reflex (Ratner, 1967; O'Brien and Dunlap, 1975; Carli, 1977; Hennig, 1978).

Duration of Restraint

In most experiments, the induction procedure involves restraint for a specified period of time, although in some studies, the subject is restrained until signs of immobility (e.g., cessation of movement, rigidity, eye closure, tremor) become apparent (O'Brien and Dunlap, 1975; Crawford, 1977; Hennig, 1978). The duration of restraint is an important consideration, since it may influence the magnitude of the immobility response. Gallup et al. (1971b) found that 15 s of restraint was optimal for eliciting and maintaining TI in chickens as compared to 5, 30, and 60 s of restraint. In rabbits, however, induction of immobility by 45 s of restraint has been reported to yield more durable TI responses than 5, 10, and 60 s of restraint (Simonov and Paikin, 1969). The duration of the immobility response is influenced not only by the duration of restraint applied during induction, but also by the number and temporal spacing of induction attempts. Potentiation of TI duration (i.e., sensitization) has been reported when repeated induction attempts occur close together in time, whereas attenuation of TI duration (i.e., habituation) occurs when repeated induction attempts are more widely spaced (Liberson, 1948; Gilman et al., 1950; Ratner, 1967; Nash and Gallup, 1976; Crawford, 1977; Lefebvre and Sabourin, 1977a,b).

Apparatus

A number of studies show that induction of TI in a device that conforms to the shape of the animal's body, such as a cloth trough (Braud and Ginsburg, 1973), a V-shaped trough (Ratner, 1958), a U-shaped trough (Klemm, 1966), or a cloth-covered U-shaped trough (Jones and Faure, 1981), enhances susceptibility to and/or duration of the immobility response. Two possible explanations for these effects have been proposed. First, such devices may prevent animals that are immobilized in a supine position from rolling to one side, which may cause the response to terminate prematurely (Ratner, 1958; Braud and Ginsburg, 1973; O'Brien and Dunlap, 1975). Alternatively, it has been suggested that the prolonged tactile stimulation provided by such devices simulates continued restraint by a predator (Klemm, 1966; Braud and Ginsburg, 1973; Ginsburg, 1975). Braud and Ginsburg (1973) noted that 1–2-d-old chicks restrained in a lateral position (so that they would not roll to one side) exhibited immobility responses in a cloth trough, but not on a flat surface. It should be noted that TI tends to be briefer and more difficult to produce in birds <2 wk of age (Ratner and Thompson, 1960; Salzen, 1963). Second, Klemm (1966) observed that rabbits immobilized in a U-shaped trough showed enhanced immobility responses compared to those immobilized in a V-shaped trough, and attributed this difference to the increased tactile stimulation provided by the contour of the U-shaped device. Finally, a number of studies have shown that tactile stimulation resulting from application of pressure (either manually or mechanically with clips or bandages) to specific body regions serves to induce and/or enhance the immobility response (Kumazawa, 1963; Ginsburg, 1975; Lefebvre and Sabourin, 1977b; de la Cruz et al., 1987; Fleischmann and Urca, 1988; Meyer, 1990). Pressure applied to the dorso-lumbar region was reported to be most effective for eliciting TI in rabbits (Buser and Viala, 1969, cited in Lefebvre and Sabourin, 1977b), whereas pressure applied to the nape of the neck was most effective for eliciting TI in mice (Fleischmann and Urca, 1988) and for enhancing TI in rats (Meyer, 1990). It is important to recognize that the response characteristics of immobility induced by pressure and/or tactile stimulation applied to various regions of the body, such as the dorsal immobility or carrying reflex observed in young rodents when they are grasped by the nape of the neck or the lordosis reflex observed in adult female rodents during behavioral estrous in response to somatosensory stimulation by a male, may differ in important ways from the response characteristics of TI. For example, in contrast to TI, which is difficult to induce in birds <2 wk of age, the carrying reflex can be elicited in young rodents until about 3 wk of age, at which time this form of immobility becomes more difficult to elicit (De La Cruz et al., 1987), and unlike TI, the lordosis reflex in female rodents is hormone-dependent (Schwartz-Giblin et al., 1989).

Measurement

Duration of the immobility episode is the most commonly utilized measure of TI, and is typically interpreted as an index of response magnitude (Ratner, 1967; Doty, 1969; Gallup et al., 1971). Susceptibility to TI has been operationally defined at least three different ways:

1. As the number or percentage of animals tested that exhibit an immobility response of a certain minimum duration (e.g., 5 or 30 s) (Gilman et al., 1950; Salzen, 1963; Braud and Ginsburg, 1973);
2. As the number of standardized induction attempts required to elicit an immobility response of a certain minimum duration (Gallup et al. 1971a; O'Brien and Dunlap, 1975; Crawford, 1977; Hennig, 1978; Jones, 1986); and
3. As duration of activity or struggling during induction in studies where the duration of the induction procedure is not specified and presumably continues until the response occurs (Prestrude and Crawford, 1970; Hennig, 1978).

Susceptibility to TI and the duration of the response are correlated, i.e., the fewer induction attempts required or the shorter the period of struggling during induction, the longer the response tends to last (Ratner, 1967; Lefebvre and Sabourin, 1977a,b), and vice versa (Boren and Gallup, 1976).

Transformation of Data

Lefebvre and Sabourin (1977a) incorporated both duration of immobility and number of induction attempts for each subject tested into a single transformed ratio score {log[(duration/ attempts) + 1]}. They argued that the transformed ratio score represents an improvement over the two individual measures (i.e., duration and number of induction attempts) for two reasons. First, since the two individual measures are negatively correlated, combining them into a ratio provides a more sensitive measure of the immobility response. Second, the duration/attempts ratio compensates for any increase in duration resulting from repeated induction attempts occurring at short intervals. On the other hand, to the extent that these measures are imperfectly correlated and differentially sensitive to various experimental manipulations (for examples, *see* Nash and Gallup, 1975b; Rager et al. 1986), one may lose important information by using such a ratio. According to Lefebvre and Sabourin (1977a), the log transformation of the ratio was then performed in order to normalize the distribution of the data. Indeed, a number of investigators have relied on logarithmic transformations for purposes of normalizing TI duration data (Gallup et al., 1970, 1971; O'Brien and Dunlap, 1975; Crawford, 1977; Hennig, 1978; Thompson and Liebreich, 1987), since these data frequently exhibit a high degree of variability (Gallup et al., 1971b) and mean variance correlations that result in heterogeneity of variance (heteroscedasticity) for various treatment groups (e.g., O'Brien and Dunlap, 1975; Hennig, 1978; Thompson and Liebreich, 1987).

Techniques for Minimizing Variability

Gallup et al. (1971b) recommended habituating subjects in order to reduce variability and minimize potential ceiling effects that might otherwise result from experimental manipulations intended to potentiate the immobility response. Specifically, their habituation procedure involved inducing TI for up to 60 s five times daily until each subject exhibited an immobility duration of 60 s or less for two consecutive trials. During the habituation procedure, intertrial intervals (ITIs) were between 2 and 5 min long in order to prevent immediate reinduction from serving as punishment for termination of TI, which would have resulted in enhanced (rather than attenuated) immobility episodes. In fact, Nash and Gallup (1976) found that in chickens, the ITI could be reduced to 15 s and the habituation procedure would remain effective, although ITIs of <15 s in duration would result in sensitization. Smith and Klemm (1977) reported similar findings for rabbits in that short ITIs (15–30 s) produced long durations of TI, whereas longer ITIs (2–3 min) produced abbreviated responses. This habituation technique has been used successfully in a number of other studies with chickens (Gallup et al., 1971; Nash and Gallup, 1975b) as well as by other investigators (Carli, 1977) to reduce variability in TI duration in rabbits.

Termination or Arousal Thresholds

Duration of immobility serves as a useful index of response magnitude provided termination of TI is allowed to occur naturally, i.e., "self-paced termination" (Ratner, 1967). Another measure of TI is the arousal or termination threshold, which is based on artificial termination of the immobility episode following exposure to tactile/nociceptive stimulation (Ratner, 1958; Klemm, 1965a; Tompkins, 1974; Rager and Gallup, 1985). Briefly, this measure involves attaching electrodes to the surface of the skin, inducing TI, and gradually increasing the intensity of electrical current delivered until response termination occurs. Thus, the termination threshold is operationally defined as the minimum intensity of current sufficient to elicit a righting response (Klemm, 1965a; Tompkins, 1974; Rager and Gallup, 1985). Whereas measures of induction and duration serve as indices of response susceptibility and magnitude (respectively), termination or arousal thresholds are believed to reflect the depth or intensity of the immobility response (Klemm, 1965a; Ratner, 1967; Rager and Gallup, 1985). Termination thresholds can be established relatively quickly and are typically less variable than measures of response duration. However, termination thresholds should probably be used in addition to, rather than instead of, duration measures, since these indices are differentially sensitive to certain experimental manipulations (see Rager and Gallup, 1985). To date, arousal thresholds have only been established using tactile/nociceptive stimulation (via delivery of electrical current). However, other forms of stimulation could be employed in a similar manner to measure the depth or intensity of the immobility response (Ratner, 1967). For example, termination of TI has also been reported to occur in response to visual (Doty, 1969) and auditory stimulation (Hatton and Thompson, 1975), although visual and auditory termination thresholds have yet to be used as dependent variables.

Other Behavioral Measures

Numerous other responses that occur during TI have also been used as dependent measures in studies of the immobility response. Gallup et al. (1971b) showed that latency to the first vocalization and the occurrence of eye closure and defecation were positively correlated with duration of the immobility episode in chickens, and Jones (1986) reported that birds that remained immobile for longer periods showed longer latencies to the first head movement and a smaller number of head movements during the immobility episode. Minor motor activity in the form of tremors has also been used as a dependent variable in studies of TI in both rabbits and chickens, although relationships between this measure and the duration of TI in these species have not been determined (Carli, 1977; Gallup et al., 1980). However, Hennig (1978) examined correlations among a variety of responses, including tremor, vocalization, defecation, eye closure, number of induction attempts, and duration of TI in squirrel monkeys, and found that duration of TI was significantly correlated only with the number of induction attempts (in an inverse manner). Thus, relationships between these various behaviors and duration of TI may vary in different species.

Physiological Measures

Physiological responses, such as respiration rate, heart rate, body temperature, and electroencephalographic (EEG) activity, also show distinct patterns during an immobility episode that may differ between species. For example, both rabbits and chickens show a reduction in heart rate (Gilman et al., 1950; Nash et al., 1976; Carli, 1977), and both chickens and iguanas exhibit an increase in respiration rate immediately following induction of TI (Gilman et al., 1950; Prestrude and Crawford, 1970; Nash et al., 1976). In contrast, humans show an increase in heart rate following restraint and rapid inversion of body position (i.e., similar to the induction procedure for TI) (Crawford, 1977), and squirrel monkeys exhibit a decrease in respiration during and immediately following induction of TI (Hennig, 1978). However, body temperature and EEG activity changes that occur during TI appear to be similar among the avian and mammalian species that have been tested. For example, body temperature decreases (i.e., hypothermia ensues) following induction of TI in chickens (Nash et al., 1976; Eddy and Gallup, 1990), rabbits (Whishaw et al., 1979a), and rats (Whishaw et al., 1979b). With regard to EEG activity, chickens, frogs, rabbits, and guinea pigs have all been shown to exhibit an increase in slow-wave activity following induction of TI similar to, but nevertheless distinct from, that observed during sleep (Schwarz and Bickford, 1956; Svorad, 1957; Kumazawa, 1963; Klemm, 1965a,b, 1977; Carli, 1977; Gentle et al., 1989).

It is important to note that TI differs from sleep in a number of ways. For example, in contrast to the reduced sensitivity to environmental stimuli that typically occurs during sleep, immobilized animals are quite responsive to a variety of environmental stimuli (Ratner, 1967), including visual (Doty, 1969), auditory (Hatton and Thompson, 1975), and tactile/nociceptive stimuli (Klemm, 1965b; Tompkins, 1974; Rager and Gallup, 1985). Indeed, Gallup et al. (1980) demonstrated that chickens will exhibit a conditioned response to a previously conditioned aversive cue when the cue is presented during TI, and that extinction of such a response occurs when the cue is presented repeatedly during an immobility episode.

A final note regarding physiological aspects of TI is that the duration of the response has been reported to exhibit a circadian rhythm. For example, in rats (which are nocturnal) maintained under a 12:12 light/dark cycle, TI durations were longest just after dark onset and shortest just before dark onset (Hennig and Dunlap, 1977). A somewhat different pattern has been reported for chickens (which are diurnal) in that the longest TI episodes occurred just prior to light onset and the shortest episodes occurred immediately following light onset (Stahlbaum et al., 1986).

Neuroanatomy

Contrary to Pavlov's suggestion that TI may be a byproduct of descending inhibitory influences arising at the level of the cortex, there is growing evidence that the cortex does not cause TI. In the first place, many animals without any naturally occurring cortex still show robust TI responses (e.g., crabs, fish, lizards). In animals with a cortex, if anything, the cortex appears to antagonize the response. In rats, for example, pups with imperfectly developed cortex remain immobile longer than do adults, and ablating the cortex in mature rats reliably prolongs the response, as does the application of potassium chloride to the surface of the cortex (Teschke et al., 1975). In chickens, ablating the anterior two-thirds of the archistriatum, which is thought to be the avian equivalent of mammalian neocortex, also prolongs the duration of TI (Maser et al., 1973).

On the basis of brain transection studies and single unit recordings taken from different brain sites, Klemm (1977) has identified an area deep in the brainstem pontine reticular formation that appears to act as a control center for TI in rats. It is interesting to note in light of the data that implicate serotonergic involvement in TI that the principal (if not sole) source of 5-HT in the brain is the raphe nuclei, which are located in that part of the brainstem identified as being essential to the response.

Neurochemical and Hormonal Changes

In addition to behavioral and physiological correlates of TI, several studies have examined changes in central neurotransmitter levels (Farabollini et al., 1986) and plasma hormone levels (Farabollini et al., 1978, 1990; Carli et al., 1979) that occur during TI in rabbits. To our knowledge, no such studies have been published using species other than rabbits. However, numerous studies have investigated effects of various pharmacological manipulations on TI in a variety of different species. Although a discussion of this extensive literature would be impractical in the context of this chapter, Gallup et al. (1983) and more recently Klemm (1989) provided a comprehensive and critical review of the psychopharmacology of TI.

Genetic Basis

It is now reasonably well established that a substantial component of the individual differences in the duration of TI has a genetic basis. In chickens, for example, there are substantial strain differences in the duration of TI that continue to obtain under conditions in which birds from different strains are raised under the same conditions. Moreover, when chickens from strains showing long durations of TI are bred with those from strains characterized by briefer reactions, the offspring show responses of intermediate duration (Gallup et al., 1976). Likewise, selective breeding applied to birds of the same strain that show differences in the duration of TI can produce responses in offspring that are highly correlated with those found in their parents. In one study, chicks derived from parents showing long durations of TI remained immobile on the average of almost a half-hour longer than those whose parents had shown brief reactions (Gallup, 1974). Effects of selective breeding on TI have also been reported in quail, mice, and rats (e.g., Jones et al., 1991; Kulikov et al., 1993; Launay et al., 1993).

Test Equipment

Equipment requirements for TI studies are minimal: a flat surface on which the subject can be tested and a stopwatch for timing the duration of the immobility episode are sufficient (Ginsburg, 1975). In cases where TI is difficult to elicit and/or maintain, increased tactile stimulation may serve to increase susceptibility and/or duration (Ginsburg, 1975; Klemm, 1977). As described previously, increased tactile stimulation can be achieved through use of a variety of simple troughs or chutes within which the subject is placed during immobility testing (Ratner, 1958; Klemm, 1966; Braud and Ginsburg, 1973; O'Brien and Dunlap, 1975; Lefebvre and Sabourin, 1977a; Hennig, 1978). Several different types of barriers or chambers designed to reduce extraneous environmental stimulation have also been employed in studies of TI (Gallup et al., 1972; Hatton and Thompson, 1975; Nash and Gallup, 1975a; Thompson and Liebreich, 1987). In our laboratory, we have used a chamber that not only minimizes environmental stimulation, but is also automated to detect response termination, record duration of the immobility episode, and allow the experimenter to monitor response termination from a remote location, i.e., outside the test room (*see* Rager et al., 1986 for a detailed description of this apparatus). These are important considerations, since, as mentioned previously, immobilized subjects respond to a variety of environmental stimuli, including the human experimenter. Indeed, Gallup et al. (1972) found that chickens immobilized in the presence of a human experimenter remained immobile significantly longer than birds immobilized with the experimenter occluded from view, and this effect was exacerbated when the experimenter maintained eye contact with subjects during testing. Thus, the use of a testing chamber that minimizes environmental stimulation and permits the experimenter to monitor subjects' responses while remaining out of view is highly recommended.

Effects of Prior Experience

We have already mentioned a number of factors that can affect susceptibility to and/or duration of the immobility response, including the duration and nature of restraint, age of the subjects, prior immobility testing (which can lead to either attenuated or enhanced responses, depending on the intertrial interval), and tactile, auditory, and/or visual stimulation (including presence of and proximity to the experimenter, and eye contact with the experimenter). Effects of prior experience are also worth mentioning for investigators preparing to embark on studies of TI. A number of experiments have demonstrated that handling or "taming" of subjects prior to testing reduces susceptibility to TI (Gilman et al., 1950; Ratner and Thompson, 1960; Salzen, 1963; Gallup et al. 1971b; Jones et al., 1991). Thus, in order to ensure a reliable and robust response, human contact with subjects prior to testing for TI should be minimized. On the other hand, deprivation of early social experience with other conspecifics actually reduces the magnitude of the immobility response (Salzen, 1963; Rovee-Collier et al., 1980). Moreover, in order to ensure a robust immobility response, subjects should be reared with conspecifics, but tested in isolation, since animals tested in the presence of a conspecific (Liberson, 1948; Salzen, 1963) exhibit shorter immobility episodes compared to animals tested in isolation. It is interesting to note that when exposed to a mirror, chickens actually show longer TI episodes, which has been attributed to increased fear resulting from eye contact with the image (Gallup, 1972).

Table 2
Human Analogs to TI

Scared stiff/frozen with fear
Shell shock
Reactions to attack by wild animals
Catatonic schizophrenia
Sleep paralysis
Drowning victims
Fainting
Rape-induced paralysis
Aircraft disasters

Natural History

Consistent with Darwin's original death-feigning hypothesis, the available evidence clearly implicates TI as being involved in the ecology of predator–prey relations. For example, Ratner (1967) noted that the behavior of many prey species often varies in a fairly systematic way as a function of the distance separating them from a potential source of danger (*see* Table 2). At appreciable distances when a predator first appears, the initial response of many prey species is to freeze as a means of minimizing detection and capitalizing on natural forms of camouflage and cryptic coloration. However, if the predator continues to approach and the distance separating the participants decreases further (i.e., detection has occurred), the next most likely reaction in this series of distant-dependent antipredator strategies is an attempt to flee. If in spite of these attempts to run or fly away the distance between the two continues to decrease and contact occurs, then fighting and struggling by the victim occurs as another means of deterring predation. However, if this form of resistance is ineffective in providing escape and contact with the predator is prolonged by more than a few seconds, then TI ensues as the terminal response in this distant related sequence of predator defenses.

Becoming immobile at close quarters with an assailant as a means of evading predation might seem counterintuitive, in the sense that the victim would appear even more vulnerable. However, it is now well established that the behavior of many predators is so rigidly programmed that they need considerable stimulus support from the victim in order to complete a predatory episode (e.g., Herzog and Burghardt, 1974). For instance, to sustain many predators in captivity requires that they be fed live food. If provided with perfectly edible meat, they will starve to death, i.e., in order for many predators to consummate a predatory episode, they need feedback from the victim. As a familiar example of the dynamics of this situation as it applies to predatory-prey relations, take the case of "playing" cat and mouse. Following initial capture, the cat will often put the mouse down, back up a few steps, crouch, and wait while keeping the mouse under careful surveillance. What does the cat wait for? It waits for the mouse to move. As soon as the mouse moves, the cat pounces on the mouse. Movement, in other words, is a provocation for further attack, and it triggers another predatory episode. However, as long as the mouse remains motionless, the probability of further attack is reduced, and in the meantime, the cat may become distracted by another mouse or a bird, and the immobilized mouse may therefore survive to pass on its genes.

There is now considerable evidence in support of this analysis. Using the presence of a stuffed hawk to simulate predation under laboratory conditions reliably prolongs the duration of TI (Gallup et al., 1971a), and a field study of predation on ducks by foxes has shown that birds that show TI often survive attacks relatively unharmed (Sargeant and Eberhardt, 1975). In addition, Thompson et al. (1981) reported a rather compelling test of the adaptive value of TI in quail under laboratory conditions using domestic cats as predators. They found that the probability of further attack was reduced and survivability was increased in quail that went immobile shortly after initial attack by cats.

Human Implications

We have all heard stories of humans who have been frozen with fear or scared stiff. Table 2 lists some of the possible human analogs to TI. There are a number of accounts of people who have survived attacks by wild animals and report having gone into a state of extreme movement inhibition or paralysis at some point during the attack. Gallup and Maser (1977) have argued that patients in catatonic stupors show many of the characteristics associated with TI, and they theorize that contained in some of the symptoms of catatonic schizophrenia may be fragments of primitive predator defenses that misfire in response to extreme environmental stress.

Suarez and Gallup (1979) surveyed the literature on sexual assault and discovered that certain reactions of victims are strikingly similar to those shown by animals during the induction of TI. At the moment of sexual assault, most of the conditions known to be conducive to the onset of TI converge, in the sense of experiencing extreme fear, contact, and being physically restrained by the assailant. As further support for this analogy, the behavior of victims also varies as a function of the distance separating them from the rapist, with attempts to escape followed by struggling and resistance at close quarters. If contact is prolonged, it is common for many victims to experience a condition described in the literature as "rape-induced paralysis." During an episode of rape-induced paralysis, which occurs at some point during the assault, victims find themselves unable to move or engage in any further resistance, and even incapable of calling out for help. Paralleling what has been found in animals, rape-induced paralysis is almost entirely restricted to efferent processes, i.e., although unable to respond, victims typically remain conscious and are often able to vividly recount the details of the attack at a later date.

Suarez and Gallup theorized that rape-induced paralysis by humans is an instance of TI triggered by some of the components of sexual assault. Indeed, just as the absence of movement and resistance can cause predators to lose interest and become distracted, Suarez and Gallup reviewed evidence that some rapists require active struggling and resistance from the victim in order to consummate the sexual attack, and in the absence of victim resistance, often find themselves incapable of even sustaining an erection. Thus, the dynamics of the relationship between the victim and the assailant appear to parallel those obtained for animals between predator and prey, and rape-induced paralysis may have adaptive significance in the sense of serving not only to minimize physical harm associated with resistance, but may even terminate an instance of sexual assault prior to consummation.

It is common for some victims of sexual assault, particularly those who experience rape-induced paralysis, to develop feelings of guilt and remorse days or weeks after the incident, feeling that if they had responded differently, the rape might not have occurred. Suarez and Gallup recommended that victims of sexual assault routinely be counseled about the natural, adaptive significance of such response inhibition and its involuntary nature. Finally, a third implication of this analysis pertains to the laws regarding sexual assault and victim recourse (which vary widely). Some states have earnest resistance statutes, which specify that in order to convict someone of rape, you have to demonstrate active resistance on the part of the victim. Rape is one of the few crimes where the behavior of the victim is taken into account. How adaptive would it be to resist handing your wallet over to a mugger as he held a pistol against your head? Why should rape victims be expected to behave any differently? Rape-induced paralysis renders earnest resistance statutes all the more ludicrous. Why should victims of sexual assault be legally penalized/handicapped for showing what may be a normal and potentially adaptive reaction over which they have no control? Suarez and Gallup concluded, therefore, from their analysis that the view of rape-induced paralysis as an instance of TI has important implications for rape prevention (i.e., assault termination), treatment implications for victims of sexual assault, and legal implications for the adjudication of rapists.

Finally, some work by Johnson (1984) implicates TI as also being involved in the behavior of people aboard airplanes that crash. Among passengers who survive the initial impact of a crash, it is often the case that their continued survival is predicated on getting out of the aircraft as soon as possible. After an airplane crash, there is a good chance that the plane may catch on fire or, in the case of aircraft that go down at sea, that it may sink. By interviewing and debriefing flight crews aboard aircraft that go down, Johnson discovered that they often encountered otherwise uninjured passengers who appear dazed, frozen with fear, and unable simply to get up out of their seats and walk to the exits. Johnson has carefully documented and studied this behavior, and he feels that it is analogous to TI in animals. Indeed, if you try to imagine what it would be like to be

aboard an aircraft that was about the crash, you would probably experience intense fear and you would be unwittingly
restrained by your seat beat. Recall that fear and restraint are
the principal features involved in the elicitation of TI. Thus,
much like instances of sexual assault, the conditions that prevail during many airline disasters would seem highly conducive to the production of TI in humans.

One fairly obvious practical implication of this analysis
would be that since TI is easily antagonized by sudden changes
in stimulation, perhaps commercial aircraft could be outfitted
with loud buzzers that are automatically activated within a short
period after the impact of a crash. TI as a response to an airline
disaster is also a classic illustration of how technological developments may be outstripping our biological ability to keep
pace. Evolution requires a considerable amount of time. Technological developments, on the other hand, although slow at
first (e.g., it took early humans thousands of years to develop
and perfect primitive arrowheads), have recently been occurring at an exponential rate, and in the process, some of our previously adaptive behavior may be rendered maladaptive.
Perhaps it is time to begin to factor some of the more salient
features of human evolutionary biology into the design of equipment and technology (see Barash, 1986).

References

Barash, D. P. (1986) The Hare and the Tortoise. Viking Penguin, New York.
Boren, J. L. and Gallup, G. G., Jr. (1976) Amphetamine attenuation of tonic
    immobility in chickens. Physiol. Psychol. 4, 429–432.
Braud, W. G. and Ginsburg, H. J. (1973) Immobility reactions in domestic
    fowl (Gallus gallus) less than 7 days old: resolution of a paradox. Anim.
    Behav. 21, 104–108.
Burghardt, G. M. and Greene, H. W. (1988) Predator simulation and duration
    of death feigning in neonate hognose snakes. Anim. Behav. 36, 1842,1843.
Carli, G. (1977) Animal hypnosis in the rabbit. Psychol. Rec. 1, 123–143.
Carli, G., Farabollini, F., and Lupo di Prisco, C. (1979) Plasma corticoster-
    one and its relation to susceptibility to animal hypnosis in rabbits. Neu-
    rosci. Lett. 11, 271–274.
Crawford, F. T. (1977) Induction and duration of tonic immobility. Psychol.
    Rec. 1, 89–107.
De La Cruz, F., Junquera, J., and Russek, M. (1987) Ontogeny of immobility
    reactions elicited by clamping, bandaging, and maternal transports in
    rats. Exp. Neurol. 97, 315–326.

Doty, R. L. (1969) The effect of environmental movement upon the duration of tonic immobility in bobwhite quail. *Psychon. Sci.* **16**, 48,49.

Eddy, T. J. and Gallup, G. G., Jr. (1990) Thermal correlates of tonic immobility and social isolation in chickens. *Physiol. Behav.* **47**, 641–646.

Farabollini, F., Lupo di Prisco, C., and Carli, G. (1978) Changes in plasma testosterone and in its hypothalamic metabolism following immobility responses in rabbits. *Physiol. Behav.* **20**, 613–618.

Farabollini, F., Lodi, L., and Lupo, C. (1986) Interaction of tonic immobility and dexamethasone in the modulation of hippocampal 5-HT activity in rabbits. *Pharm. Biochem. Behav.* **25**, 781–784.

Farabollini, F., Facchinetti, F., Lupo, C., and Carli, G. (1990) Time-course of opioid and pituitary-adrenal hormone modifications during the immobility reaction in rabbits. *Physiol. Behav.* **47**, 337–341.

Fleischmann, A. and Urca, G. (1988) Clip induced analgesia and immobility in the mouse: activation by different sensory modalities. *Physiol. Behav.* **44**, 39–45.

Gallup, G. G., Jr. (1972) Mirror-image stimulation and tonic immobility in chickens. *Psychon. Sci.* **28**, 257–259.

Gallup, G. G., Jr. (1974a) Animal hypnosis: factual status of a fictional concept. *Psychol. Bull.* **81**, 836–853.

Gallup, G. G., Jr. (1974b) Genetic influence on tonic immobility in chickens. *Anim. Learn. Behav.* **2**, 145–147.

Gallup, G. G., Jr. (1977) Tonic immobility: the role of fear and predation. *Psychol. Rec.* **27**, 41–61.

Gallup, G. G., Jr. and Maser, J. D. (1977) Tonic immobility: evolutionary underpinnings of human catalepsy and catatonia, in *Psychopathology: Experimental Models* (Maser, J. D. and Seligman, M. E. P., eds.), Freeman, San Francisco, pp. 334–357.

Gallup, G. G., Jr., Nash, R. F., Potter, R. J., and Donegan, N. H. (1970) Effect of varying conditions of fear on immobility reactions in domestic chickens (*Gallus gallus*) *J. Comp. Physiol. Psychol.* **73**, 442–445.

Gallup, G. G., Jr., Nash, R. F., Donegan, N. H., and McClure, M. K. (1971a) The immobility response: a predator-induced reaction in chickens. *Psychol. Rec.* **21**, 513–519.

Gallup, G. G., Jr., Nash, R. F., and Wagner, A. M. (1971b) The tonic immobility reaction in chickens: response characteristics and methodology. *Behav. Res. Meth. Instrument.* **3**, 237–239.

Gallup, G. G., Jr., Cummings, W. H., and Nash, R. F. (1972) The experimenter as an independent variable in studies of animal hypnosis in chickens (*Gallus gallus*) *Anim. Behav.* **20**, 166–169.

Gallup, G. G., Jr., Ledbetter, D. H., and Maser, J. D. (1976) Strain differences among chickens in tonic immobility: evidence for an emotionality component. *J. Comp. Physiol. Psychol.* **90**, 1075–1081.

Gallup, G. G., Jr., Boren, J. L., Suarez, S. D., Wallnau, L. B., and Gagliardi, G. J. (1980) Evidence for the integrity of central processing during tonic immobility. *Physiol. Behav.* **25**, 189–194.

Gallup, G. G., Jr., Boren, J. L., Suarez, S. D., and Wallnau, L. B. (1983) The psychopharmacology of tonic immobility in chickens, in *The Brain and*

Behavior of the Fowl (Ookawa, R., ed.), Japan Scientific Societies Press, Tokyo, pp. 43–59.

Gentle, M. J., Jones, R. B., and Wooley, S. C. (1989) Physiological changes during tonic immobility in *Gallus gallus* var domesticus. *Physiol. Behav.* **46,** 843–847.

Gilman, T. T., Marcuse, F. L., and Moore, A. U. (1950) Animal hypnosis: a study in the induction of tonic immobility in chickens. *J. Comp. Physiol. Psychol.* **43,** 99–111.

Ginsburg, H. J. (1975) Defensive distance and immobility in young precocial birds (*Gallus gallus*). *Dev. Psychobiol.* **8,** 281–285.

Goleman, D. (1985) *Vital Lies, Simple Truths.* Simon and Schuster, New York.

Hatton, D. C. and Thompson, R. W. (1975) Termination of tonic immobility in chickens by auditory stimulation. *Bull. Psychon. Soc.* **5,** 61,62.

Hennig, C. W. (1978) Tonic immobility in the squirrel monkey (*Saimiri sciureus*). *Primates* **19,** 333–342.

Hennig, C. W. and Dunlap, W. P. (1977) Circadian rhythms of tonic immobility in the rat: evidence of an endogenous mechanism. *Anim. Learn. Behav.* **5,** 253–258.

Herzog, H. A., Jr. and Burghardt, G. M. (1974) Prey movement and predatory behavior of juvenile western yellow-bellied racers, *Coluber constrictor mormon. Herpetologica* **30,** 285–289.

Hicks, L. E., Maser, J. D., Gallup, G. G., Jr., and Edson, P. H. (1975) Possible serotonergic mediation of tonic immobility: effects of morphine and serotonin blockade. *Psychopharmacologia* **42,** 51–56.

Johnson, D. (1984) *Just in Case: A Passenger's Guide to Safety and Survival.* Plenum, New York.

Jones, R. B. (1986) Conspecific vocalisations, tonic immobility and fearfulness in the domestic fowl. *Behav. Proc.* **13,** 217–225.

Jones, R. B. and Faure, J. M. (1981) Sex and strain comparisons of tonic immobility ("righting time") in the domestic fowl and the effects of various methods of induction. *Behav. Proc.* **6,** 47–55.

Jones, R. B., Mills, A. D., and Faure, J. M. (1991) Genetic and experiential manipulation of fear-related behavior in japanese quail chicks (*Coturnix coturnix japonica*) *J. Comp. Psychol.* **105,** 15–24.

Klemm, W. R. (1965a) Drug potentiation of hypnotic restraint of rabbits, as indicated by behavior and brain electrical activity. *Lab. Anim. Care* **15,** 163–167.

Klemm, W. R. (1965b) Potentiation of animal "hypnosis" with low levels of electrical current. *Anim. Behav.* **13,** 571–574.

Klemm, W. R. (1966) A method to encourage extensive study of animal hypnotic behavior. *J. Exp. Anal. Behav.* **9,** 63,64.

Klemm, W. R. (1971) EEG and multiple-unit activity in limbic and motor systems during movement and immobility. *Physiol. Behav.* **7,** 337–343.

Klemm, W. R. (1977) Identity of sensory and motor systems that are critical to the immobility reflex ("animal hypnosis"). *Psychol. Rec.* **1,** 145–159.

Klemm, W. R. (1989) Drug effects on active immobility responses: what they tell us about neurotransmitter systems and motor functions. *Prog. Neurobiol.* **32,** 403–422.

Kulikov, A. V., Kozlachkova, E. Y., Maslova, G. B., and Popova, N. K. (1993) Inheritance of predisposition to catalepsy in mice. *Behav. Genet.* **23**, 379–384.

Kumazawa, T. (1963) "Deactivation" of the rabbit's brain by pressure application to the skin. *Electroencephalog. Clin. Neurophysiol.* **15**, 660–671.

Launay, F., Mills, A. D., and Faure, J. M. (1993) Effects of age, line and sex on tonic immobility responses and social reinstatement behaviour in Japanese quail *Coturnix japonica. Behav. Proc.* **29**, 1–16.

Lefebvre, L. and Sabourin, M. (1977a) Effects of spaced and massed repeated elicitation on tonic immobility in the goldfish (*Carassius auratus*). *Behav. Biol.* **21**, 300–305.

Lefebvre, L. and Sabourin, M. (1977b) Response differences in animal hypnosis: a hypothesis. *Psychol. Rec.* **1**, 77–87.

Liberson, W. T. (1948) Prolonged hypnotic states with "local signs" induced in guinea pigs. *Science* **108**, 40,41.

Maser, J. D., Klara, J. W., and Gallup, G. G., Jr. (1973) Archistriatal lesions enhance tonic immobility in the chicken (*Gallus gallus*). *Physiol. Behav.* **11**, 729–734.

Maser, J. D. and Gallup, G. G., Jr. (1977) Tonic immobility and related phenomena: a partially annotated, tricentennial bibliography, 1936 to 1976. *Psychol. Rec.* **27**, 177–217.Meyer, M. E. (1990) Dorsal pressure potentiates the duration of tonic immobility and catalepsy in rats. *Physiol. Behav.* **47**, 531–533.

Nash, R. F. and Gallup, G. G., Jr. (1975a) Aversiveness of the induction of tonic immobility in chickens (*Gallus gallus*). *J. Comp. Physiol. Psychol.* **88**, 935–939.

Nash, R. F. and Gallup, G. G., Jr. (1975b) Effect of different parameters of shock on tonic immobility. *Behav. Res. Meth. Instrum.* **7**, 361–364.

Nash, R. F. and Gallup, G. G., Jr. (1976) Habituation and tonic immobility in domestic chickens. *J. Comp. Physiol. Psychol.* **90**, 870–876.

Nash, R. F., Gallup, G. G., Jr., and Czech, D. A. (1976) Psychophysiological correlates of tonic immobility in the domestic chicken (*Gallus gallus*). *Physiol. Behav.* **17**, 413–418.

O'Brien, T. J. and Dunlap, W. P. (1975) Tonic immobility in the blue crab (*Callinectes sapidus, Rathbun*): its relation to threat of predation. *J. Comp. Physiol. Psychol.* **89**, 86–94.

Prestrude, A. M. and Crawford, F. T. (1970) Tonic immobility in the lizard, Iguana iguana. *Anim. Behav.* **18**, 391–395.

Rager, D. R. and Gallup, G. G., Jr. (1985) Tonic immobility in chickens (*Gallus gallus*): shock-termination thresholds. *J. Comp. Psychol.* **99**, 350–356.

Rager, D. R., Gallup, G. G., Jr., and Beckstead, J. W. (1986) Chlordiazepoxide and tonic immobility in chickens: a paradoxical enhancement. *Pharmacol. Biochem. Behav.* **25**, 1237–1243.

Ratner, S. C. (1958) Hypnotic reactions of rabbits. *Psycholog. Rep.* **4**, 209,210.

Ratner, S. C. (1967) Comparative aspects of hypnosis, in *Handbook of Clinical and Experimental Hypnosis* (Gordon, J. E., ed.), The Macmillan Company, New York, pp. 550–587.

Ratner, S. C. and Thompson, R. W. (1960) Immobility reactions (fear) of domestic fowl as a function of age and prior experiences. *Anim. Behav.* **8**, 186–191.

Rovee-Collier, C. K., Kaufman, L. W., and Farina, P. (1980) The critical cues for diurnal death feigning in young chicks: a functional analysis. *Am. J. Psychol.* **93**, 259–268.

Salzen, E. A. (1963) Imprinting and the immobility reactions of domestic fowl. *Anim. Behav.* **11**, 66–71.

Sargeant, A. B. and Eberhardt, L. E. (1975) Death feigning by ducks in response to predation by red foxes (*Vulpes fulva*). *Am. Midland Naturalist* **94**, 108–119.

Schwartz-Giblin, S., McEwen, B. S., and Pfaff, D. W. (1989) Mechanisms of female reproductive behavior, in *Psychoendocrinology* (Brush, F. R. and Levine, S., eds.), Academic, San Diego, pp. 41–104.

Schwarz, B. E. and Bickford, R. G. (1956) Encephalographic changes in animals under the influence of hypnosis. *J. Nervous Mental Dis.* **124**, 433–439.

Simonov, P. V. and Paikin, D. I. (1969) The role of emotional stress in the hypnotization of animals and man, in *Psychophysiological Mechanisms of Hypnosis* (Chertok, L., ed.), Springer-Verlag, New York, pp. 67–87.

Smith, G. W. and Klemm, W. R. (1977) The fear hypothesis revisited: other variables affecting duration of the immobility reflex (animal hypnosis) *Behav. Biol.* **36**, 751–758.

Stahlbaum, C. C., Rovee-Collier, C., Fagen, J. W., and Collier, G. (1986) Twi-light activity and antipredator behavior of young fowl housed in artifi-cial or natural light. *Physiol. Behav.* **36**, 751–758.

Suarez, S. D. and Gallup, G. G., Jr. (1979) Tonic immobility as a response to rape in humans: a theoretical note. *Psychol. Rec.* **29**, 315–320.

Svorad, D. (1957) Reticular activating system of brain stems and animal hypnosis. *Sci.* **125**, 156.

Teschke, E. J., Maser, J. D., and Gallup, G. G., Jr. (1975) Cortical involve-ment in tonic immobility ("Animal hypnosis"): effect of spreading corti-cal depression. *Behav. Biol.* **13**, 139–143.

Thompson, R. K. R. and Liebreich, M. (1987) Adult chicken alarm calls xenhance tonic immobility in chicks. *Behav. Proc.* **14**, 49–61.

Thompson, R. K. R., Foltin, R. W., Boylan, R. J., Sweet, A., Graves, C. A. and Lowitz, C. E. (1981) Tonic immobility in Japanese quail can reduce the prob-ability of sustained attack by cats. *Anim. Learn. Behav.* **9**, 145–149.

Tompkins, E. C. (1974) The use of the immobility reflex (animal hypnosis) as a possible procedure for detecting sedative activity. *Life Sci.* **15**, 671–684.

Whishaw, I. Q., Flannigan, K. P., and Barnsley, R. H. (1979a) Development of tonic immobility in the rabbit: relation to body temperature. *Dev. Psychobiol.* **12**, 595–605.

Whishaw, I. Q., Schallert, T., and Kolb, B. (1979b) The thermal control of immobility in developing infant rats: is the neocortex involved? *Physiol. Behav.* **23**, 757–762.

CHAPTER 3

Circadian Organization of Locomotor Activity in Mammals

Ralph E. Mistlberger

Introduction

Daily rhythms of form and function are ubiquitous across eukaryotic taxa, and are manifest at all levels of biological organization. In mammalian species, most behavioral and physiological systems display at least some evidence of a 24-h temporal structure, with daily maxima and minima exhibiting stable differences between systems (Szabo et al., 1978). Although some daily rhythms may be directly driven by exposure to environmental cycles of light and dark (LD), temperature, humidity, and so forth, most reflect an innate temporal program provided by internal 24-h biological clocks. Primary evidence for this is the observation that daily rhythms can persist ("free-run") indefinitely in constant environments devoid of 24-h time cues, displaying periodicities that are "circadian" (approximately a day) rather than precisely 24 h. Moreover, this circadian periodicity, although affected by various internal and external factors, is clearly genetically determined, as demonstrated by selective breeding studies (Bunning, 1973) and gene mutations (Hall and Rosbash, 1988).

Although a broad range of rhythmic variables has been measured in various species, for matters of convenience, much of what is known about mammalian circadian organization has

From: *Motor Activity and Movement Disorders*
P. R. Sanberg, K. P. Ossenkopp, M. Kavaliers, Eds. Humana Press Inc., Totowa, NJ

derived from studies of locomotor activity, primarily wheel-running, in a small number of nocturnal (night active) rodent species, principally hamsters, rats, mice and flying squirrels. Important contributions have also been made using diurnal (day active) mammals, particularly tree shrews (Meijer et al., 1990), squirrel monkeys, and humans (*see* Moore-Ede et al., 1982). For these few species, an extensive library of data has accumulated describing the formal properties and physiological mechanisms of their locomotor activity rhythms. For other common laboratory animals, such as cats (Randall et al., 1985) and guinea pigs (Kurumiya and Kawamura, 1988), relatively little work has been done, primarily because these species exhibit very weak or labile activity rhythms in captivity.

The goal of this chapter is to provide an overview of circadian rhythms of mammalian activity and of some of the rules by which these rhythms are influenced by photic and nonphotic time cues. A central concept to be elaborated is that the mammalian circadian system consists of multiple coupled oscillators that are subject to phase and period control by periodic environmental stimuli ("Zeitgebers") and by feedback from the organism's own activity and state of arousal. The following practical questions are addressed:

1. How are activity rhythms monitored and measured?
2. How are they synchronized to LD cycles and what factors determine rates of resynchronization to shifted LD cycles?
3. How are they affected by constant light and dark, daily feeding schedules, and acute or chronic exposure to arousing or stressful stimuli?
4. How are they influenced by measurement devices?
5. What differences might one expect to find as a function of sex, species, and ontogeny?

Circadian Rhythms in Constant Conditions

Figures 1–5 illustrate the circadian rhythms of wheel-running in typical male representatives of three rodent species, the rat (*Rattus norvegicus*; Wistar albino), the hamster (*Mesocricetus auratus*, Syrian or Golden) and the mouse (*Mus musculus*; DBA). The animals represented in Fig. 1, 4, and 5 were housed in stan-

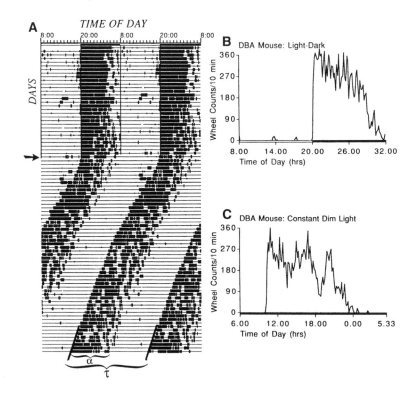

Fig. 1. Wheel-running activity of a mouse recorded under a 12:12 light–dark (LD) cycle, and in constant dim light (LLdim; <.1 lx, red). **(A)** Standard actogram format for plotting activity data. Each line represents 2 d, or 288 consecutive 10-min time bins plotted left to right. Time bins during which activity counts were recorded are represented by a vertical deflection; bins during which no activity occurred are represented by a point. Consecutive days are aligned left to right and top to bottom, i.e., the chart is double-plotted to facilitate inspection of non-24-h rhythms. Under LD, activity is largely restricted to the dark period and has a period of 24 h, reflecting entrainment to LD. Under LLdim, beginning on the day indicated by the arrow, the wheel-running rhythm persists (free-runs) with a period (τ) that is <24 h. τ can be determined by the slope of a regression line fit to activity onsets over a set of consecutive circadian cycles, as illustrated. Alpha (α) represents the daily active phase. **(B)** Waveform of wheel-running rhythm averaged over 7 d in LD. **(C)** Educed waveform of wheel-running averaged over seven circadian cycles, with the averaging interval set at τ of the free-running rhythm.

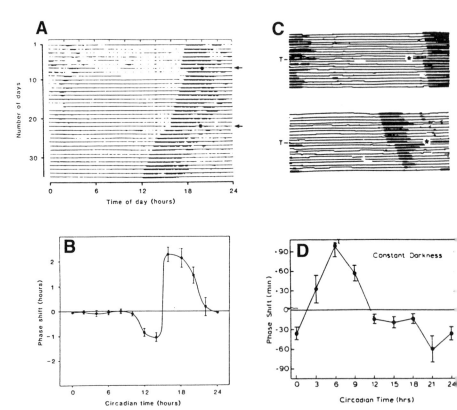

Fig. 2. **(A)** Event record of hamster wheel-running activity in constant darkness (DD). Each line represents 24 h, with time reading left to right and successive d aligned top to bottom. Each revolution of the running wheel caused a single vertical deflection of the event record pen, producing a solid dark line during continuous running. In this chart, the onset of continuous running began earlier each day, i.e., τ of the free-running rhythm was <24 h. A 15-min light pulse (marked by the stars on the day indicated by the first arrow), presented several hours after the beginning of the daily running phase, caused the rhythm to shift to a later time on subsequent days, i.e., "phase delay." The second light pulse, indicated by the other arrow, was presented near the end of the daily running phase and produced a "phase advance" shift, whereby activity began and ended earlier on subsequent days. **(B)** The phase-response curve for hamsters summarizes the relationship between the time at which a light pulse is presented and the direction and magnitude of the phase-shift produced. To construct this curve, τ of the circadian rhythm is first normalized to a 24-h time scale, referred to on the abscissa as circadian

dard Wahmann wheel cages with 34-cm activity wheels or wire-floored plastic cages with 17-cm activity wheels. Wheel revolutions detected by switch closures were monitored by Apple computer and displayed with 10-min resolution in a standard double-plotted raster format using commercially available software (Behavioral Cybernetics, Cambridge, MA) for the MacIntosh. This system represents a standard incarnation of modern microcomputer-controlled laboratories designed for continuous measurement of circadian rhythms in freely moving rodents. Similar systems are available from Mini-mitter Co. (Sunriver, OR) and Stanford Software Systems (Palo Alto, CA).

The individual records chosen illustrate species' typical rhythms for continuous recordings taken in LD, constant light (LL), and constant dark or dim (DD, usually < 1 lx). The period (duration of one daily cycle, designated "τ") and precision (standard deviation of activity onset times around the slope of the regression line through activity onsets on successive days; *see* Fig. 1A) of activity rhythms free-running in LL or DD exhibit continuous, normally distributed variation within strains (Pittendrigh and Daan, 1976a). Period and precision are correlated across species, with greater precision associated with average τ closer to 24 h (Pittendrigh and Daan, 1976a). Hamsters are generally found to have as a group the most precise and robust

---

time (CT). For nocturnal rodents, CT12 is, by convention, the time when running activity begins. When a light pulse is presented around CT12, a delay phase-shift of 1 h occurs on average. This is indicated on the ordinate by a –1. The largest delay shifts occur at about CT13–14. The largest advance shifts (positive sign) occur at about CT16–18. No significant shifts occur when light pulses are presented between CT0 and CT10, i.e., the hamsters' daily rest phase. (Reprinted with permission from Takahashii and Zatz [1982].) **(C)** Event records from two hamsters recorded in constant dark. At the time marked by the stars, the animals received 2.5 mg TRZ (ip injection). Injections during the inactive phase induced phase advances, whereas injections during the active phase usually induced phase delays, as summarized in the phase-response curve for the group **(D)**. TRZ also triggered a bout of wheel running lasting several hours. If the animal is prevented from running by confinement, no phase-shift occurs, indicating that behavioral effects of the drug may mediate the phase-shifting response.

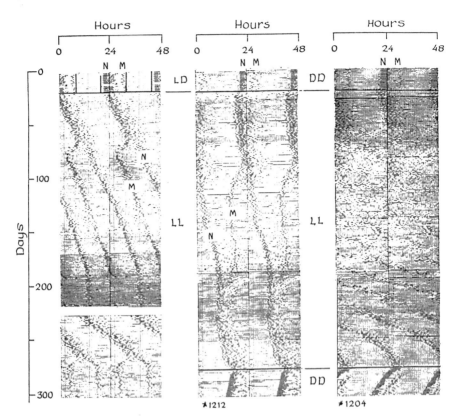

Fig. 3. Event records from three hamsters illustrating different forms of rhythm dissociation commonly observed in constant light (LL). **(Left)** This animal exhibited typical nocturnal running in an LD cycle, and free-ran with τ > 24 in LL (140 lx) until the rhythm "split" into two components (labeled N and M), which free-ran independently until coupling 180° out of phase. **(Middle)** The split components in the activity rhythm of this animal free-ran independently until recoupling at their original phase relation. **(Right)** Wheel-running in this animal became disorganized and aperiodic after about 30 d in LL. A long τ free-run eventually "nucleated" from the noise. The latter two cases also illustrate how τ is lengthened in LL as compared to constant dark (DD).

wheel-running rhythms; this, combined with a high threshold for arousal during their daily inactive phase, has made hamsters the most commonly used species in studies where circadian phase (for example, activity onset time) is the dependent variable.

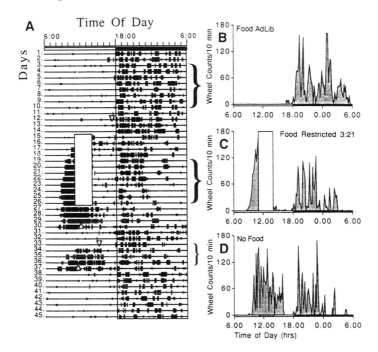

Fig. 4. (A) Actogram of wheel-running from a rat recorded under LD. Inverted triangles represents the beginning of total food deprivation. Triangle represent the end of food deprivation. The hollow vertical bar represents 3-h daily feeding time during food restriction. Activity is nocturnal during ad lib feeding and total food deprivation prior to food restriction. During food restriction, a prominent bout of running precedes mealtime each day; the timing of this bout is maintained during four subsequent days of total food deprivation. During a second deprivation test, the bout reappears. (B–D) Average waveforms of activity from the set of days indicated by the brackets on the right of (A).

The distribution of wheel-running within the daily active phase (α) also exhibits strain and individual variation; typically, mice (e.g., Fig. 1B) and hamsters free-run with wheel-running concentrated early in α, whereas wheel-running in rats is generally dispersed more evenly throughout α (e.g., Fig. 4B). In mice, this pattern has been shown to correlate with τ, with earlier concentrations of wheel activity associated with a shorter τ (Edgar et al., 1991). Some species, such as the meadow vole, exhibit particularly striking 2–5 h "ultradian" rhythms of

Time Of Day (hrs)

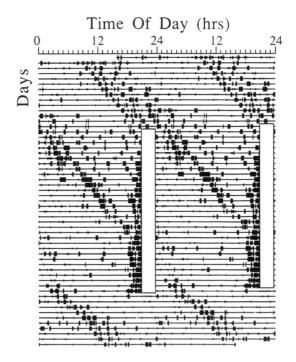

Fig. 5. Double-plotted actogram of wheel-running from a rat in constant dim light during ad lib and restricted feeding. Mealtime during restriction is indicated by the vertical hollow bar. A free-running rhythm with τ > 24 h is evident prior to food restriction. This rhythm persists during food restriction and appears to free-run with entraining to mealtime. A second daily bout of activity is evident in anticipation of mealtime each day.

activity within α (Rowsemitt, 1986). Ultradian rhythms of wheel-running are less prominent in mice, hamsters, and rats, although strain differences within species have been described (e.g., Wollnik et al., 1987).

Activity rhythms measured by running wheels are tightly synchronized with activity rhythms measured by other devices, such as tilt cages, photocells, food-bin monitors, drinkometers, and so forth; we know of no clear case in which circadian rhythms in different measures of activity become desynchronized and free-run with different τ. However, rhythm parameters other than τ may differ; for example, the amplitude of the wheel-running rhythm is usually greater than that of general cage activity (e.g., hamsters may show very weak circadian organization of feeding, drinking, and general cage activity despite showing very robust daily rhythms of wheel-running), and the onset of the daily active phase as measured by general cage activity may be advanced (occur earlier) relative to the onset as measured by wheel running. This phase difference varies systematically with τ (Aschoff et al., 1973).

Acute Effects of Light: Entrainment

Under natural photoperiods, circadian activity rhythms in visually intact animals do not free-run, but are synchronized to the LD cycle (Fig. 1A). This synchrony reflects phase and period control (i.e., entrainment) of the circadian pacemaker. Early in the daily active period of nocturnal rodents, light decelerates the motion of the pacemaker, whereas late in the active period, light accelerates the pacemaker. This circadian rhythm of pacemaker sensitivity to light can be represented graphically in the form of a phase–response curve (PRC; Fig. 2B; *see* Pittendrigh, 1981, for review). In a nocturnal rodent free-running in DD, exposure to a single light pulse (e.g., 15 min at 50 lx) early in α (in DD, α is designated the "subjective night," when the rodent would usually not see light) "phase delays" the activity rhythm (activity now begins and ends later on subsequent days), whereas exposure to an identical pulse of light late in the subjective night "phase advances" the rhythm (activity begins and ends earlier). During most of the subjective day (when nocturnal rodents are inactive), light has little or no effect on the pacemaker. The cumulative effect of these phase–dependent responses to light exposure under natural photoperiods is a daily phase adjustment that is sufficient to offset the difference between τ and the period of the LD cycle (designated T). The degree to which acute phase-shifts (discrete or nonparametric effects of light) or continuous velocity modulation (continuous or parametric effects of light) contribute to stable entrainment in various nocturnal and diurnal species is unknown, but discrete effects alone are sufficient in nocturnal rodents. Thus, rodents exhibit stable entrainment to "skeleton" photoperiods that consist of 2 daily light pulses separated by 6–18 h (Pittendrigh and Daan, 1976b). In a semi-natural enclosure, activity rhythms in flying squirrels can be entrained by as little as a few seconds of light exposure occurring at the same external clock time every few days (Decoursey, 1986).

The shape of the PRC to light, combined with the "circadian" value of τ, is adaptive in that it ensures stable entrainment of activity rhythms to LD, with an appropriate phase (e.g., activity beginning after dusk) that adjusts as needed to seasonal variations in the duration of the photoperiod (Pittendrigh and Daan, 1976c). PRC shape and τ also set the range of T values within which stable entrainment can occur, and determine the rate at which re-entrainment to a shifted LD cycle is accomplished. Since the maximal light-induced phase-shift in nocturnal rodents is, under standard conditions, on the order of a few hours, circadian rhythms can generally entrain to fixed LD cycles only in the range of about 22–26 h. This range can be extended somewhat if the LD cycle is lengthened or shortened gradually. At the upper and lower limits of this range, "relative coordination" may be observed, with the free-running rhythm alternately accelerating and decelerating as the lights-on portion of the LD cycle passes over photosensitive phases of the pacemaker (Pittendrigh and Daan, 1976b). When a LD cycle within the range of entrainment is acutely phase-shifted, re-entrainment occurs, usually in a series of daily phase-shifts ("transient" cycles), the direction and size of which depend on several factors, including τ, PRC shape, the size and direction of the LD shift, and the intensity of light (Aschoff et al., 1975). Also, re-entrainment rates may be affected by housing conditions; for example, wheel-running rhythms in hamsters re-entrained to an LD reversal in 6–9 d, whereas the drinking rhythm in hamsters housed without wheels required 25–40 d before the preshift light-to-dark drinking ratio was reestablished (Zucker and Stephan, 1973; *see* Feedback Effects of Activity on the Clock for further discussion of the effects of wheel-running on circadian timing).

In addition to pacemaker-mediated effects of light (e.g., entrainment), light can also exert direct effects on behavioral states. In nocturnal rodents, these so-called masking effects include acute or tonic suppression of activity by light (negative masking) and stimulation of activity by dark (positive masking; Aschoff et al., 1982). The waveform of activity rhythms under LD thus reflects both pacemaker-mediated and direct effects of LD (Fig. 1B,C).

Tonic Effects of Light

In years past, although much less so today, it was common to encounter laboratories in which LD schedules in animal colonies and test chambers were unknown, unregulated, or absent. For reasons of convenience, LL may appear to have some advantages in many studies. However, because τ varies across individuals, single-housed animals in LL free-run independently and do not retain the group synchrony observed under LD. Moreover, as light intensity increases, τ increases in nocturnal animals and decreases in most diurnal animals, thus hastening desynchrony of individual rhythms from external (laboratory) time (Aschoff, 1981). Presumably, this dependence of τ on light intensity reflects PRC shape; if the area under the delay portion of the PRC exceeds that under the advance portion, the pacemaker will necessarily undergo greater total daily deceleration than acceleration as the intensity of photic stimulation of photosensitive phases increases. Moreover, slowing of the pacemaker as light falls over its delay phase will prolong exposure to the delay phase. The duration of the daily active phase also changes systematically with τ, decreasing as τ increases.

Under prolonged LL, free-running rhythms are often disrupted. In hamsters, a common form of disruption is "splitting," whereby the daily activity rhythm divides (splits) into two components that free-run independently for a few (e.g., Fig. 3A) or many (e.g., Fig. 3B) cycles, usually stably recoupling at about 180° out of phase (Fig. 3A; Pittendrigh and Daan, 1976c). "Colliding" patterns have also been described, in which the end component of α delays relative to α onset, inducing a phase-shift or transient arrhythmia when it collides and fuses with α onset (Mrosovsky and Hallonquist, 1986). Instances of colliding-like phenomena have also been observed in mouse and squirrel monkey activity rhythms in LL (Mistlberger, unpublished observations).

Rats exposed to bright LL rarely show coherent splitting or colliding patterns; rather, there is usually a progressive decline in the amplitude of the activity rhythm until no significant circadian periodicity can be detected. The manner in which arrhythmia develops is suggestive of uncoupling within a population of oscillators; activity components appear to break away from α, a process dubbed "threading" (Eastman and Rechtschaffen, 1983), until no primary α remains. Mouse and hamster rhythms are also susceptible to arrhythmia under LL (e.g., Fig. 3C). Diurnal species, by contrast, do not usually show arrhythmia under LL, but may show splitting under DD or very dim LL (Meijer et al., 1990). None of these states are permanent; when LD is restored, so too is a unitary circadian organization.

Structure of the Circadian System: Mark I

These manifestations of circadian disorder under artificial lighting conditions reveal underlying structure in the mammalian circadian system. Splitting suggests that there are two circadian pacemakers with two stable coupling modes (0° and 180°). Other observations support this model (Pittendrigh and Daan, 1976c). Threading suggests a population of oscillators. The two models can be combined in the proposal that a master pacemaker (consisting of two coupled oscillators) gates and/ or couples the output of a population of secondary oscillators (Davis and Menaker, 1980; Rusak, 1989).

For primates, a different multioscillator organization has been proposed based on the observation in humans that temperature and sleep–wake rhythms may become uncoupled under prolonged temporal isolation, with the sleep–wake cycle slowing to 30–50 h durations, despite a 25-h free-running temperature cycle. This "spontaneous internal desynchronization" suggests that the human circadian system may consist of two pacemakers that control different sets of rhythms (Aschoff and Wever, 1976). However, the degree to which apparent desynchronization represents biological structure as opposed to procedural or volitional factors unique to human studies has been questioned (Eastman, 1982; Zulley and Campbell, 1985). Nonetheless, marked slowing of some components of internal timing during apparent desynchronization seems quite robust, as indicated by measures of subjective time estimation and meal frequency (Aschoff, 1985; Aschoff et al., 1986). To date, comparative studies have revealed at best questionable evidence for spontaneous internal desynchronization in squirrel monkeys (Sulzman et al., 1977) and no evidence in other species.

Nonphotic Effects on Circadian Rhythms

Restricted Feeding Schedules

Additional layers of complexity in the mammalian circadian system have been revealed by manipulations of nonphotic stimuli. Rats maintained on feeding schedules in which a restricted amount or duration of food is provided at the same time each day exhibit anticipatory wheel-running, food-bin activity, or unreinforced operant responding (e.g., lever pressing) initiated 1–3 h before mealtime (Fig. 4). This food anticipatory activity (FAA) appears analogous to the "time sense" by which forager bees associate time of day with food availability at particular places. However, traditional learning interpretations of rat FAA fall short of capturing its properties under various feeding challenges (Mistlberger, 1994). Rather, the available evidence, as summarized here, indicates that FAA represents the output of a food-entrainable circadian pacemaker that is separate from the master light-entrainable pacemaker that regulates day–night cycles of activity:

1. In DD or LL, the light-entrainable rhythm, in rats, free-runs with τ usually longer than 24 h. FAA, however, remains stably coupled to a fixed daily feeding recurring exactly every 24 h (Fig. 5; Boulos et al., 1980).
2. During total food deprivation, FAA recurs each day at its usual time, for up to five cycles (the limit of most food deprivation studies; Fig. 4; Coleman et al., 1982; Rosenwasser et al., 1984).
3. If feeding is shifted to a new time, FAA often resynchronizes in a series of transient cycles, taking several days (Stephan, 1984).
4. If the feeding cycle differs appreciably from 24 h, no FAA develops; the "limits to anticipation" are in the 22–31 h (i.e., circadian) range (Stephan, 1981).
5. Feeding cycles at the limits to anticipation may result in free run of the FAA component, with τ around 25–27 h in the few cases reported (e.g., Stephan, 1981).

These and other observations (for review, see Mistlberger, 1994) indicate that FAA represents a clock-controlled bout of activity that is entrained by feeding time and not by light. This interpretation is supported by transplant (Lehman et al., 1987) and in vitro electrophysiology studies (Gillette, 1991), which demonstrate that the master circadian pacemaker mediating photic entrainment resides in the hypothalamic suprachiasmatic nuclei (SCN). Ablation of the SCN eliminates or disrupts light-entrainable, free-running rhythms but does not disturb FAA; indeed, many of the properties of FAA reviewed above were demonstrated in SCN-ablated rats that were arrhythmic prior to restricted feeding. In vivo electrophysiology studies show that circadian rhythms of neural activity in the SCN of rats are not affected by scheduled feeding (Inouye, 1982). Neural sites critical for FAA are unknown (see Mistlberger, 1994).

When restricted feeding schedules are replaced by ad lib food access, the daily rhythm of FAA usually disappears within a few days. However, if the rat is subsequently food deprived, the daily bout of FAA reappears, often very close to the time of day or circadian phase at which it originally occurred (Fig. 4; Coleman et al., 1982; Rosenwasser et al., 1984). This has been observed in intact and SCN-ablated rats, although the temporal precision appears to be greater in intact rats. When FAA reappears, activity at other phases is often attenuated or even absent. These observations suggest a multioscillator organization in which two anatomically separate circadian pacemakers mediate photic and food entrainment, respectively. The two pacemakers appear to be mutually, but weakly coupled, and can be conceived as competing for expression in behavior, with the photically entrainable pacemaker dominant under ad lib feeding, and the food-entrainable pacemaker dominant when food is absent or restricted.

The photically-entrainable pacemaker appears to have a compound organization, as indicated by observations of splitting in LL. The food-entrainable pacemaker may also have a compound organization, since intact rats exhibit FAA to two meals each day, provided that the intermeal interval exceeds about 5 h (optimally, 7 h or more; Stephan, 1989a). Moreover, SCN-ablated rats can anticipate 2 meals/d even when the meals are presented on different T cycles, e.g., meals recurring once every 25 and 26 h (Stephan, 1989b). Intact and SCN-ablated rats do not anticipate 3 meals/d; under such schedules, FAA is usually expressed to at most two of the meals, even if all intermeal intervals are identical, and may "jump" or shift in a series of transient cycles from one meal to another (Stephan, 1989a). These responses to multiple meals suggest that the food-entrainable pacemaker can "split" into a maximum of two oscillatory components that can couple independently to separate feeding events for at least short durations. A formal description of the PRC for feeding zeitgebers is not yet available, so it is not yet clear how two oscillators can couple to separate meals in the face of competing coupling forces from the other meal and the other oscillator.

Although traditional associative or computational learning processes appear inadequate to account for many of the oscillatory properties of FAA, such processes must play a role in time–place learning. Rats, like bees, can learn to associate particular feeding places with particular times of food availability (Boulos and Logothetis, 1990). For example, in an open-field 2 meal/d paradigm, rats can exhibit temporally appropriate place preferences conditioned by pairing each of two feeding places with a unique feeding time (Mistlberger, unpublished observations). Locomotor activity under restricted feeding schedules may thus reflect a memory for feeding time and place combined with circadian pacemaker entrainment.

Locomotor rhythms in other species are also affected by restricted feeding schedules. In some species, including C57BL/6J mice (Abe et al., 1989) and two marsupial carnivores (O'Reilly et al., 1986; Kennedy et al., 1990), restricted feeding appears to entrain a food-sensitive pacemaker that is readily dissociated from the light-entrainable pacemaker. In these animals, as in rats, LD-entrained activity retains its nocturnal phase even if feeding time is restricted to the daytime, and free-runs independently of feeding time in LD or DD. In a minority of cases, the free-running rhythm may entrain to feeding time; in rats, for example, this occurs in <20% of individuals tested. Such entrainment could be mediated by coupling between the food- and light-entrainable pacemakers, or perhaps might reflect an entraining influence of FAA *per se* on the light-entrainable, free-running pacemaker (*see* Feedback Effects of Activity).

Other species, such as the CS mouse (Abe et al., 1989) and Syrian hamster (Abe and Rusak, 1992; Mistlberger, 1993b), generally show little evidence of independent food- and light-entrainable oscillators. In these animals, activity rhythms are more likely to either ignore or entrain completely to feeding time, rather than dissociate into separate free-running and FAA components. However, in intact hamsters, instances of FAA concurrent with a free-running rhythm have occasionally been observed (Abe and Rusak, 1992). Moreover, FAA is not disrupted by SCN ablation, and exhibits both persistence during deprivation and circadian limits to entrainment (Abe and Rusak, 1992; Mistlberger, 1992a, 1993c). Hamsters thus also appear to possess a multioscillator circadian organization involving separate food- and light-entrainable pacemakers. Species differences in intact animals may thus reflect differences not in the pacemaker components of the circadian system, but in the coupling strengths among these components. Weak coupling would permit rapid dissociation into separate food- and light-entrained rhythms when the animal is exposed simultaneously to both zeitgebers, whereas strong coupling would tend to prevent such dissociation. In the latter case, either light or food serves as a dominant zeitgeber. In hamsters, light is clearly dominant (Mistlberger, 1993b), whereas in the rabbit, for example, food would appear to be dominant in many instances (Jilge and Stähle, 1993).

The food-entrainable component of the mammalian circadian system is clearly adaptive in providing a mechanism for coordinating foraging behavior with temporal windows of food availability. This coordination also extends to metabolic function; exocrine and endocrine secretions that regulate food digestion, absorption, and distribution all exhibit circadian rhythms synchronized with and in many cases anticipatory to feeding time (reviewed in Boulos and Terman, 1980).

Restriction of Food and Other Resources: Effects of Measurement Device and Procedure

FAA detected in wheel running and food-bin measures of activity usually develops rapidly and may be evident by the second day of food restriction, although as a rule a single feeding event is usually not sufficient to establish or shift FAA

(Stephan, 1992). FAA usually increases in duration until stabilizing within 7–14 d. Other measures of activity may be less useful for detecting FAA. For example, tilt-cage, photobeam, or "animex" measures of general cage activity may increase, remain unchanged, or decrease prior to daily feedings (e.g., Nagai et al., 1982); this may be contingent on the availability of competing activities, such as running in a wheel or waiting at a food access window (place preference), since in some studies we have observed significant levels of anticipatory locomotion in tilt-floor cages lacking food hoppers or wheels. In one instructive instance, we found in some rats that ablation of the hypothalamic paraventricular nucleus eliminated anticipation in general cage activity, but not in food-bin activity (Mistlberger and Rusak, 1988).

Rats may exhibit anticipatory activity to resources other than food, but this appears to be strikingly dependent on procedural variables. Several early studies reported that rats can anticipate a daily opportunity to drink water, but these studies were confounded by unregulated food intake, since rats (and hamsters) take large meals during water access. When this confound was controlled, little anticipation was observed in general cage activity (Mistlberger and Rechtschaffen, 1985) and none at all in wheel-running (Mistlberger, 1992b). However, in some rats, reasonable levels of anticipation were evident in activity directed at a water access bin (Mistlberger, 1992b). Similar results were reported for salt-hungry rats; anticipation of a daily salt access period was absent in wheel-running, but evident, at least under some conditions, in unreinforced lever pressing for salt (Rosenwasser et al., 1985, 1988).

This latter study identified another procedural factor that may affect anticipatory activity. When salt access time was signaled by external cues, no anticipatory lever pressing was evident. A similar result was obtained previously in a study of pigeons (Abe and Sugimoto, 1987). When food access was signaled by a light, anticipatory pecking at a food key was absent. When the light was then left on continuously, temporally precise anticipatory pecking was evident the next day. This suggests that some animals may attend to discriminative stimuli and use these in lieu of internal cues to ensure that daily feeding opportunities are exploited. However, a thorough analysis of interactions between such external cues and spontaneous FAA in rats showed that although the rise in anticipatory lever pressing before each meal can be delayed by predictive cues in the environment, in no case was anticipation eliminated (Terman et al., 1984). This is consistent with the fact that food anticipation in rats is observed in all laboratories despite the widespread use of food delivery systems that signal food access by auditory, visual, and olfactory stimuli, including those associated with manual delivery of food cups. Thus, in food-restricted rats, anticipatory activity programs manifest as behaviors, such as wheel-running, appear relatively unaffected by the availability of discriminative environmental cues.

Feedback Effects of Activity on the Clock: Circadian Structure Mark II

The traditional model of the circadian system can be conceived as a top-down hierarchy with master pacemakers coordinating or gating secondary or slave oscillators, which in turn directly drive or modulate effector systems for specific behavioral and physiological rhythms. At each level of the hierarchy, mutual coupling interactions are permissible, for example between compound pacemakers for food and light entrainment, or among a population of secondary oscillators. Bottom-up (i.e., feedback) coupling has not been a feature in this scheme, because phase control of the clock by effector systems for behavior might result in repeated phase-shifts caused by unpredictable, arousing daily events. This might be incompatible with stable photic entrainment. In principle, the gears of the clock should drive the hands of the clock (e.g., wheel-running behavior), and not vice versa.

However, as early as 1960, Aschoff noted that the positive correlation between light intensity and τ could be related to the "level of excitement" (or activity) of the animal, since light intensity also directly affects activity levels. This suggestion that activity level (a hand of the clock) could affect pacemaker motion was hinted at empirically in Aschoff's subsequent report that hamsters with access to running wheels exhibit longer free-running τ's than hamsters without wheels (Aschoff et al., 1973). A thorough examination of this issue did not begin for another decade. First, Yamada et al. (1986) reported that free-running τ

in rats was shortened by free access to a running wheel. Mrosovsky then demonstrated in hamsters that a bout of wheel activity, stimulated by cage cleaning, exposure to a conspecific, or transfer to a novel wheel cage, could phase shift free-running rhythms of home cage wheel activity (Mrosovsky, 1988). The PRC obtained was roughly a mirror image of that for light, with phase advances in response to activity induced in the middle of the subjective day (when the hamster is usually inactive) and small phase delays in response to activity induced during the latter half of the subjective night (Fig. 2C,D). Consistent with this PRC, free-running rhythms in hamsters can be entrained by a scheduled daily bout of activity induced by transfer to a novel running wheel (Mrosovsky, 1988) or by restricted access to a seminatural foraging area containing seeds (Rusak et al., 1988). Subsequent studies in mice have yielded similar results; if the home cage activity wheel is disabled for 12–18 h each day, C57 mice exhibit robust spontaneous wheel running during the 6–12 h period of wheel availability. This scheduled yet voluntary activity entrains free-running rhythms of sleep–wake and drinking (Edgar and Dement, 1992).

The implications of this apparent feedback of activity to the pacemaker are wide-ranging. It suggests that any event or manipulation that even transiently alters activity or arousal levels may acutely alter the motion of the clock, possibly changing its free-running phase and period, or its phase of entrainment to some other zeitgeber, such as LD. It also suggests that locomotor activity (or "exercise") may be a useful clinical or experimental tool for manipulating the phase of mammalian rhythms. These implications have already received strong empirical confirmation, as reviewed below, in just a few years of work.

A variety of drugs and hormones are known to alter circadian phase and period. Many of these also affect behavioral states directly. For example, benzodiazepines (BZDs), including triazolam (TRZ), midazolam, diazepam, and chlordiazepoxide (CDZ), can phase-shift activity rhythms of hamsters free-running in DD or LL, with a PRC essentially the same as that for novel cage-induced activity (Fig. 2C,D; Turek and Losee-Olson, 1986). TRZ, the first of these to be examined thoroughly, also stimulates activity in hamsters at doses that induce phase-shifts. If activity is prevented by restraint (Van Reeth and Turek,

1989) or confinement to a small nest box for several hours after drug administration (Mrosovsky and Salmon, 1990), no phase shift occurs. Restraint alone has no effect on circadian phase, unless it occurs during the first 1–2 h of the subjective night, at which time it can produce small phase delays (Van Reeth et al., 1991). The results indicate that, at least for TRZ, phase-shifts are mediated by behavioral effects of the drug, rather than a direct action of the drug on the pacemaker. The same may not be true of all BZDs, since there is evidence that the phase-shifting action of CDZ, which is essentially the same as TRZ, is not dependent on induction of activity (Biello and Mrosovsky, 1992). Also, in squirrel monkeys, TRZ injections induce sedation at doses that cause phase shifts (Mistlberger et al., 1991).

Hormones that alter the phase or period of circadian rhythms include testosterone and estrogen, both of which affect activity levels. In female rodents, activity is increased and its onset phase advanced on the day of proestrous, when estradial secretion is maximal (Zucker, 1979). Ovariectomy (OVX) results in reduced levels of wheel-running and a delay in the entrained phase of activity rhythms in LD. Chronic estradial treatment increases wheel-running, advances the phase of LD entrainment, and shortens free-running τ (Gerall et al., 1973; Morin et al., 1977) In DD, OVX rats exhibit a longer τ than intact rats, but only in populations of animals with access to running wheels (Ruis de Elvira et al., 1992). These results suggest that estradial influences pacemaker period by virtue of its effects on activity levels. Testosterone may alter τ of male rats by a similar behavioral mechanism; τ is lengthened by castration and shortened by chronic testosterone treatment, and activity levels correlate with testosterone level (Daan et al., 1975).

Most of the work on activity feedback effects on the circadian system has been done using animals housed in LL or DD. However, a few studies have demonstrated that these effects are sufficiently robust to alter the phase of entrainment under natural photoperiods. In photically entrained hamsters, the phase of activity onset can be delayed by up to 2 h by 3 h of novel cage-induced wheel-running scheduled each day at the end of the dark period (Mistlberger, 1991a). This is consistent with predictions based on the PRC to novel cage-induced activity, since phase delays are normally observed at the end of the subjective night in free-running hamsters. Another prediction from the PRC is that phase advances of photically entrained activity onset would be produced by novelty-induced running scheduled in the middle of the light period (*see* Fig. 2D). However, instead, more complex responses have been observed that suggest that a daily bout of novelty-induced activity can, given appropriate timing, split the photically entrained activity rhythm, advancing one component toward the scheduled activity time and phase delaying the other component (Mrosovsky and Janik, 1993; Mistlberger, unpublished observations).

There is some evidence that photic and nonphotic zeitgebers combine arithmetically in their effects on circadian phase. In one study, an 8-h permanent advance of the LD cycle (dark onset now 8 h earlier), coupled with a 3-h bout of novelty-induced wheel running beginning at the new time of dark onset on the first day of the LD shift, resulted in greatly accelerated re-entrainment; hamsters receiving the "exercise" treatment adjusted to the new LD cycle in just 1.6 d, compared to about 8.5 d in control hamsters left undisturbed during the LD shift (Mrosovsky and Salmon, 1987). Subsequent analysis of the contributions of photic and nonphotic cues suggested a simple additive effect, although the possibility of synergism between the two zeitgebers was not ruled out (Reebs and Mrosovsky, 1989).

More recent data, however, suggest that, in some cases, photic and nonphotic zeitgebers exhibit nonlinear interactions; wheel activity induced at certain phases of the subjective night can attenuate the phase-shifting effects of concurrent light exposure, while having little phase-shifting effect on its own (Ralph and Mrosovsky, 1992). More dramatic effects were previously reported in an arthopod; scorpions that were active during a light pulse showed a shift consistent with a nonphotic zeitgeber, whereas those that were inactive showed shifts consistent with a photic zeitgeber (Hohmann et al., 1990). Behavioral activity, or some correlate, may block or modulate the effects of light on the clock. More work is needed to substantiate and understand this surprising effect better.

Novelty-induced wheel-running has to date been the primary probe used to explore feedback effects on the clock. However, it is not clear that wheel running or nonspecific arousal *per se* is sufficient for such effects to occur. Dose–response studies indicate that hamsters must run about 5000 wheel revolutions in 3 h to undergo a maximal phase shift of their free-running rhythms (Janik and Mrosovsky, 1993). However, some hamsters induced to run twice this amount by cold challenge did not exhibit phase-shifts (Janik and Mrosovsky, 1993). There are other reports that amphetamine-induced activity does not shift the clock in mice (Edgar, personal communication), and that a daily scheduled mealtime, which induces anticipatory activity, a strong feeding response, and obvious arousal, does not mimic novelty-induced or forced running in its effects on photically entrained rhythms in hamsters and rats (Mistlberger, 1991a,b). These studies suggest that all manifestations of arousal and activity are not equivalent in their effects on the circadian system, but the critical underlying (motivational?) variables remain to be clarified. Possible species differences also need to be explored; for example, forced treadmill running or restricted daily feeding schedules can entrain free-running rhythms in rats, but by this and other measures, feedback effects appear to be weaker in rats than in hamsters or mice (Mistlberger, 1991b).

Physiological mechanisms of activity feedback to the circadian pacemaker also remain to be clarified. Neither adrenalectomy nor naloxone pretreatments, which block components of sympathetic arousal, alter nonphotic phase-shifts induced by TRZ injections (Van Reeth and Turek, cited in Van Reeth et al., 1991). Increases in body temperature that accompany intense wheel running also do not appear to be a mediating variable (Wickland and Turek, 1991).

One set of recent studies indicate that nonphotic phase–shifts in hamsters are mediated by neuropeptide Y release from a thalamic intergeniculate leaflet projection to the SCN pacemaker (Biello and Mrosovsky, 1994; Janik and Mrosovsky, 1994). Another set of studies suggest that nonphotic entrainment in mice and suppression of photic phase-shifts by concurrent behavior may be dependent on serotonin release from a raphe projection to the SCN (Edgar, personal communication; Rea et al., 1994). Species and paradigm generality of these mechanisms remain to be established.

Other Nonphotic Stimuli

Daily temperature cycles can entrain activity rhythms in many poikilothermic and some homeothermic species, including the pocket mouse (Lindberg and Hayden, 1974), Long Evans rat (Francis and Coleman, 1988), and pig-tailed macaques (Tokura and Aschoff, 1983). However, in these homeothermic species the effective temperature range is generally quite large. Entrainment may be mediated by behavioral responses to heat and cold.

Social interactions also may entrain activity rhythms. Group-housed animals can exhibit group synchrony in DD, whereas single-housed animals in adjacent cages do not. Brief interactions, which in some cases are associated with fighting, can phase-shift or entrain free-running rhythms in hamsters (Mrosovsky, 1988), although in other studies, these social stimuli were inneffective. Cohousing two hamsters, for example, did not result in mutual synchrony of their activity rhythms (Refinetti et al., 1992). Social interactions involve obvious changes in arousal and activity, but whether these changes constitute the entrainment signal to a nonphotically-entrainable pacemaker remains to be determined.

Life-Span Changes in Circadian Activity Rhythms

Circadian activity rhythms in rodents are evident from weaning to old age, but show prominent changes across the life-span, including a progressive shortening of τ, an advance of activity onset in LD, and a reduction in amplitude (reviewed in Bliwise, 1993). Overall levels of activity also decline, which may play a causal role in the period and amplitude changes noted. Similar life-span changes characterize human circadian function.

Recent studies indicate that the circadian system in aged hamsters also shows changes in its responses to photic and nonphotic stimuli. For example, aged hamsters show markedly enhanced phase-shifts in response to light pulses at certain phases (Rosenberg et al., 1991), but reduced phase-shifts to nonphotic events (Van Reeth et al., 1992), although this latter result may reflect reduced behavioral arousal to such events, rather than any decline in the sensitivity of the pacemaker to behavioral arousal (Mrosovsky and Biello, 1994). In aged rats, FAA is slow to emerge and reduced in amplitude (Mistlberger et al., 1990).

Conclusion

Recent years have witnessed substantial progress toward comprehensive formal and physiological explanations of daily biological rhythms (cf. Menaker, 1993). Beneficiaries include not only the circadian biologists immersed in this task, but all students of behavior who must deal with time dependencies in the expression and responsivity of activity in the laboratory, the field, or the clinic. An appreciation of these time dependencies is obviously crucial for optimal experimental design in the behavioral and neural sciences (cf., Mistlberger, 1990).

Acknowledgments

This work was supported by the National Science and Engineering Research Council, Canada.

References

Abe, H. and Rusak, B. (1992) Anticipatory activity and entrainment of circadian rhythms in Syrian hamsters exposed to restricted palatable diets. *Am. J. Physiol.* **263**, R116–R124.

Abe, H. and Sugimoto, S. (1987) Food anticipatory response to restricted food access based on the pigeon's biological clock. *Anim. Learning Behav.* **15**, 353–359.

Abe, H., Kida, M., Tsuji, K., and Mano, T. (1989) Feeding cycles entrain circadian rhythms of locomotor activity in CS mice but not in C57BL/6J mice. *Physiol. Behav.* **45**, 397–401.

Aschoff, J. (1960) Exogenous and endogenous components in circadian rhythms. *Cold Spring Harbor Symp. Quant Biol.* **25**, 11–28.

Aschoff, J. (1981) Freerunning and entrained circadian rhythms, in *Handbook of Behavioral Neurobiology. IV. Biological Rhythms* (Aschoff, J., ed.), Plenum, New York, pp. 81–93.

Aschoff, J. (1985) On the perception of time during prolonged temporal isolation. *Hum. Neurobiol.* **4**, 41–52.

Aschoff, J. and Wever, R. (1976) Human circadian rhythms: a multioscillator system. *Fed. Proc.* **35**, 2326–2332.

Aschoff, J., Figala, J., and Poppel, E. (1973) Circadian rhythms of locomotor activity in the golden hamster measured with two different techniques. *J. Comp. Physiol. Psych.* **85**, 20–28.

Aschoff, J., Hoffman, K., Pohl, H., and Wever, R. (1975) Re-entrainment of circadian rhythms after phase–shifts of the zeitgeber. *Chronobiologia* **2**, 23–78.

Aschoff, J., Daan, S., and Honma, K. I. (1982) Zeitgebers, entrainment and masking: some unsettled questions, in *Vertebrate Circadian Systems* (Aschoff, J., Daan, S., and Groos, G. A., eds.), Springer-Verlag, New York, pp. 13–24.

Aschoff, J., von Goetz, C., Wildgruber, C., and Wever, R. A. (1986) Meal timing in humans during temporal isolation without time cues. *J. Biol. Rhythms* **1(2)**, 151–162.

Biello, S. and Mrosovsky, N. (1992) Nest box restriction does not block phase advances to the benzodiazepine chlordiazepoxide at CT5. *Soc. Res. Biol. Rhythms Abstract* **3**, 169.

Biello, S. M., Janik, D., and Mrosovsky, N. (1994) Neuropeptide Y and behaviorally induced phase-shifts. *Neuroscience*, in press.

Bliwise, D. (1993) Sleep in normal aging and dementia. *Sleep* **16**, 40–81.

Boulos, Z. and Logothetis, D. E. (1990) Rats anticipate and discriminate between two daily feeding times. *Physiol. Behav.* **48**, 523–529.

Boulos, Z. and Terman, M. (1980) Food availability and daily biological rhythms. *Neurosci. Biobehav. Rev.* **4**, 119–131.

Boulos, Z., Rosenwasser, A. M., and Terman, M. (1980) Feeding schedules and the circadian organization of behaviour in the rat. *Behav. Brain Res.* **1**, 39–65.

Bunning, E. (1973) *The Physiological Clock.* Springer-Verlag, New York.

Coleman, G. J., Harper, S., Clarke, J. D., and Armstrong, S. (1982) Evidence for a separate meal-associated oscillator in the rat. *Physiol. Behav.* **29**, 107–115.

Daan, S., Damassa, D., Pittendrigh, C. S., and Smith, E. R. (1975) An effect of castration and testosterone replacement on circadian pacemaker in mice. *Proc. Natl. Acad. Sci. USA* **72**, 3744–3747.

Davis, F. C. and Menaker, M. (1980) Hamsters through time's window: temporal structure of hamster locomotor rhythmicity. *Am. J. Physiol.* **239**, R149–R155.

DeCoursey, P. J. (1986) Light-sampling behavior in photoentrainment of a rodent circadian rhythm. *J. Comp. Physiol.* **A159**, 161–169.

Eastman, C. I. (1982) The phase–shift model of spontaneous internal desynchronization, in *Vertebrate Circadian Systems* (Aschoff, J., Daan, S., and Groos, G. A., eds.), Springer-Verlag, New York, pp. 262–267.

Eastman, C. and Rechtschaffen, A. (1983) Circadian temperature and wake rhythms of rats exposed to prolonged continuous illumination. *Physiol. Behav.* **31**, 417–427.

Edgar, D. M. and Dement, W. C. (1992) Regularly scheduled voluntary exercise synchronizes the mouse circadian clock. *Am. J. Physiol.* **261**, R928–R933.

Edgar, D. M., Martin, C. E., and Dement, W. C. (1991) Activity feedback to the mammalian circadian pacemaker: influence on observed measures of rhythm period length. *J. Biol. Rhythms* **6**, 185–199.

Francis, A. J. P. and Coleman, G. J. (1988) The effect of ambient temperature cycles upon circadian running and drinking activity in male and female laboratory rats. *Physiol. Behav.* **43**, 471–477.

Gerall, A. A., Napoli, A. M., and Cooper, U. C. (1973) Daily and hourly estrous running in intact, spayed and estrone implanted rats. *Physiol. Behav.* **10**, 225–229.

Gillette, M. U. (1991) SCN electrophysiology in vitro, rhythmic activity and endogenous clock properties, in *Suprachiasmatic Nucleus: The Mind's Clock* (Klein, D. C., Moore, R. Y., and Reppert, S. M., eds.), Oxford University, New York, pp. 125–143.

Hall, J. C. and Rosbash, M. (1988) Mutations and molecules influencing biological rhythms. *Annu. Rev. Neurosci.* **11**, 373–393.

Hohmann, W., Michel, S., and Fleisnner, G. (1990) Locomotor activity and light pulses as competing zeitgeber stimuli in the scorpion circadian system. *Soc. Res. Biol. Rhythms Abstract* **2**, 52.

Inouye, S. T. (1982) Restricted daily feeding does not entrain circadian rhythms of the suprachiasmatic nucleus in the rat. *Brain Res.* **232**, 193–199.

Janik, D. and Mrosovsky, N. (1993) Nonphotically induced phase-shifts of circadian rhythms in the golden hamster: activity response curves at different ambient temperatures. *Physiol. Behav.* **53**, 431–436.

Janik, D. and Mrosovsky, N. (1994) Intergeniculate leaflet lesions and behaviorally-induced shifts of circadian rhythms. *Brain Res.*, in press.

Jilge, B. and Stähle, H. (1993) Restricted food access and light-dark: impact of conficting zeitgebers on circadian rhythms of the rabbit. *Am. J. Physiol.* **264**, R708–R715.

Kennedy, G. A., Coleman, G. J., and Armstrong, S. M. (1990) The effect of restricted feeding on the wheel-running activity rhythms of the predatory marsupial Dasyurus viverrinus. *J. Comp. Physiol.* **166**, 607–618.

Kurumiya, S. and Kawamura, H. (1988) Circadian oscillation of the multiple unit activity in the guinea pig suprachiasmatic nucleus. *J. Comp. Physiol.* **A162**, 301–308.

Lindberg, R. G. and Hayden, P. (1974) Thermoperiodic entrainment of arousal from torpor in the little pocket mouse, *Perognathus longimembris. Chronobiologia* **1**, 356–361.

Meijer, J. H., Daan, S., Overkamp, G. F. J., and Herman, P. M. (1990) The two-oscillator circadian system of tree shrews (Tupaia belangeri) and its response to light and dark pulses. *J. Biol. Rhythms* **5**, 1–16.

Menaker, M. (1993) Special topic: circadian rhythms. *Annu. Rev. Physiol.* **55**, 657–659.

Mistlberger, R. E. (1990) Circadian pitfalls in experimental designs employing food restriction. *Psychobiology* **18**, 23–29.

Mistlberger, R. E. (1991a) Scheduled daily exercise or feeding alters the phase of photic entrainment in Syrian hamsters. *Physiol. Behav.* **50**, 1257–1260.

Mistlberger, R. E. (1991b) Effects of daily schedules of forced activity on free-running rhythms in the rat. *J. Biol. Rhythms* **6**, 71–80.

Mistlberger, R. E. (1992a) Non-photic entrainment of circadian activity rhythms in suprachiasmatic nuclei ablated hamsters. *Behav. Neurosci.* **102(1)**, 192–202.

Mistlberger, R. E. (1992b) Anticipatory activity rhythms under daily schedules of water access in the rat. *J. Biol. Rhythms* **7**, 149–160.

Mistlberger, R. E. (1993a) Circadian properties of anticipatory activity to restricted water access in suprachiasmatic-ablated hamsters. *Am. J. Physiol.* **264**, R22–R29.

Mistlberger, R. E. (1993b) Effects of scheduled food and water access on circadian rhythms of hamsters in constant light, dark, and light-dark. *Physiol. Behav.* **53**, 509–516.

Mistlberger, R. E. (1994) Circadian food-anticipatory activity: formal models and physiological mechanisms. *Neurosci. Biobehav. Rev.* **18**, 171–195.

Mistlberger, R. E. and Rechtschaffen, A. (1985) Periodic water availability is not a potent zeitgeber for entrainment of circadian locomotor rhythms in rats. *Physiol. Behav.* **334**, 17–22.

Mistlberger, R. E. and Rusak, B. (1988) Food anticipatory circadian rhythms in rats with paraventricular and lateral hypothalamic lesions. *J. Biol. Rhythms* **3**, 277–291.

Mistlberger, R. E., Houpt, T. A., and Moore-Ede, M. C. (1990) The effects of aging on food-entrained circadian rhythms in the rat. *Neurobiol. Aging* **11**, 619–624.

Mistlberger, R. E., Houpt, T. A., and Moore-Ede, M. C (1991) The benzodiazepine triazolam phase–shifts circadian activity rhythms in a diurnal primate, the squirrel monkey *(Saimiri sciureus)*. *Neurosci. Lett.* **124**, 27–30.

Moore-Ede, M. C., Sulzman, F. M., and Fuller, C. A. (1982) *The Clocks that Time Us*. Harvard University Press, Cambridge, MA.

Morin, L. P., Fitzgerald, K. M., and Zucker, I. (1977) Estradiol shortens the period of hamster circadian rhythms. *Science* **196**, 305–307.

Mrosovsky, N. (1988) Phase response curves for social entrainment. *J. Comp. Physiol.* **A162**, 35–46.

Mrosovsky, N. and Biello, S. M. (1994) Nonphotic phase-shifting in the old and the cold. *Chronobiol. Int.* **11**, 232–252.

Mrosovsky, N. and Hallonquist, J. D. (1986) Colliding of activity onset and offset: evidence for multiple circadian oscillators. *J. Comp. Physiol.* **A159**, 187–190.

Mrosovsky, N. and Janik, D. (1993) Behavioral decoupling of circadian rhythms. *J. Biol. Rhythms* **8**, 57–65.

Mrosovsky, N. and Salmon, P. A. (1987) A behavioural method for accelerating re-entrainment of rhythms to new light–dark cycles. *Nature* **330**, 372,373.

Mrosovsky, N. and Salmon, P. A. (1990) Triazolam and phase–shifting acceleration re-evaluated. *Chronobiol. Int.* **7**, 35–41.

Nagai, K., Mori, T., and Nakagawa, H. (1982) Different responses of running wheel and Animex activity to restricted feeding and drug-induced anorexia in rats. *Biomed. Res.* **3**, 333.

O'Reilly, H., Armstrong, S., and Coleman, G. J. (1986) Restricted feeding and circadian activity rhythms of a predatory marsupial, *Dasyuroides byrnei*. *Physiol. Behav.* **38**, 471–476.

Pittendrigh, C. S. (1981) Circadian systems, entrainment, in *Handbook of Behavioral Biology, vol. 4, Biological Rhythms* (Aschoff, J., ed.), Plenum, New York, pp. 95–124.

Pittendrigh, C. S. and Daan, S. (1976a) A functional analysis of circadian pacemakers in nocturnal rodents. V. Pacemaker structure: a clock for all seasons. *J. Comp. Physiol.* **106**, 333–355.

Pittendrigh, C. S. and Daan, S. (1976b) A functional analysis of circadian pacemakers in nocturnal rodents, IV. Entrainment: pacemaker as clock. *J. Comp. Physiol.* **A106**, 291–331.

Pittendrigh, C. S. and Daan, S. (1976c) A functional analysis of circadian pacemakers in nocturnal rodents. I. The stability and lability of spontaneous frequency. *J. Comp. Physiol.* **106**, 223–252.

Ralph, M. R. and Mrosovsky, N. (1992) Behavioral inhibition of circadian responses to light. *J. Biol. Rhythms* **7**, 353–360.

Randall, W., Johnson, R. F., Randall, S., and Cunningham, J. T. (1985) Circadian rhythms in food intake and activity in domestic cats. *Behav. Neurosci.* **99**, 1162–1175.

Rea, M. A., Glass, J. D., and Colwell, C. S. (1994) Serotonin modulates photic responses in the hamster suprachiasmatic nuclei. *J. Neurosci.* 14, 3635–3642.

Reebs, S. G. and Mrosovsky, N. (1989) Large phase–shifts of circadian rhythms caused by induced running in a re-entrainment paradigm: the role of pulse duration and light. *J. Comp. Physiol.* A165, 819–825.

Refinetti, R., Nelson, D. E., and Menaker, M. (1992) Social stimuli fail to act as entraining agents of circadian rhythms in the golden hamster. *J. Comp. Physiol.* 170, 181–187.

Rosenberg, R., Zee, P. C., and Turek, F. W. (1993) Phase response curves to light in young and old hamsters. *Am. J. Physiol.* 261, R491–495.

Rosenwasser, A. M. and Adler, N. T. (1986) Structure and function in the circadian timing systems: evidence for multiple coupled circadian oscillators. *Neurosci. Biobehav. Rev.* 10, 431–448.

Rosenwasser, A. M., Pelchat, R. J., and Adler, N. T. (1984) Memory for feeding time: possible dependence on coupled circadian oscillators. *Physiol. Behav.* 32, 25–30.

Rosenwasser, A. M., Schulkin, J., and Adler, N. T. (1985) Circadian wheel-running activity of rats under schedules of limited daily access to salt. *Chronobiol. Int.* 2(2), 115–119.

Rosenwasser, A. M., Shulkin, J., and Adler, N. T. (1988) Anticipatory appetitive behaviour of adrenalectomized rats under circadian salt-access schedules. *Anim. Learning Behav.* 16, 324–329.

Rowsemitt, C. N. (1986) Seasonal variations in activity rhythms of male voles: mediation by gonadal hormones. *Physiol. Behav.* 37, 797–803.

Ruiz de Elvira, M. C., Persaud, R., and Coen, C. W. (1992) Use of running wheels regulates the effects of the ovaries on circadian rhythms. *Physiol. Behav.* 52, 277–284.

Rusak, B. (1989) The mammalian circadian system: models and physiology. *J. Biol. Rhythms* 4, 121–134.

Rusak, B., Mistlberger, R. E., Losier, B., and Jones, C. H. (1988) Daily hoarding opportunity entrains the pacemaker for hamster activity rhythms. *J. Comp. Physiol.* A164, 165–171.

Stephan, F. K. (1981) Limits of entrainment to periodic feeding in rats with suprachiasmatic lesions. *J. Comp. Physiol.* 143, 401–410.

Stephan, F. K. (1984) Phase-shifts of circadian rhythms in activity entrained to food access. *Physiol. Behav.* 32, 663–671.

Stephan, F. K. (1989a) Forced dissociation of activity entrained to T cycles of food access in rats with suprachiasmatic lesions. *J. Biol. Rhythms* 4, 467–479.

Stephan, F. K. (1989b) Entrainment of activity to multiple feeding times in rats with suprachiasmatic lesions. *Physiol. Behav.* 46, 489–497.

Stephan, F. K. (1992) Resetting of a circadian clock by food pulses. *Physiol. Behav.* 52, 997–1008.

Sulzman, F. M., Fuller, C. A., and Moore-Ede, M. C. (1977) Spontaneous internal desynchronization of circadian rhythms in the squirrel monkey. *Comp. Biochem. Physiol.* 58A, 63–67.

Szabo, I., Kovats, T. G., and Halberg, F. (1978) Circadian rhythm in murine reticuloendothelial function. *Chronobiologia* 5, 137–143.

Takahashi, J. S. and Zatz, M. (1982) Regulation of circadian rhythmicity. *Science* **2178**, 1104–1111.

Terman, M., Gibbon, J., Fairhurst, S., and Waring, A. (1984) Daily meal anticipation: interaction of circadian and interval timing. *Ann. NY Acad. Sci.* **423**, 470–487.

Tokura, H. and Aschoff, J. (1983) Effects of temperature on the circadian rhythm of pig-tailed macaques, Macaca nemestrina. *Am. J. Physiol.* **245**, R800–R804.

Turek, F. W. and Losee-Olson, S. H. (1986) A benzodiazepine used in the treatment of insomnia phase–shifts the mammalian circadian clock. *Nature* **321**, 167,168.

Van Reeth, O. and Turek, F. W. (1989) Stimulated activity mediates phase-shifts in the hamster circadian clock induced by dark pulses or benzodiazepines. *Nature* **339**, 49–51.

Van Reeth, O., Hinch, D., Tecco, J. M., and Turek, F. W. (1991) The effects of short periods of immobilization on the hamster circadian clock. *Brain Res.* **545**, 208–214.

Van Reeth, O., Zhand, Y., Zee, P. C., and Turek, F. W. (1992) Aging alters feedback effects of the activity-rest cycle on the circadian clock. *Am. J. Physiol.* **263**, R981–R986.

Wickland, C. and Turek, F. W. (1991) Phase-shifting of triazolam on the hamster's circadian rhythm of activity is not mediated by a change in body temperature. *Brain Res.* **560**, 12–16.

Wollnik, F., Gartner, K., and Buttner, D. (1987) Genetic analysis of ultradian and circadian locomotor activity rhythms in LEW/Ztm and ACI/Ztm rats. *Behav. Genet.* **17**, 167–178.

Yamada, N., Shimoda, K., Takahashi, K., and Takahashi, S. (1986) Change in period of free-running rhythms determined by two different tools in blinded rats. *Physiol. Behav.* **36**, 357–362.

Zucker, I. (1979) Hormones and hamster circadian organization, in *Biological Rhythms and Their Central Mechanism* (Suda, M., Hayaishi, O., and Nakagawa, H., eds.), Elsevier, New York, pp. 369–382.

Zucker, I. and Stephan, F. K. (1973) Light-dark rhythms in hamster eating, drinking and locomotor behaviors. *Physiol. Behav.* **11**, 239–250.

Zulley, J. and Campbell, S. S. (1985) Napping behavior during "spontaneous internal desynchronization": sleep remains in synchrony with body temperature. *Hum. Neurobiol.* **4**, 123–126.

CHAPTER 4

Sex Differences in Rodent Spontaneous Activity Levels

Larissa A. Mead,
Eric L. Hargreaves, and Liisa A. M. Galea

Introduction

The spontaneous activity of rodents has been studied extensively since the early part of this century. This interest reflects the desire to understand the underlying mechanisms and patterns of spontaneous behavior, as well as to characterize the internal and external factors influencing motor activity levels. Many of the latter factors are discussed elsewhere in this book, with the focus of the present chapter being on sex differences in the activity levels and patterns of common laboratory rodents.

The study of sex differences in such basic phenomena as activity levels is not only theoretically interesting, but may also provide insight into the evolutionary mechanisms of sexual selection (Darwin, 1871). For example, sex differences are found in the home range sizes and spatial abilities of polygynous species of voles that are not found for monogamous species, suggesting that such differences evolved to improve the odds of finding a mate (Gaulin and FitzGerald, 1986). Greater reproductive success is clearly the mechanism through which sexually dimorphic abilities and activity levels have evolved, in order to best utilize the species-specific distribution of sexual resources (Trivers, 1972).

From: *Motor Activity and Movement Disorders*
P. R. Sanberg, K. P. Ossenkopp, M. Kavaliers, Eds. Humana Press Inc., Totowa, NJ

Although reports of sex differences in activity levels or patterns are widespread, they are not necessarily uniform in either direction or degree. The literature is complicated by the wide variety of methodologies and species used to examine activity. For simplicity, the present chapter is organized according to species. A selective overview of the literature examining activity patterns of males and females is presented, along with a brief discussion of factors influencing sex differences in activity patterns. Organizational and activational influences of gonadal hormones are emphasized. Where appropriate, more detailed attention is given to studies from our laboratories.

Rats

In 1925, Hitchcock reported a striking sex difference in the spontaneous activity of rats (*Rattus norvegicus*): In rats housed for 6 mo in revolving wheels, males exhibited running wheel activity levels only 56% that of females. Although earlier studies had shown that the activity of female rats varied with the estrous cycle (Wang, 1923; Slonaker, 1924), the report by Hitchcock was the first study demonstrating a sex difference in the spontaneous activity of rats.

This finding of greater female activity levels has been replicated many times using several different paradigms, and is now considered one of the most consistent findings in the rat activity literature (*see* Archer, 1975; Beatty, 1979, 1992 for reviews). Notably, however, females are not more active than males at birth, but instead become more active as they mature. Sex differences typically become evident following puberty (Richter, 1933), which occurs at approx 40–50 d of age (Bennett and Vickery, 1970). This was illustrated by Blizard et al. (1975), who observed the behavior of male and female rats at 41, 52, or 62 d of age. Activity was measured by the number of segments traversed during the first 2 min in a circular open field. In naive animals, a sex difference in activity was not found for 41-d-old rats, but females were significantly more active than males at both 52 and 62 d of age. The pattern was similar in animals originally tested at 41 d of age and retested at 52 and 62 d, with sex differences found only at the latter two ages. These results suggested that the sex difference in activity could not be solely accounted for by a stronger reaction to a novel environment by naive adult females relative to naive males. Novelty of the environment did, however, have an influence on activity levels, as overall activity levels generally decreased with repeated exposure to the open field.

This sex difference appears to be caused by a developmental change in the activity of females, but not of males. An open field study by Beatty and Fessler (1976) found that, in females, both locomotor activity and rearing increased progressively with age, beginning around the time of weaning; in males, however, there was no consistent change in activity with age, and only a modest increase in rearing. Consequently, sex differences in activity were not evident until about 50 d of age, confirming the results of Blizard et al. (1975). Valle and Bols (1976) also found a significant sex difference in open-field activity favoring female rats at 120 d of age, but not at 30 d of age. Hyde and Jerussi (1983), using an automated open field, measured activity at 28, 42, 56, 77, 91, and 105 d of age, and found that male activity increased slightly, but not significantly, over this time, whereas female activity increased dramatically. Sex differences in this sample became evident at 56 d of age. A more recent study (Renner et al., 1992) failed to find a significant sex difference in open-field activity among rats aged 30, 60, and 90 d, but did show a greater age-related increase in activity for females than for males. Thus, one may tentatively conclude that sex differences in activity levels in the rat appear around the time of puberty, and are a function of increasing female activity.

The vast majority of experiments in this area have measured the overall activity levels of the subjects in some type of open field. This often consisted of a large square testing area with an outer wall, in which smaller square areas were demarcated. The traditional measures used with open fields have been the number of squares entered or the number of lines crossed. Short time periods were the norm, to prevent experimenter fatigue. These methods, although generally effective at the time, provided somewhat crude measures of gross activity levels; the newer automated open fields now allow experimenters to describe the actual behavior of the animals, characterizing their activity in more detailed and discrete terms, without the necessity of labor-intensive observation. These automated systems also allow the continuous sampling of activity over time periods as long as several days. This is advantageous in that "basal" activity levels can be distinguished from "exploratory" activity levels. Since most rodent species show a decreasing pattern of activity during the first 30 min in the activity monitors, after which activity reaches basal levels (e.g., Hargreaves et al., 1994), the traditional short measurement periods have not been adequate to fully describe activity patterns.

One of the automated measurement systems that has gained wide acceptance in this field is the Omnitech Digiscan Animal Activity Monitoring System (Ossenkopp and Kavaliers, in press). The Digiscan Animal Activity Monitor is a 40 × 40 × 30.5-cm open field with a grid of infrared beams mounted horizontally every 2.5 cm, and a second tier of beams mounted 15 cm above the floor. When in operation, the pattern of beam interruptions produced as the animal moves around is analyzed and recorded by an analysis unit, then routed to a microcomputer where the data is stored on disk.

This system was used by Hargreaves et al. (1990) and Hargreaves and Cain (1990) to examine the developmental aspects of sex differences in rat activity patterns, and to characterize the qualitative differences in the activity of males and females. They examined a number of locomotor activity variables:

1. Total distance traveled;
2. Average distance per movement;
3. Average speed;
4. Number of horizontal movements;
5. Time in horizontal movement;
6. Time per horizontal movement;
7. Number of vertical movements;
8. Time in vertical movement; and
9. Time per vertical movement.

Subjects consisted of eight male and eight female rats from two matched litters. Developmental testing began when the rats were weaned at 25 d of age and continued for 20 consecutive days. Each rat was tested daily in the automated open field for three consecutive samples of 10 min each. Sex differences were not evident for any of the activity measures, with activity levels of males and females increasing with age at approximately the same rate.

The rats were then retested in a similar fashion at 60 d of age. Three 10-min samples were collected from the animals on four consecutive days. No sex differences were found for the horizontal and vertical measures of number of movements, time in movement, or time per movement. These results indicated that the male and female rats executed the same number of movements, spent the same amount of time in movement, and spent the same amount of time per movement, regardless of whether these movements were locomotory or rearing behaviors.

There were, however, significant sex differences in the total distance traveled, average distance, and average speed. Female rats traveled farther overall than males, farther per individual movement, and at a faster speed (Fig. 1). Since an increase in average speed would result in concomitant increases in each of the other measures, these results suggested that sex differences in open field tests resulted from the female rats moving faster than male rats, rather than from simply making more movements (Hargreaves and Cain, 1990). There were also interactions between sex and sample for total distance and average distance, indicating that the sex differences observed for these variables faded over the three 10-min samples, whereas the difference observed for average speed remained constant. These data suggest that some sex differences in rat activity are caused by differences in *exploratory* activity—that is, activity occurring when an animal is initially placed in an unfamiliar environment—rather than differences in basal activity levels. This apparent effect of novelty of the environment is supported by the observation that sex differences were considerably reduced on the fourth day of testing as compared to the first day of testing.

To further examine this initial exploratory period, this experiment was repeated using the same rats at 122 d of age, using three 3-min sample periods. Although results were essentially the same as at 60 d of age, females also spent significantly more time per vertical movement, and tended to spend more total time in vertical movement. These results are consistent with other reports that female rats spend more time rearing than do males (e.g., Beatty and Fessler, 1976; Valle and

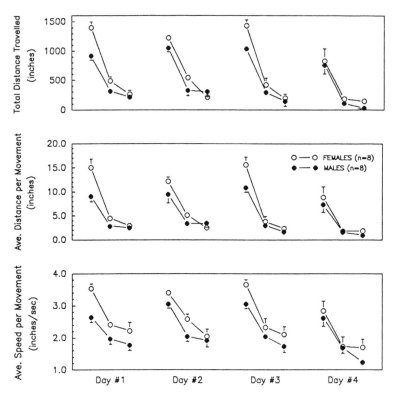

Fig. 1. Activity measures obtained from the Digiscan automated open field that revealed sex differences favoring female rats. Each point represents a 10-min sample period. Note that sex differences evident during the first sample period for total distance and average distance have disappeared by the third sample period. Differences between males and females also fade across the 4-d testing period.

Bols, 1976). Since these earlier reports were based on observation periods of only a few minutes in duration, it is not surprising that a sex difference in rearing, intuitively an exploratory behavior, is not detectable over longer time samples. In fact, this result supports the notion that sex differences in activity are strongly influenced by novelty and the exploratory demands of the environment.

Hormonal Influences

The mechanisms underlying these sex differences in activity appear to be under the influence of gonadal hormones

(Beatty, 1979, 1992). Experimental manipulations of circulating hormone levels have been especially useful in revealing the organizational effects of gonadal hormones on activity. For example, Blizard et al. (1975) demonstrated that female rats given testosterone propionate injections perinatally displayed activity levels equivalent to those of males, and significantly less than control females, when tested in an open field at 50 and 60 d of age. In addition, Beatty (1979, 1992) has discussed several studies in which neonatal castration resulted in increased activity levels in adulthood, and neonatal ovariectomy reduced open-field activity in adulthood.

A related line of evidence involves the effects of endogenous fluctuations of hormones, which occur across the estrous cycle, on the spontaneous activity of female rats. A good deal of research in the first half of this century examined these effects using running wheels and rotating cages (e.g., Wang, 1923; Slonaker, 1924; Brobeck et al., 1947), and consistently found a 4–5 d cycle in activity levels, such that activity was greatest during the period of high estrogen (late proestrus/estrus phases) and lowest during the period of low estrogen (diestrus phase). A later study (Hitt et al., 1968) confirmed these findings using a method in which activity in the running wheel was measured during daily 1-h recording periods. These observations suggested that the difference in running activity between the stages of the estrous cycle is very reliable, and is also relatively consistent throughout the course of the day.

Differences in activity across the estrous cycle have also been reported for rats using the open field paradigm (Burke and Broadhurst, 1966; Quadagno et al., 1972; Birke and Archer, 1975). However, Beatty (1979, 1992) discussed a number of studies involving hormonal manipulations that failed to significantly influence open-field activity and concluded that although activational effects of estrogen greatly influence activity in the running wheel, their influences in the open-field are more modest.

This issue was addressed in a study by Sauvé et al. (1990) that examined the effect of the estrous cycle on daily samples of activity in an automated open field. The activity of eight female rats was examined across a period of 31 d. Animals were 145 d old at the start of testing, which followed 10 d of habituation to the experimental procedures. Each day the rats were tested in the Digiscan Activity system for three consecutive samples of 3 min each. This study did not reveal any cyclic pattern of activity corresponding to the different phases of the estrous cycle, even though there was an effect of the estrous cycle on body weight in the predicted direction (Brobeck et al., 1947; Gentry and Wade, 1976), indicating that the estrous cycles were normal and had been identified accurately. As such, these results are in apparent conflict with those of Burke and Broadhurst (1966), Quadagno et al. (1972), and Birke and Archer (1975).

However, in contrast to the repeated-measures design used by Sauvé et al. (1990), Birke and Archer (1975) used a between-subjects design, in which each animal was tested only once for a 4 min period. Quadagno et al. (1972) tested their rats across 3 d, for 2 min at a time, and found activity to be greater for females at the estrus phase compared to the diestrus phase, but not significantly so. Subsequently, in a separate study using different animals, Quadagno et al. (1972) tested an unspecified number of females once each at their estrus and diestrus phases, and found them to be significantly more active at the estrus phase than the diestrus phase.

Comparison of the results of the preceding studies to those of Sauvé et al. (1990), whose animals were highly experienced in the open field, suggests that a familar testing environment may moderate activational hormone influences on activity. This suggestion is supported by the results of Burke and Broadhurst (1966), who tested their rats in an open field on five consecutive days for 2 min periods each day. Across this one estrous cycle time period, of three strains of rat studied, one displayed significantly greater activity at estrus than diestrus and one displayed greater activity, which was not statistically significant. The third strain had had previous exposure to the open field, a fact that is given by the authors as a possible explanation for the lack of an estrous cycle effect in this group.

Based on the above evidence, one may conclude that there is clearly a consistent effect of novelty of the environment on rat activity patterns. Data from our laboratories have revealed that sex differences in activity are most prominent during the initial exploration of a new environment, as compared to later basal activity levels, and that cyclic activity patterns across the estrous cycle disappear with experience in the open field.

Mice

The literature concerning sex differences in the activity of laboratory mice (*Mus musculus*) is much less straightforward than that for rats. Whereas some studies report greater female spontaneous activity, others report greater male activity, and even more report no sex difference in activity. A large part of the variation in this field may be caused by the number of different strains of mice used, since strain differences in locomotor activity have been demonstrated (Peeler, 1990). However, if strain was an important factor in the determination of sex differences, then one would expect to find sex by strain interactions in activity levels, and this has not been the case (Peeler, 1990).

In his 1975 review of the rodent sex difference literature, Archer concluded that female mice, in general, were more active than males in an open field, although he acknowledged that the variation in this finding was much greater than that for rats. This conclusion has been supported by a number of more recent studies. For example, Goodrich and Lange (1986), using a 3-min "headpoke test," found female mice to be more active than males. In addition, a series of experiments by Broida and Svare (1984) found repeatedly that female mice were more active than males, when measured by the mean number of photobeam interruptions during home cage movement.

However, the balance of the literature suggests that greater female activity is not necessarily the norm. For example, Kvist and Selander (1987) showed that males locomoted more than females during a 2 min period in an open field. Similarly, Buhot (1986) found that, although there were very few significant differences between male and female mice in nest-box exploration measures, females were consistently less active than males. Furthermore, numerous studies, published both before and since Archer's 1975 review, have failed to demonstrate any sex difference in activity levels in mice. These include studies that have used traditional open field measures (Candland and Nagy, 1969; Sayler and Salmon, 1971; Simmel et al., 1976), as well as studies using alternate methods, such as activity cages (Nagy and Ritter, 1976), runway activity measures (Peeler, 1990) and automated open fields (Lamberty and Gower, 1988).

Kemble and Enger (1984) extended these investigations of common laboratory mice to northern grasshopper mice (*Onychomys leucogaster*). When tested in an open field for three daily 5-min sessions, no reliable sex differences emerged. In addition, these mice were tested in activity wheels for 12 consecutive days, and no difference between males and females was revealed. We have also conducted a preliminary investigation of the spontaneous activity of the deer mouse (*Peromyscus maniculatus*), using the Digiscan automated open field. Aged adult deer mice ($N = 16$) were tested for 12 consecutive 5-min sample periods on each of 5 d, where testing sessions occurred once every 2 d. Nine standard activity measures were analyzed (*see* Rats), with no significant differences emerging between males and females on any of these measures.

These varying results do not lend themselves to a neat synopsis of the mouse literature. In fact, at present it seems that it would be unwise to attempt to make generalizations about the sex differences (or lack thereof) in activity across the many different strains, conditions, and measures that have been used.

Hormonal Influences

Despite the disarray of the findings for mouse spontaneous activity levels, early gonadal hormones have been shown to have consistent influences on adult behavior. For example, Broida and Svare (1984) demonstrated that gonadectomy at 60–70 d of age significantly reduced home-cage activity in both male and female mice (Rockland-Swiss albino), with ovariectomized females remaining significantly more active than castrated males. These results suggested that the perinatal hormone environment is influential in the development of adult activity patterns. Broida and Svare (1984) went on to show that neonatally castrated male mice were more active as adults than were sham-operated controls. Additionally, in an experiment in which female mice were ovariectomized at birth, then immediately injected with testosterone propionate (TP) or oil, it was found that the TP-treated females were less active than the oil-treated controls. These results strongly suggested that, at least in this strain of mouse, androgens have an organizational effect of lowering activity levels.

Further evidence of the substantial influence of early hormonal exposure comes from research demonstrating that intrauterine position plays a significant role in adult activity levels, as measured in an automated open field (Kinsley et al., 1986) and in home-range size (Zielinski et al., 1992). Female mice developing *in utero* between two males exhibit higher levels of testosterone in both amniotic fluid samples and fetal blood titers than females developing between two females (vom Saal and Bronson, 1980). Kinsley et al. (1986) reported that in a ranking from most to least active, females situated between two females were most active, followed by females situated between two males, then by males situated between two females, and finally, males situated between two males. Interestingly, when home range size was examined for mice whose intrauterine position was known, larger ranges were found for females that had been situated between two males than for females situated between two females (Zielinski et al., 1992). Together, these sets of data suggest that the role of early androgens in adult activity levels is not as clearcut as Broida and Svare's (1984) work would indicate. It is likely that early exposure to high levels of androgens has a masculinizing effect on numerous measures of activity, which cannot be totally accounted for by a gross decrease in basal activity levels.

Gerbils

Activity patterns in male and female Mongolian gerbils (*Meriones unguiculatus*) are relatively straightforward; most experiments have indicated that there is not a significant sex difference in the activity of these animals. There are, however, a number of reports of greater female activity. Again, differing outcomes appear to depend largely on methodological factors, such as the specific activity measure, experience with the testing procedure, and housing conditions.

Studies showing greater female activity have included both running wheel and open field paradigms. Roper (1976) tested gerbils for 21 d in home cages containing a running wheel, and found that females ran more overall and ran faster than did males.

There were also qualitative differences between the activity patterns of males and females: In general, females ran very quickly and consistently for up to 8 h/night, making only brief departures from the wheel, whereas males ran more slowly, ran for only a few revolutions at a time, and frequently left the wheel for long periods. This study found that females ran more during the dark phase than did males, whereas males ran more during the light phase than did females.

Two later studies examined the behavior of gerbils of varying ages during the first minute in an open field (Cheal and Foley, 1985; Cheal, 1987). Groups of gerbils were tested either repeatedly (in a longitudinal design) or singly (in a cross-sectional design). Cheal and Foley (1985) reported that in gerbils aged 0–5 mo, there were no sex differences in locomotor activity. However, Cheal (1987) reported that, in gerbils aged 6–18 mo, females tested repeatedly were more active in the open field than males, but females tested only once were less active than males. These studies seem to suggest that habituation is important in the demonstration of sex differences favoring females in gerbils. In Roper's study, the gerbils had been habituated for some time to their environment before any measures were taken, and in this circumstance, females were more active. Similarly, in Cheal's studies, only adult animals that were well accustomed to their environment demonstrated a sex difference favoring females. This is opposite to the finding in rats that greater female activity is best demonstrated during the initial exploratory period in a new environment (Hargreaves and Cain, 1990).

In contrast, numerous other open field studies have failed to demonstrate significant sex differences in locomotor activity in the gerbil. For example, Thiessen et al. (1969) did not find a sex difference in the number of lines crossed in an open field when gerbils were tested for 11 daily 5-min test sessions. In fact, by the end of the 11 test days, the male gerbils were slightly more active, contrary to the suggestion that females tend to be more active after habituation. Bols and Wong (1973) used eight daily 5-min open field samples during the dark phase. They found no sex difference in number of squares entered in their control animals, nor was there a trend toward greater female activity with increasing experience in the open field. Similarly,

Hull et al. (1973) failed to find a sex difference using single 10-min open field samples.

Work done in our laboratory has confirmed the absence of a sex difference in the open field (Mead et al., 1995). Male and female gerbils (N = 58) were tested using the Digiscan automated open field system over 12 consecutive 5-min sample periods, just prior to the dark phase. The gerbils ranged in age from 24–197 d, and each animal was tested only once. No sex differences were found in any of the horizontal, vertical, or movement characteristic measures, even when analyses were restricted to adult animals (Fig. 2). This finding indicates that increases in gonadal hormone levels at puberty do not result in sexually dimorphic activity levels in gerbils, as is the case for rats. In addition, no sex by time interactions were found, indicating that female activity did not increase relative to males' across the 1-h time period (i.e., as the animals habituated to the environment).

Hormonal Influences

In 1968, Barfield and Beeman established that normal adult female gerbils have an estrous cycle of 4–6 d. Shortly thereafter, Vick and Banks (1969) demonstrated a 5–6-d cycle in running wheel activity in the gerbil, corresponding to the phases of the estrous cycle; the subjects ran most at estrus, and least at diestrus, consistent with the pattern seen in rats. Thus, running wheel activity in gerbils appears to be modulated by endogenously cycling estrogen levels.

Hamsters

Work with golden hamsters (*Mesocricetus auratus*) has revealed sex differences of a nature similar to that of rats; females are generally more active than males, and this difference is under the control of gonadal hormones. Early studies with golden hamsters were among the first experiments designed to determine whether the sex differences in activity observed in rats could be generalized to other species, and whether any such differences were under the influence of circulating gonadal hormones. Swanson, in 1966, tested male and female hamsters in an open field during four daily 2-min testing sessions, where

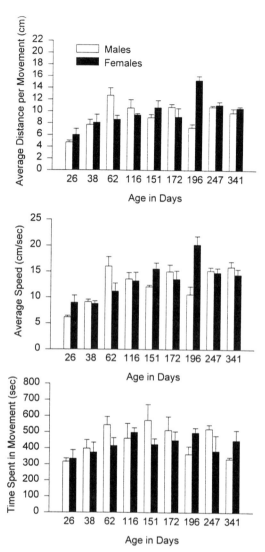

Fig. 2. Three activity measures obtained from the Digiscan automated open field demonstrating a lack of sex differences in gerbils of varying ages tested over a 1 h period. Although there are some age groups that appear to show greater male or female activity for some activity measures, these do not conform to any recognizable pattern, and clearly do not show a pattern of increasing sex differences following puberty.

the total number of floor units traversed constituted the measure of activity. When tested at the age of 65 d (postpuberty), females showed greater open-field ambulation than males. In fact, females were reported to ambulate "about twice as much as males" (p. 523). Within each testing session, females displayed significantly greater ambulation during the first 30 s than in subsequent intervals, whereas the activity of males remained relatively constant throughout the sessions. This is consistent with the demonstration of a stronger sex difference in rats, favoring females, during the initial exploratory period.

Hormonal Influences

Swanson (1966, Exp. 2) also attempted to determine whether the pubertal hormone surge was responsible for the observed sex differences. This experiment demonstrated that prepubertal (d 17) castration of males increased ambulation significantly, relative to intact males, when tested at 80 d of age. The activity of these castrated males did not differ from the activity of intact or ovariectomized females. The activity of males castrated postpubertally (d 95) and retested at 115 d of age was not different from male controls. Conversely, females were relatively unaffected by gonadectomy, irrespective of when it occurred. Additionally, when activity was examined across the estrous cycle, no significant effects were found. These results suggested that estrogens are not necessary either for the establishment or the maintenance of the observed activity patterns; however, androgens may have an early organizational effect resulting in lowered adult activity levels.

We now know, however, that activational effects of estrogen are important in modulating activity levels and rhythms in female hamsters (e.g., Richards, 1966; Zucker et al., 1980). For example, Richards (1966) made observations of golden hamsters in the home cage and in running wheels, across the stages of the estrous cycle, and found that there was more overall movement in the home cage and greater wheel-running at the estrus and proestrus phases than during the other 2 d.

To further examine the organizational effects of hormones, Swanson (1967) injected 2-d-old hamsters with testosterone propionate (TP), estradiol benzoate (EB), or an oil control, and then tested them in the open field between 60 and 90 d of age. TP-treatment in females reduced the ambulation score to that of males, a result consistent with the findings of the Swanson (1966) study. There was no difference between the activity of

TP-treated males and control males. EB treatment produced effects similar to those of TP, with early estrogen treatment reducing female scores to that of males, and also slightly reducing ambulation scores in males compared to male controls. The similarity of these hormone effects suggests that the TP was aromatized to 17-β-estradiol, and that both TP and EB treatments acted on the brain as estrogens.

Voles

Evidence for sex differences in the spontaneous locomotor activity of voles has favored greater activity in males, a pattern that is different than for most rodents. One of the few studies that have examined sex differences in the activity levels of voles in the laboratory is that of Gaulin et al. (1990). This study was designed to determine whether sex differences in activity levels were sufficient to explain sex differences in spatial abilities. Several symmetrical mazes were outfitted with infrared photocells, and male and female meadow voles (*Microtus pennsylvanicus*) and prairie voles (*M. ochrogaster*) were each tested in these mazes. This paradigm did not reveal any significant sex differences in activity across the two species. However, adult male meadow voles tended to be slightly more active than females, whereas male prairie voles were slightly less active than females.

Work in our laboratories, using the Digiscan automated open field, has confirmed the finding of greater male activity in adult meadow voles for the measures of total distance traveled, time spent in horizontal movement, and overall horizontal activity (Fig. 3) (Tysdale, 1990; Tysdale et al. 1990). Average speed was not affected by sex, indicating that this is not an important determinant of sexually dimorphic activity levels in voles, as is the case for rats.

Hormonal Influences

Hormone levels vary by season in voles (Rowsemitt and Berger, 1983; Seabloom, 1985; Bronson, 1989; Galea, 1994), with increased levels of gonadal hormones during the breeding season relative to the nonbreeding season. In keeping with this, a study by Splinter et al. (1993) found that greater male activity

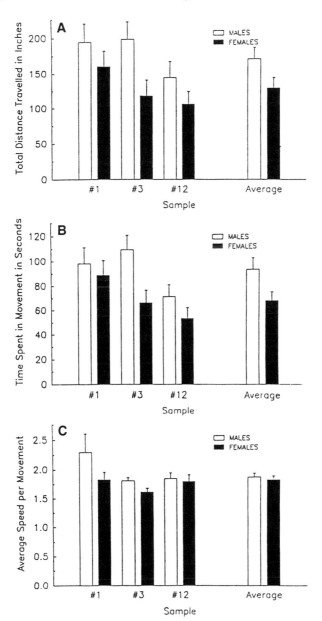

Fig. 3. Activity measures of adult meadow voles obtained over 1 h (12 consecutive 5-min sample periods) of testing. There were significant sex differences, favoring males, for total distance **(A)** and time spent in movement **(B)**. Unlike rats, these voles did not evidence a sex difference in average speed **(C)**. Note the decline in activity levels between the first and twelfth sample periods for total distance and time spent in movement.

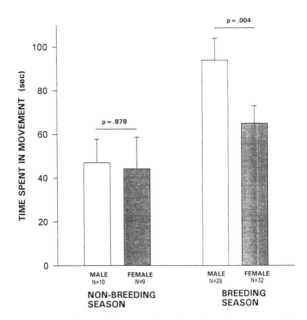

Fig. 4. Activity of meadow voles during simulated breeding and nonbreeding seasons. Mean time spent in movement over 1 h of testing. There was a significant sex difference during the "breeding season" (maintained under a 16:8 h light/dark cycle), with males displaying greater activity than females. No difference in activity was evident during the "nonbreeding season" (maintained under a 8:16 h light/dark cycle). These voles were housed in mixed-sex pairs.

in adult meadow voles was observed only when male and female voles were housed together under a long-day light–dark cycle to simulate the breeding season (Fig. 4). Sex differences in activity were not observed when:

1. The voles were housed under a short-day light–dark cycle to simulate the nonbreeding season;
2. When males and females were housed separately (unpublished data); or
3. In young, nonreproductive animals.

Although the meadow voles in Gaulin et al.'s 1990 study were maintained on a long-day light–dark cycle, they were housed singly, which may account for the lack of a significant sex difference in activity in their data.

In addition, at least two studies have directly manipulated androgen levels in adult male voles and examined the influence on subsequent activity levels in the laboratory. Rowsemitt (1986) reported that among male montane voles (*M. montanus*) castrated at 12–16 wk of age, animals implanted with Silastic tubing containing testosterone displayed more total activity on a running wheel over a period of 6 mo than animals that received empty implants. A complementary result has been found for meadow voles (Turner et al., 1980). Turner et al. (1980) showed that castration of adult male meadow voles decreased activity in an open field. They also found that male meadow voles exhibited a greater number of movements in a circular open field during the breeding season relative to the nonbreeding season (Turner et al., 1983).

Results obtained in the field are similiar to those obtained in the laboratory. A number of studies have found that there are sex differences favoring males in home range size during the breeding season, but this difference disappears during the nonbreeding season (for review *see* Madison, 1985; Madison and McShea, 1987). These studies suggest a picture of endogenously fluctuating hormone levels across seasons that influence activity levels in a sexually dimorphic fashion. Activity levels, as measured by home range size, also appear to be directly related to reproductive status in the monogamous male prairie vole; reproductive males (with scrotal testes) display larger home ranges than nonreproductive males (with abdominal testes) (Swihart and Slade, 1989). Data from our laboratories have also suggested that pubertal hormone factors may be influential in the development of adult sex differences, since juvenile meadow voles, aged 7–36 d, do not display sex differences in spontaneous activity (Fig. 5) (Kavaliers et al., 1990).

These studies indicate that androgens play an activational role in the modulation of activity levels in male voles. The effect is contrary to the typical effect of androgens, which, in other species, appear to play a strictly organizational role in lowering adult activity levels (e.g., Swanson, 1966; Blizard et al., 1975; Broida and Svare, 1984). The androgen-induced facilitation of activity in male voles is, however, consistent with the increased activity of male voles compared to female voles, a pattern that is also contrary to that of most other rodent species studied in the laboratory.

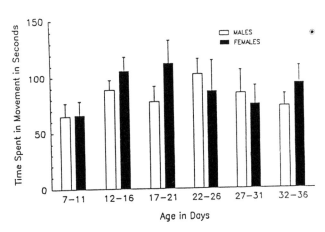

Fig. 5. Mean time spent in movement over 1 h of testing. There were no sex differences in activity level for these young, nonreproductive meadow voles.

There is less evidence supporting a role of hormones in the activity of the female vole. Housing conditions may be an important factor when examining activity in females, since exposure to a male serves to increase estradiol levels in both the female meadow and prairie vole (Cohen-Parsons and Carter, 1987; Galea et al., 1995). Indirect evidence suggests that greater levels of estradiol in the female vole may be associated with a decrease in activity levels since there are sex differences in activity favoring males when males and females are housed together, and no evident sex differences in activity when males and females are housed separately. Corroborating evidence comes from data on wild meadow voles that suggests that lactating females are less active and have smaller home ranges than nonlactating females (Madison, 1985; Sheridan and Tamarin, 1988). Furthermore, when adult female voles (*M. townsendii*) were given subcutaneous implants of testosterone in the field, an increase in home range size was found (Taitt and Krebs, 1982).

Thus, it appears from both laboratory and field data that sex differences in activity in voles are dependent on reproductive status, such that increased androgen levels appear to increase activity levels, and increased estrogens appear to decrease activity levels. In addition, it is possible there are sex differences in the activity rhythms of voles, with females being somewhat more diurnal than males (Madison, 1985), therefore time of day is a factor that should also be taken into consideration when interpreting vole activity data.

Summary of Sex Differences Literature

In summary, the evidence collected to date suggests that among rats and hamsters, females are more active; among voles, males are generally more active; among mice, there are reports of sex differences in both directions, but the literature is too mixed to attempt a generalization with confidence; and among gerbils, most studies indicate no differences, however there are isolated reports of sex differences, with activity levels reported higher for both males and females.

Issues to Consider
Regarding Sex Differences in Activity Levels

It is clear from the above, brief discussions that the finding of a sex difference in activity, even within a species, cannot be considered a straightforward phenomenon, and that the list of factors that may contribute to such a finding is likely to be lengthy. Some of these factors will be discussed in this section.

Reproductive Status

There is an abundance of evidence indicating that reproductive status, and the associated hormonal condition, is a major determinant of activity level. This has been found to be true for both males and females, for every species that has been tested. Since the focus of this chapter is on spontaneous activity, the present discussion will be limited to natural hormonal variations, such as occur during the estrous cycle or across seasons.

The Estrous Cycle

The estrous cycle has been shown to influence activity levels in every species for which it has been tested. However, it appears that some methods of measuring activity are more sensitive to the activational effects of the estrous cycle than others. For example, female rats, gerbils, and hamsters all show greater levels of activity during the estrus phase when tested in running wheels (e.g., Richards, 1966; Hitt et al., 1968; Vick and Banks, 1969). When rats are tested in the open field, however, the results are more varied, and in many cases no differences across the cycle can be detected (see Beatty, 1979, 1992; Sauvé et al., 1990). This effect of activity measure must be attributed to the fact that the running wheel and the open-field are not measuring precisely the same behavior (Mather, 1981).

In general, the influence of the estrous cycle on the observation of sex differences is small. This is provided that the females are tested in a random order to prevent any systematic variance in estrous cycle phase. On the other hand, if a group of females is tested within a short time period, especially if they are housed together, this may have an undue influence on the group mean for females, because the cycles of these females may be synchronized. In such a situation, if many of the females are tested during the estrus phase, a sex difference may be more evident than if many of the females are tested at phases associated with lower activity levels.

Housing

Housing of laboratory animals may also affect reproductive status. This has been shown to be especially true for female voles, which are induced ovulators (Sawrey and Dewsbury, 1985). Exposure to males for this species results in increased estradiol levels, which may in turn decrease activity levels (see Voles). This is supported by the finding of sex differences in activity favoring males when voles are housed together (Splinter et al., 1993), but not when they are housed singly (Gaulin et al., 1990). An additional factor related to housing is the possibility of social suppression of sexual maturity, which may occur when same-sex littermates are housed together. Such suppression has been observed among wild gerbils, both male and female, resulting in suppressed activity levels and range sizes (Agren et al., 1989). Since many experimental animals are housed in same-sex littermate groups, and typically the animals are not examined for their reproductive status, the possibility that social suppression has a silent influence on the activity of animals in many experiments cannot be ruled out.

Seasonal Influences

Seasonal influences on reproductive status refer to patterns of breeding where breeding occurs only (or at a higher rate) during a certain time of year, typically the spring and summer. These breeding patterns may be associated with altered hormonal states that may in turn influence activity levels. Seasonal influences have been found to be especially pronounced for voles; whether tested in the laboratory or in the field, male meadow voles have been found to be consistently more active than females, but only during the breeding season, or during a simulated breeding condition (*see* Voles). For this reason, it is imperative that researchers using wild-caught animals know at what time of year their subjects were captured, and have an adequate understanding of the influence that laboratory conditions (e.g., photoperiod) may have on the behavior of their subjects.

Age and Experience

The influence of age is most clearly seen in the transition between juvenile and adult animals, and is perhaps more accurately called the influence of puberty. In species for which sex differences are typically observed, it is nearly always the case that the differences in activity are not present before puberty, but develop during or shortly following puberty. This has been demonstrated in rats (e.g., Blizard et al., 1975) and in meadow voles (Kavaliers et al., 1990; Tysdale et al., 1990). Such changes are clearly caused by the effects of pubertal surges of gonadal hormones.

To our knowledge, there have not been any reports of an attenuation of sex differences in activity levels as adult animals age into senescence; however, sex differences clearly may be affected by experience with the testing situation. This was tested explicitly by Swanson (1969), who found that hamsters that were initially tested shortly after weaning did not demonstrate sex differences in activity when retested at 100 d of age. However, among hamsters tested for the first time at 100 d, females had significantly greater activity levels than males, the expected difference. This is an important issue that should be addressed when designing experiments, and is particularly relevant when deciding whether a longitudinal or a cross-sectional design is most appropriate to assess activity across various ages. An illustration of the difficulties that may arise in interpretation was provided by Cheal and Foley (1985) who examined the behavior of male and female gerbils using both longitudinal and cross-sectional protocols. At the age of 5 mo, among gerbils that had been tested only once, males showed greater activity than females; however, among gerbils that had been tested repeatedly since birth, 5 mo-old males demonstrated less activity than females.

On a smaller scale, the importance of experience in the situation is emphasized by the fact that sex differences are often most robust at the very outset of a testing session, during the initial exploratory phase, and tend to fade away after the first 10–20 min (Hargreaves and Cain, 1990). This effect also holds true across testing sessions, where activity assessed daily tends to become less sexually dimorphic, even during the exploratory period. An excellent example of the effect this may have on results was provided, inadvertantly, by Sauvé et al. (1990) who found absolutely no effect of the estrous cycle on open-field activity in female rats that were highly experienced in the apparatus. In contrast, open field experiments that employed between-subjects designs, or in which animals were tested for very few consecutive days, have reliably found activity differences between the estrus and diestrus phases (*see* Rats).

These issues comprise some of the more common methodological issues that experimenters should be aware of when assessing the spontaneous activity of rodents, particularly where sex is a relevant factor. Although species is unquestionably the strongest predictive variable of activity patterns, a good deal of evidence also indicates that gonadal hormones have common influences across most rodent species. Early exposure to androgens appears to lower adult activity levels in all species studied except voles, and estrogen fluctuations across the estrous cycle are usually positively correlated with fluctuations in activity levels. Information such as this is important, because knowledge of the factors contributing to sex differences will likely provide further insight into the factors underlying other normal variations in rodent activity patterns, which may in turn influence a wide variety of behaviors studied in behavioral neuroscience. As stated at the outset of this chapter, the study of sex differences in activity levels can also be a valuable tool in the exploration of some of the selective forces that have acted on a species' behavioral characteristics.

References

Agren, G., Zhou, Q., and Zhong, W. (1989) Ecology and social behavior of Mongolian gerbils, *Meriones unguiculatus*, at Xilinhot, Inner Mongolia, China. *Anim. Behav.* **37**, 11–27.

Archer, J. (1975) Rodent sex differences in emotional and related behavior. *Behav. Biol.* **14**, 451–479.

Barfield, M. A. and Beeman, E. A. (1968) The oestrous cycle in the Mongolian gerbil, *Meriones unguiculatus*. *J. Reprod. Fert.* **17**, 247–251.

Beatty, W. W. (1979) Gonadal hormones and sex differences in nonreproductive behaviors in rodents: organizational and activational influences. *Horm. Behav.* **12**, 112–163.

Beatty, W. W. (1992) Gonadal hormones and sex differences in nonreproductive behaviors, in *Handbook of Behavioral Neurobiology, vol. 11: Sexual Differentiation* (Gerall, A. A., Moltz, H., and Ward, I. L., eds.), Plenum, New York, pp. 85–128.

Beatty, W. W. and Fessler, R. G. (1976) Ontogeny of sex differences in open-field behavior and sensitivity to electric shock in the rat. *Physiol. Behav.* **16**, 413–417.

Bennett, J. P. and Vickery, B. H. (1970) Rats and mice, in *Reproduction and Breeding Techniques for Laboratory Animals* (Hafez, E. S. E., ed.), Lea and Febiger, Philadelphia, pp. 299–315.

Birke, L. I. A. and Archer, J. (1975) Open-field behaviour of oestrous and dioestrous rats: evidence against an "emotionality" interpretation. *Anim. Behav.* **23**, 509–512.

Blizard, D. A., Lippman, H. R., and Chen, J. J. (1975) Sex differences in open-field behavior in the rat: the inductive and activational role of gonadal hormones. *Physiol. Behav.* **14**, 601–608.

Bols, R. J. and Wong, R. (1973) Gerbils reared by rats: effects on adult open-field and ventral marking activity. *Behav. Biol.* **9**, 741–748.

Brobeck, J. R., Wheatland, M., and Strominger, J. L. (1947) Variations in regulation of energy exchange associated with estrus, diestrus and pseudopregnancy in rats. *Endocrinology* **40**, 65–72.

Broida, J. and Svare, B. (1984) Sex differences in the activity of mice: modulation by postnatal gonadal hormones. *Horm. Behav.* **18**, 65–78.

Bronson, C. H. (1989) *Mammalian Reproductive Biology*. The University of Chicago Press, Chicago, IL.

Buhot, M.-C. (1986) Nest-box exploration and choice in male and female mice tested under individual and social conditions. *Behav. Processes* **13**, 119–148.

Burke, A. W. and Broadhurst, P. L. (1966) Behavioural correlates of the oestrous cycle in the rat. *Nature* **209**, 223–224.

Candland, D. K. and Nagy, Z. M. (1969) The open field: some comparative data. *Ann. NY Acad. Sci.* **159**, 831–851.

Cheal, M. (1987) Adult development: plasticity of stable behavior. *Exp. Aging Res.* **13**, 29–37.

Cheal, M. and Foley, K. (1985) Developmental and experiential influences on ontogeny: the gerbil (*Meriones unguiculatus*) as a model. *J. Comp. Psychol.* **99**, 289–305.

Cohen-Parsons, M. and Carter, C. S. (1987) Males increase serum estrogen and estrogen receptor binding in brain of female voles. *Physiol. Behav.* **39**, 309–314.

Darwin, C. R. (1871) *The Descent of Man and Selection in Relation to Sex.* John Murray, London.

Galea, L. A. M. (1994) *Developmental, Hormonal, and Neural Aspects of Spatial Learning in Rodents.* PhD. Thesis. The University of Western Ontario, London, Ontario.

Galea, L. A. M., Kavaliers, M., Ossenkopp, K.-P. and Hampson, E. (1995) Gonadal hormone levels and spatial learning performance in the Morris water-maze in the male and female meadow vole, *Microtus pennsylvanicus*. *Horm. Behav.*, **29**, 106–125.

Gaulin, S. J. C. and FitzGerald, R. W. (1986) Sex differences in spatial ability: an evolutionary hypothesis and test. *Am. Naturalist* **127**, 74–88.

Gaulin, S. J. C., FitzGerald, R. W., and Wartell, M. S. (1990) Sex differences in spatial ability and activity in two vole species (*Microtus ochrogaster* and *M. pennsylvanicus*) *J. Comp. Psychol.* **104**, 88–93.

Gentry, R. T. and Wade, G. N. (1976) Sex differences in sensitivity of food intake, body weight, and running-wheel activity to ovarian steroids in rats. *J. Comp. Physiol. Psychol.* **90**, 747–754.

Goodrich, C. and Lange, J. (1986) A differential sex effect of amphetamine on exploratory behavior in maturing mice. *Physiol. Behav.* **38**, 663–666.

Hargreaves, E. L. and Cain, D. P. (1990) Speed of movements and not number of movements characterize sexually dimorphic open-field behavior of adult rats. *Can. Psychol. Assoc. Abstr.* **31(2a)**, 359.

Hargreaves, E. L., Ossenkopp, K.-P., Kavaliers, M., and Cain, D. P. (1994) Weight/age of adult male rats predict activity levels in an automated open-field. *Soc. Neurosci. Abstr.* **20**, 169.

Hargreaves, E. L., Tysdale, D. M., Cain, D. P., Ossenkopp, K.-P., and Kavaliers, M. (1990) Sex differences in the spontaneous locomotor activity of the laboratory rat: multivariate and temporal patterns. *Soc. Neurosci. Abstr.* **16**, 742.

Hitchcock, F. A. (1925) Studies in vigor. V. The comparative activity of male and female albino rats. *Am. J. Physiol.* **75**, 205–210.

Hitt, J. C., Gerall, A. A., and Giantonio, G. W. (1968) Detection of estrous activity cycle by 1-hour samples of running behavior. *Psychonom. Sci.* **10**, 159,160.

Hull, E. M., Langan, C. J., and Rosselli, L. (1973) Population density and social, territorial, and physiological measures in the gerbil (*Meriones unguiculatus*) *J. Comp. Physiol. Psychol.* **84**, 414–422.

Hyde, J. F. and Jerussi, T. P. (1983) Sexual dimorphism in rats with respect to locomotor activity and circling behavior. *Pharmacol. Biochem. Behav.* **18**, 725–729.

Kavaliers, M., Tysdale, D. M., Hargreaves, E. L., Ossenkopp, K.-P., and Shivers, R. R. (1990) Developmental changes in the spontaneous locomotor activity of juvenile male and female meadow voles. *Soc. Neurosci. Abstr.* **16**, 742.

Kemble, E. D. and Enger, J. M. (1984) Sex differences in shock motivated behaviors, activity, and discrimination learning of northern grasshopper mice. *Physiol. Behav.* **32**, 375–380.

Kinsley, C., Miele, J., Konen, C., Ghiraldi, L., and Svare, B. (1986) Intrauterine contiguity influences regulatory activity in adult female and male mice. *Horm. Behav.* **20**, 7–12.

Kvist, B. and Selander, R.-K. (1987) Sex difference in open-field activity after learning in mice. *Scandinav. J. Psychol.* **28**, 88–91.

Lamberty, Y. and Gower, A. J. (1988) Investigation into sex-related differences in locomotor activity, place learning and passive avoidance responding in NMRI mice. *Physiol. Zool.* **44**, 787–790.

Madison, D. M. (1985) Activity rhythms and spacing, in *Biology of New World Microtus* (Tamarin, R. H., ed.), Special Publication No. 8, The American Society of Mammologists, pp. 373–419.

Madison, D. M. and McShea, W. J. (1987) Seasonal changes in reproductive tolerance, spacing, and social organization in meadow voles: a microtine model. *Am. Zool.* **27**, 899–908.

Mather, J. G. (1981) Wheel-running activity: a new interpretation. *Mammal. Rev.* **11**, 41–51.

Mead, L. A., Hargreaves, E. L., Ossenkopp, K.-P., and Kavaliers, M. A. (1995) Multivariate assessment of spontaneous locomotor activity in the Mongolian gerbil *(Meriones unguiculatus):* influences of age and sex. *Physiol. Behav.* **57**, 893–899.

Nagy, Z. M. and Ritter, M. (1976) Ontogeny of behavioral arousal in the mouse: effect of prior testing upon age of peak activity. *Bull. Psychonom. Soc.* **7**, 285–288.

Ossenkopp, K.-P. and Kavaliers, M. Measuring spontaneous locomotor activity in small mammals, in *Measuring Movement and Locomotion: From Invertebrates to Humans* (Ossenkopp, K.-P., Kavaliers, M., and Sanberg, P. R., eds.), R. G. Landes, Austin, TX, in press.

Peeler, D. F. (1990) Two measures of activity in genetically defined mice as a function of strain, time of day, and previous experience. *Psychobiology* **18**, 327–338.

Quadagno, D. M., Shryne, J., Anderson, C., and Gorski, R. A. (1972) Influence of gonadal hormones on social, sexual, emergence, and open field behaviour in the rat *(Rattus norvegicus) Anim. Behav.* **20**, 732–740.

Renner, M. J., Bennett, A. J., and White, J. C. (1992) Age and sex as factors influencing spontaneous exploration and object investigation by preadult rats *(Rattus norvegicus) J. Comp. Psychol.* **106**, 217–227.

Richards, M. P. M. (1966) Activity measured by running wheels and observation during the oestrous cycle, pregnancy and pseudopregnancy in the Golden hamster. *Anim. Behav.* **14**, 450–458.

Richter, C. P. (1933) The effect of early gonadectomy on the gross body activity of rats. *Endocrinology* **17**, 445–450.

Roper, T. J. (1976) Sex differences in circadian wheel running rhythms in the Mongolian gerbil. *Physiol. Behav.* **17**, 549–551.

Rowsemitt, C. N. (1986) Seasonal variations in activity rhythms of male voles: mediation by gonadal hormones. *Physiol. Behav.* **37**, 797–803.

Rowsemitt, C. N. and Berger, P. J. (1983) Diel plasma testosterone rhythms in male *Microtus montanus,* the montane vole, under long and short photoperiods. *Gen. Comp. Endocrinology* **50**, 354–358.

Sauvé, D., Hargreaves, E. L., Stewart, D. J., and Cain, D. P. (1990) Multivariate analyses of spontaneous locomotor activity of the laboratory rat across the estrous cycle: a preliminary report. *Soc. Neurosci. Abstr.* **16**, 743.

Sawrey, D. K. and Dewsbury, D. A. (1985) Control of ovulation, vaginal estrus, and behavioral receptivity in voles (*Microtus*). *Neurosci. Biobehav. Rev.* **9**, 563–571.

Sayler, A. and Salmon, M. (1971) An ethological analysis of communal nursing by the house mouse (*Mus musculus*) *Behavior* **40**, 62–85.

Seabloom, R. W. (1985) Endocrinology, in *Biology of New World Microtus* (Tamarin, R. H., ed.), Special Publication No. 8, The American Society of Mammologists, pp. 685–724.

Sheridan, M. and Tamarin, R. H. (1988) Space use, longevity, and reproductive success in meadow voles. *Behav. Ecol. Sociobiol.* **22**, 85–90.

Simmel, E. C., Haber, S. B., and Harshfield, G. (1976) Age, sex, and genotype effects on stimulus exploration and locomotor activity in young mice. *Exp. Aging. Res.* **2**, 253–269.

Slonaker, J. R. (1924) The effect of pubescence, oestruation and menopause on the voluntary activity in the albino rat. *Am. J. Physiol.* **68**, 294–315.

Splinter, A. L., Galea, L. A. M., Kavaliers, M., and Ossenkopp, K.-P. (1993) Effects of live-trap restraint on motor activity levels and scentmarking in meadow voles. *Can. Soc. Brain. Behav. Cognit. Sci. Abstr.* 71.

Swanson, H. H. (1966) Sex differences in behaviour of hamsters in open-field and emergence tests: effects of pre and postpubertal gonadectomy. *Anim. Behav.* **14**, 522–529.

Swanson, H. H. (1967) Alteration of sex-typical behaviour of hamsters in open-field and emergence tests by neo-natal administration of androgen or oestrogen. *Anim. Behav.* **15**, 209–216.

Swanson, H. H. (1969) Interaction of experience with adrenal and sex hormones on the behaviour of hamsters in the open field test. *Anim. Behav.* **17**, 148–154.

Swihart, R. K. and Slade, N. A. (1989) Differences in home-range size between sexes of *Microtus ochrogaster*. *J. Mammol.* **70**, 816–820.

Taitt, M. J. and Krebs, C. J. (1982) Manipulation of female behaviour in field populations of *Microtus townsendii*. *J. Anim. Ecol.* **51**, 681–690.

Thiessen, D. D., Blum, S. L., and Lindzey, G. (1969) A scent marking response associated with the ventral sebaceous gland of the Mongolian gerbil (*Meriones unguiculatus*). *Anim. Behav.* **18**, 26–30.

Trivers, R. L. (1972) Parental investment and sexual selection, in *Sexual Selection and the Descent of Man* (Campbell, B., ed.), Aldine, Chicago, pp. 136–179.

Turner, B. N., Iverson, S. L., and Severson, K. L. (1980) Effects of castration on open-field behavior and aggression in male meadow voles (*Microtus pennsylvanicus*). *Can. J. Zool.* **58**, 1927–1932.

Turner, B. N., Iverson, S. L., and Severson, K. L. (1983) Seasonal changes in open-field behavior in wild male meadow voles (*Microtus pennsylvanicus*). *Behav. Neural. Biol.* **39**, 60–77.

Tysdale, D. M. (1990) *Sex and Developmental Effects on the Locomotor Activity of the Meadow Vole*, Microtus pennsylvanicus: *A Multivariate Analy-*

sis. Unpublished Honours Thesis, University of Western Ontario, London, Ontario.

Tysdale, D. M., Hargreaves, E. L., Kavaliers, M., Ossenkopp, K.-P., and Cain, D. P. (1990) Sex differences in the spontaneous locomotor activity of the meadow vole (*Microtinae*): reversal of the usual laboratory rodent pattern. *Soc. Neurosci. Abstr.* **16**, 742.

Valle, F. P. and Bols, R. J. (1976) Age factors in sex differences in open-field activity of rats. *Animal. Learning. Behav.* **4**, 457–460.

Vick, L. H. and Banks, E. M. (1969) The estrous cycle and related behavior in the Mongolian gerbil, *Meriones unguiculatus* Milne Edwards. *Comm. Behav. Biol. Part A.* **3**, 117–124.

vom Saal, F. S. and Bronson, F. H. (1980) Sexual characteristics of adult female mice are correlated with their blood testosterone levels during prenatal development. *Science (Washington, DC)* **208**, 597–599.

Wang, G. H. (1923) Relation between "spontaneous" activity and oestrous cycle in the white rat. *Comp. Psychol. Mono.* **2**, 1–27.

Zielinski, W. J., vom Saal, F. S., and Vandenbergh, J. G. (1992) The effect of intrauterine position on the survival, reproduction and home range size of female house mice (*Mus musculus*). *Behav. Ecol. Sociobiol.* **30**, 185–191.

Zucker, I., FitzGerald, K. M., and Morin, L. P. (1980) Sex differentiation of the circadian system in the golden hamster. *Am. J. Physiol.* **238** (Regulatory Integrative Comparative Physiology 7): R97–R107.

CHAPTER 5

Automated Video-Image Analysis of Behavioral Asymmetries

R. K. W. Schwarting, J. Fornaguera, and J. P. Huston

Introduction

The analysis of behavioral asymmetries is a widely employed approach to investigate behavior in relation to functions of the brain. Out of the repertory of presently available measures, turning behavior is probably the one that has been examined the most. Spontaneous or conditioned turning may be used either as an independent variable to look for related asymmetries in the brain, or turning may be analyzed dependent on stimulation (electrical, chemical) or lesion of the nervous system (for reviews, *see* Pycock, 1980; Pycock and Kilpatrick, 1989; Miller and Beninger, 1991). One of these lesion techniques is almost inevitably associated with the analysis of turning behavior, namely the unilateral 6-hydroxydopamine (6-OHDA) lesion of the nigrostriatal dopamine (DA) system, which serves as a unilateral model of Parkinson's disease. Here, the study of turning behavior has provided substantial information about the role of the basal ganglia, the neurotransmitters involved, and mechanisms of functional recovery. Turning behavior has also been used for screening of potential therapeutic drugs and the study of possible neurotrophic factors or

From: *Motor Activity and Movement Disorders*
P. R. Sanberg, K. P. Ossenkopp, M. Kavaliers, Eds. Humana Press Inc., Totowa, NJ

of brain grafts. Different kinds of methods (e.g., rotometers) are available to measure turning, and out of these, the automated analysis of video images is the most recent and probably the most promising. With this approach, not only turning, but also thigmotactic scanning, another potential measure of behavioral asymmetry, can be evaluated automatically. This method will be explained in the following, and its applicability will be demonstrated by several experimental examples.

Some Considerations on the Behaviors to Be Measured

With respect to the study of behavioral asymmetries, the measure of turning has been investigated most often; however, there are different definitions of the kind of behavior that is to be described by the term "turning" (*see,* for example, Pycock and Kilpatrick, 1989). Thus, a delimiting statement is required before going into detail regarding the measurement of turning. Here, turning is defined as follows: First, there is a locomotor component, that is, the animal (usually a rat or mouse) is moving with its body in the horizontal position; therefore, behavior during rears is not considered. Second, the animal is moving to one side; that is, it continuously or discontinuously produces a circular path about a vertical axis. This movement may be performed within a very narrow diameter, apparently directed at the animal's own body (*see* Fig. 1). Such behavior is sometimes described as "pivoting" or "head-to-tail" turning, and usually includes an asymmetric posture, that is, a curvature of the rat's longitudinal axis. However, a turning movement may also be performed within a wider diameter not pointing at its own body and without an asymmetric posture. Furthermore, narrow or wide turning movements may be performed as full circles (360°) or as partial turns, for example, quarter or half turns.

In contrast, there are other behaviors that are not included in this definition, especially head-turns in an otherwise stationary animal. Furthermore, and especially in an open field situation, rats will show a behavior that could falsely be measured as turning, but should be considered separately: The animal may locomote along walls or edges within or surrounding the environment. This behavior, when performed within a distance

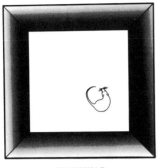 

TURNING   THIGMOTACTIC SCANNING

Fig. 1. A rat showing turning behavior **(Left)** or thigmotactic scanning **(Right)** in an open field. The turn shown is performed within a narrow diameter with the snout almost pointing at the animal's tail. However, the diameter of turning may be wider or even narrower than in this example. Thigmotactic scanning is defined as a locomotion along walls or edges, in this case, the wall of the environment. This behavior is performed within a distance that ensures that the vibrissae on one side of the rat's face can have contact with the wall.

that ensures that the vibrissae of the animal can have contact with the environmental border (*see* Fig. 1) is termed thigmotactic scanning (Huston et al., 1990), and has been functionally related to exploration and emotionality (Grossen and Kelley, 1972; Sanberg et al., 1987; Treit and Fundytus, 1989).

Thigmotactic scanning might resemble turning, particularly when the animals is continuously following the walls of a small circular testing arena in one direction. However, it is not included in the definition of turning, since in contrast to turning itself, it is apparently determined by the presence of an external stimulus, in this case, the wall. Such variables have often been neglected because turning behavior (especially after unilateral lesions in the basal ganglia) is often considered to reflect an asymmetry in postural or in motor mechanisms, which are thought to be largely independent of sensory stimuli. However, several studies using unilateral sensory manipulations or even brain-damaged animals have pointed to the importance of sensory mechanisms determining the direction of asymmetry in an animal, and it has been repeatedly demonstrated that the direction of behavioral asymmetry can be dependent on the kind of testing environment and the presence of walls and edges within or surrounding it (e.g., Szechtman, 1983; Pisa and Szechtman, 1986; Ziegler and Szechtman, 1988). Furthermore, the investigation of thigmotactic scanning has shown that it can be lateralized by several unilateral manipulations, and that the direction and time-course of asymmetry depend on the kind of treatment, which includes sensory manipulation (e.g., vibrissae removal; Steiner et al., 1986), lesion of the barrel cortex (Adams et al., 1992), trephination of the overlying skull (Adams et al., 1994), chemical stimulation in the brain (Wise and Holmes, 1986), and 6-OHDA lesion of the nigrostriatal DA system (e.g., Steiner et al., 1988; Fornaguera et al., 1993, 1994; Sullivan et al., 1993). Thus, scanning can also serve as a useful and sensitive indicator of functional asymmetry.

Taken together, even in a simple testing environment, such as the open field, there are at least two measures available that can be used to investigate behavioral asymmetry. Out of these, turning behavior has been examined the most. Accordingly, the following short overview of methods deals mainly with the measurement of turning.

Conventional Behavioral Methods

Behavioral asymmetries in turning are usually measured either mechanically with rotometers (e.g., Ungerstedt and Arbuthnott, 1970), by observation (e.g., Adani et al., 1991), or with automated video-image analysis (Bonatz et al., 1987; Schwarting et al., 1993). Furthermore, infrared systems for the measurement of turning were reported (Hodge and Butcher, 1979; Young et al., 1993).

Out of these, the most common method to measure turning in rodents is the rotometer. Here, rotation of the animal is detected mechanically by devices, including a harness put around the animal's chest and a wire fixed to the harness, which transmits the rotation of the animal to a microswitch or photocell system mounted above the animal. There are several variations of such rotometers that differ mainly in the kind of testing environment, such as bowls (Ungerstedt and Arbuthnott, 1970; Barber et al., 1973; Jerussi, 1982; Etemadzadeh et al., 1989), spheres (Greenstein and Glick, 1975; Schwarz et al., 1978; Brann et al., 1982), flat-bottomed cylinders (Walsh and Silbergeld, 1979), or rectangular boxes (Guilleux and Peterfalvi, 1974; Schwarting and Huston, 1987). Undoubtly, the rotometer technique has contributed substantially to the study of behavioral asymmetries. It has been used in hundreds of studies dealing with spontaneous or conditioned turning in intact animals, to investigate the effects of brain lesions or pharmacological treatments, and it has become a standard pharmacological test for the screening of potential therapeutical drugs against Parkinson's disease.

The wide dissemination of this method is certainly determined by the fact that rotometers are relatively simple, easy to use, and usually inexpensive; furthermore, they produce quantitative data and allow time-dependent analyses. However, they have some drawbacks that limit their application, particularly when qualitative aspects of behavior are of importance. First, a harness has to be fixed to the animal. This requires interaction between the experimenter and the animal, and can irritate the animal during the testing session, thereby artificially influencing its behavior. Second, behavior is reduced to "turns" of a wire, which may lead to a loss of important behavioral information or even to a misinterpretation of data. For example, a rotometer cannot discriminate between different diameters of turns, and it cannot detect where a turn was performed in a certain testing environment. Furthermore, rotometers cannot discriminate turning from thigmotactic scanning, the importance of which is elucidated by the following example: In a testing area of 40 × 40 cm side length, an animal has been continuously scanning the enclosing walls with the right side of the body; however, a rotometer would have measured this behavior as turns to the left. Based on the rotometer data, one would interpret the animal's behavior as an asymmetry to the left, although it actually had an asymmetry for the right side, since it was following the wall with that side of the body.

Thus, if a differential and qualitative analysis of behavioral asymmetries is intended, other methods have to be considered. The most obvious way to measure "actual" behavior is the observational method, which, in combination with rating scales, has also been used to measure behavioral asymmetries in simple environments like an open field (Naylor and Olley,

1972). This method has the advantage that many different behavioral items can be differentiated and considered (*see* for example, Adani et al., 1991). However, it is dependent on the skills of the observer, his or her motivation, and state. Thus, the observational method is not useful for long testing sessions or for large numbers of tests, such as in drug screening.

Methods for Video-Image Analysis of Behavior

Probably the best approximation to a human observer can be provided by the automated analysis of video images, and there are several methodical variations and experimental applications for behavioral studies in animals. Regarding the species under investigation, most methods work with small rodents, such as rats or mice, but systems for fish and insects have also been reported (Lubinski et al., 1977; Dusenbery, 1985; Ye and Bell, 1991; Nilsson et al., 1993).

Usually, the animal is monitored by a videocamera positioned either above or below the testing environment; a side view is reported in Kernan et al. (1987,1989). The easiest way to detect an animal within a testing environment is to use the contrast between the two (e.g., Tanger et al., 1978; Spruijt and Gispen, 1983); thus, for example, an albino rat is tested in a dark apparatus. However, conditions with low contrasts can also be used, especially when using marked animals (e.g., Crawley et al., 1982). The analog video image is then converted into digitized data by means of a video-image digitizer and is then transferred to a computer. The computer can work either with the whole two-dimensional view of the animal or can reduce it to a spot, which reflects the center of the rat's image. The second approach is often preferred, since image analysis requires large memory space and calculation time.

There are several behavioral parameters that can be analyzed by the present video systems. The most common are: locomotion, usually as distance traversed or time spent locomoting (e.g., Livesey and Leppard, 1981; Renner et al., 1990; Vorhees et al., 1992), resting time (Ye and Bell, 1991), immobility or catalepsy (Martin et al., 1992), spatial location within the testing apparatus, for example, peripheral vs central activity (e.g., Tanger et al., 1978; Vorhees et al., 1992), social interaction (approach, avoidance) between animals (Crawley et al., 1982; Höglund et al., 1983; Spruijt et al., 1992), swimming behavior in the Morris water maze (Spruijt et al., 1990; Wolfer and Lipp, 1992; Spooner et al., 1994), or footprints as a measure of functional recovery (Tanger et al., 1984; Walker et al., 1994). Furthermore, a system with two cameras, one taking a vertical and one a horizontal view, has been presented (Spruijt and Gispen, 1983; Kernan et al., 1987, 1989; Hopper et al., 1990) which is constructed to measure standing, sitting, lying, walking, grooming, and so forth. Finally, there are now several video systems commercially available (Videomex-V, SMART, The Observer, Etho Vision, Poly-Track, Video Tracking Activity System). These systems can also be used for the study of open field behavior and different kinds of mazes (e.g., radial, T, Y, water) and they yield a spectrum of behavioral measures that goes beyond the scope of this review.

However, with respect to the measurement of turning or scanning, these methods are very limited. Thus, some systems can measure turning behavior and can differentiate between left and right turns (for example, Torello et al., 1983); however, they may require placing a mark on the animal's dorsal body surface and do not discriminate between different diameters of turns. With regard to thigmotactic scanning, there are systems that can measure the time spent in the periphery (e.g., Tanger et al., 1978; Ploeger et al., 1992; Vorhees et al., 1992), the distance to a wall (Crawley et al., 1982), or the distance traveled and time spent close to the walls (Simon et al., 1994). However, most of these systems do not differentiate this parameter between a stationary and a moving animal, and do not detect which side of the body was pointing at the wall; thus, they are not suitable for the analysis of behavioral asymmetry.

The Video-Image Analyzing System (VIAS)

Several years ago, we presented a VIAS (Bonatz et al., 1987) that does not require marking the animal and that can measure turning, can discriminate between different diameters of turning (such as tight and wide turns), can detect partial (quarter, half, three-quarters) and full turns, and can measure locomotion as distance traveled. Meanwhile, this VIAS system, which

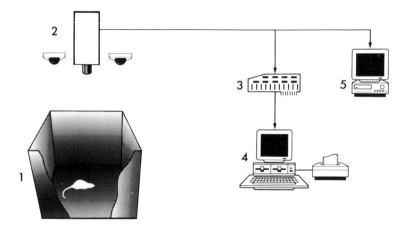

Fig. 2. Diagram of the hardware components of the VIAS, including animal chamber (1), videocamera and red lights (2), video digitizer (3), personal computer and printer (4), and an optional video recorder (5).

is used in combination with an open field serving as the testing environment, has proven to be worthwhile in several studies, including the study of functional recovery and the behavioral effects of drugs in the 6-OHDA model (Steiner et al., 1988; Schwarting et al., 1991a; Fornaguera et al., 1993), the role of substance P as a recovery-promoting factor (Mattioli et al., 1992), the analysis of conditioned drug effects (e.g., Carey, 1989), or the relationship between the vibrissae and the basal ganglia (e.g., Milani et al., 1990; Schwarting et al., 1990, 1991b). Furthermore, a new system (Schwarting et al., 1993) has been developed that incorporates the features of the former version and adding, among others, the feature of measuring thigmotactic scanning. This system, together with some of its potential applications, will be explained in the following sections.

The Hardware

The VIAS consists of four basic hardware components (*see* Fig. 2): the animal chamber, a camera, a video interface, and the computer. The animal chamber currently used is an open field made from black wooden walls and a black cloth floor.

The animal (albino rat) is monitored by a camera positioned at about 150 cm above the center of the open field. We use either a tube camera (Philips LDH 402 with a Newvicon tube XQ 1277) in combination with a 25 mm/F 1.8 lens or a CCD camera (type 90 100; Himmelreich GmbH, Schwaigern, Germany), from which the infrared filter has been removed. Lighting is provided by four dim red lights, which are arranged around the camera, and produce a luminous density of about 0.005–0.007 cd/m^2. However, any other lighting condition can be used, provided that the rat contrasts sufficiently with the background to provide a detectable signal for the computer.

The analog video signal is digitized by means of a video-digitizer (VIDEO-1000 I, Ing. Büro Manfred Fricke, Berlin, Germany), which is installed in an AT-386 (25 MHz) personal computer with 4 MB RAM and a mathematical coprocessor (US83c87, 25 MHz, USLI Systems, CA). Furthermore, the computer system includes a floppy disk (1.2 MB), a hard disk (20 MB), a conventional b/w monitor, and a matrix printer.

In addition, one can also connect a video recorder to the camera for the registration of supplementary behavioral data. Alternatively, the VIAS can also analyze turning, scanning, and locomotion from video records. Therefore, instead of directly analyzing behavior, one may record behavior by means of the video recorder and analyze (or reanalyze) it later.

The Software

The video digitizer provides the computer with a digitized image that is composed of 640 by 200 pixels (bits). This input results in a corresponding monochrome image on the screen of the computer monitor. No graduation is made. Only switched-on bits produce a signal (white dots on the screen). Thus, a rat is depicted as a pattern of white dots (*see* Fig. 3).

The software is designed to analyze the bit image representing the analog video-image of the animal chamber. To analyze this bit image, an x-y coordinate system is overlaid. Two parameters provide the main information for all further evaluations: (1) the central point of the switched-on image in relation to the x-y coordinate system and (2) the angle α of its longitudinal axis (orthogonal regression line). Based on these

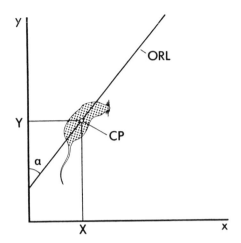

Fig. 3. The x-y coordinate system overlaid on the bit image. The dots represent the coordinate points generated by the switched-on pixels of the rat's image. The rat's contour is also shown here for clarification. CP is the central point of the x- and y-coordinates of the rat's image; α is the angle of the longitudinal axis (orthogonal regression line, ORL) through the rat's image with the y-axis.

two parameters, turning, scanning, and locomotion can be analyzed by the program, which is written in Turbo Pascal 6.0 and 8086 Assembler.

Measurement of Turning and Locomotion

The principal operations for the evaluation of turning and locomotion have been described in detail previously (Bonatz et al., 1987). In short, a digitized video-image is taken about every 150 ms. The position of the central point and the angle α of its longitudinal axis are evaluated from every image and compared between images. Locomotion is defined as the movement of the central point and is expressed as cm/min. Movements of the central point of <0.7 cm between two images are discarded to exclude behavior that does not result in locomotion (such as grooming), but that might change the position of the central point. Turning behavior is analyzed using the change of the angle α of the longitudinal axis over time. If the shape of the animal (i.e., the longitudinal axis) rotates for a certain angle "rot" in a time interval $dt = t_2 - t_1$, then rot is calculated by:

$$\text{rot} = \alpha_2 - \alpha_1 \qquad (1)$$

if $-90° < \alpha_2 - \alpha_1 < 90°$

$$\text{rot} = \alpha_2 - \alpha_1 + 180° \qquad (2)$$

if $-180° \leq \alpha_2 - \alpha_1 \leq -90°$

$$\text{rot} = \alpha_2 - \alpha_1 - 180° \qquad (3)$$

if $90° \leq \alpha_2 - \alpha_1 \leq 180°$

Thus:

$$-90° \leq \text{rot} \leq 90° \qquad (4)$$

If rot $< 0°$, then the rotation is counterclockwise. Otherwise (rot $> 0°$), the rotation is clockwise. Since the animal is not marked, the program cannot discriminate between its head and tail ends. Therefore, a rotation of $90° + x°$ during one measurement cycle cannot be distinguished from a $90° - x°$ rotation in the opposite direction. As a solution to this problem the program has to trace rotation within very short time intervals (about every 150 ms), during which the rat is unlikely to turn more than 90°. Additionally, all rotations of more than 80° within one measurement cycle are excluded. This approach has proven to be an accurate solution to measuring the direction of turning (Bonatz et al., 1987).

The direction of rotation and its degree performed in one cycle (150 ms) is memorized. If the rotation measured in the next cycle is in the same direction as in the former cycle, its angle is added to that of the first. In contrast, if the second rotation is in the opposite direction, its value is subtracted from that of the first (negative results are set as 0) and added to that of the other direction. This cumulation is continued until 90° (i.e., a quarter rotation) are accumulated in one direction. The direction of this quarter turn is stored together with the distance the central point has moved since the previous quarter turn. This distance is taken as the perimeter of the rotation and serves to calculate the diameter. A sequence of quarter turns to the left or to the right represents the basis for all the following calculations of 1/4, 1/2, 3/4, or full turns.

The following criteria had to be set in order to exclude possible artifacts:

1. Criterion of size: The rat's image must consist of at least 20 points. Thereby, we minimize artifacts resulting from

behaviors that may lead to a smaller shape in the plane (e.g., rearing).

2. Criterion of shape: The rat's image must differ from a circular area to allow a reliable calculation of the longitudinal axis. Thus, all shapes with a longitudinal axis shorter than 1.5 times that of the traverse axis are excluded. Thereby, we eliminate artifacts arising out of behaviors that lead to a circular shape (e.g., grooming behavior).

3. Criterion of locomotion: Turning behavior is defined as an asymmetry of the animal when ambulating. Thus, for example, head turns of an otherwise stationary animal are not registered. To fulfill this definition, the same criterion was used as for locomotion, that is, similar to the evaluation of locomotion itself, turning is evaluated only when the central point of the image is moving at least 0.7 cm between two consecutive images.

If one of these criteria is not reached, a "dropout" results, and calculation continues with the next image. The total number of dropouts is indicated in the final printout and can then be related to the detection frequency of about 5–6 images/s to control for the quality of the measurement.

Measurement of Thigmotactic Scanning

Thigmotactic scanning is defined as locomotion along the walls of the environment so that the vibrissae of one side can have contact with the wall (*see* Figs. 1 and 4). To measure this behavior, the same criteria have to be fulfilled as in the case of turning behavior. Thus, the rat's image has to have a certain size (>20 points), has to move (criterion of locomotion), and has to have an ellipsoid shape (criterion of shape). Furthermore, and most importantly, the image has to locomote within a certain distance and almost in parallel with respect to one of the walls of the environment. Therefore, the exact position of the four walls has to be indicated on the monitor by the experimenter. Any rectangular box of at least 20 × 20 cm up to about 100 × 100 cm can be used. The program then automatically defines a virtual inner zone of 5-cm width along each wall. If the central point of the rat's image is locomoting within this zone and if the longitudinal axis (α) does not deviate more than ±15° from the corresponding wall, then scanning behavior is

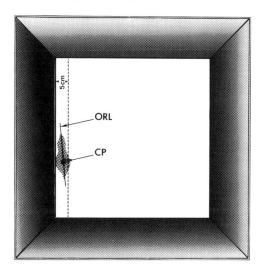

Fig. 4. The detection of thigmotactic scanning. The rat (i.e., its central point CP) has to locomote within a virtual corridor (5-cm wide) along one of the walls, the shape of its video-image has to be ellipsoid, and its longitudinal axis (ORL) must not deviate more than ±15° from the direction of the corresponding wall. Scanning is analyzed as distance/time or duration/time and is expressed separately for the left and the right side of the body.

recorded. This record is expressed as distance/time (cm/min) or duration/time (s/min). Clockwise movements are recorded as scanning with the left side of the body, and counterclockwise movements reflect scanning with the right side.

All these behavioral variables can be evaluated during any time interval of the whole measurement. The minimal interval currently used is 1 min. The printout chosen for the time interval to be analyzed shows the following data:

1. Locomotion (in cm);
2. Turning toward the left and the right (1/4, 1/2, 3/4, and full turns) in different diameters (<20, 20–30, <30, 30–55, >55 cm); and
3. Thigmotactic scanning with the left and the right side of the body (as distance/time or as duration/time). In addition, or alternatively to the printout, the data can also be directly transferred in the ASCII format to a spreadsheet or statistical software.

Furthermore, the "path of locomotion" may be printed out for any time interval of interest. In this case, the VIAS prints out the position of the center point for every image taken, that is 5–6 times/s. This results in a line of dots, which reflects the locomotion of the animal. Finally, the program, which usually only stores calculated data from the individual images, can also store the images themselves. These images can later be retrieved almost in real time and may be used (similar to a video record) for the purpose of teaching, for example.

Examples for Applications

Measurement of Behavioral Asymmetries in Animals with Unilateral Brain Lesions

Spontaneous Behavior and Effects of Dopaminergic Challenges

As indicated in the introduction, behavioral asymmetries are most often analyzed in animals with unilateral brain lesions. Here, the following questions may be of interest: What is the relationship between the degree of brain damage and the occurrence or expression of behavioral deficits? What is the time-course of functional recovery? How do drugs affect behavior? How may possible therapeutic treatments (drugs, transplants) induce or improve recovery?

In the case of the well-known unilateral 6-OHDA lesion, which is taken as a unilateral model of Parkinson's disease, DA neurons are destroyed in the nigrostriatal DA system on one side of the brain. This unilateral lesion is known to induce asymmetries in behavior, that is, turning to the side of DA loss, and a multimodal contralateral sensory neglect (e.g., Ungerstedt and Arbuthnott, 1970; Marshall et al., 1974; Ljungberg and Ungerstedt, 1976). This sensory neglect is probably reflected in the lateralized reduction of thigmotactic scanning, which is observed contralateral to the side of DA loss (Steiner et al., 1988; Mattioli et al., 1992; Fornaguera et al., 1993).

Both measures, obtained in the same animal by the automated video-image analysis, can be used to quantitate behavioral deficits in relation to the degree of neostriatal DA loss, to determine the occurrence and time-course of behavioral recovery, and to relate these behavioral with neural measures (*see*

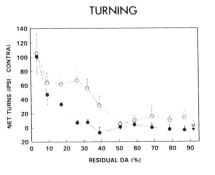

Fig. 5. Asymmetries in turning **(Left)** and thigmotactic scanning **(Right)** in groups of animals with different degrees of unilateral neostriatal DA lesion (expressed as percent of the contralateral hemisphere). Turning was measured as quarter turns, and scanning was measured as time spent during 15 min either on d 1 (open circles) or on d 7 (full circles). Both measures are expressed as net behavior, that is, ipsi- minus contraversive quarter turns or ipsi- minus contralateral scanning. Therefore, positive values (group means ± SEM) represent asymmetries for the side of DA lesion, and the dotted horizontal line represents symmetry. Animals with 6-OHDA lesions were assigned in ascending order of neostriatal DA denervation to 11 groups (9 animals/group). The behavioral data from each group are plotted on the *x*-axis according to the mean residual DA level of the respective group. Control animals, which had received unilateral vehicle injections, are plotted in the same way (open triangle, d 1; full triangle, d 7).

also Morgan et al., 1991, 1993; Schwarting et al., 1991a). Thus, in a recent study (Fornaguera et al., 1994), behavioral asymmetries in undrugged behavior were monitored during the first week after a unilateral 6-OHDA lesion of the substantia nigra and were related to the degree of neostriatal DA loss (determined as residual DA levels, percent of the contralateral hemisphere). This analysis showed that initially (d 1) after 6-OHDA administration, lateralized deficits can occur over a wide range of neostriatal DA lesions and that animals recovered to symmetry (on d 7) unless the lesion exceeded a certain degree (Fig. 5): On the first day, tight ipsiversive turning was observed in animals with a mean residual DA level of 32% or less. Out of these, the strongest asymmetries were found in animals with the most severe lesions. After 1 wk, these asymmetries had totally recovered, except in animals with mean residual DA levels of 17% or less. In these animals, partial recovery occurred with mean residual DA of 9 or 17%, whereas no recovery was found in animals with the most severe lesions (3% residual DA). With respect to scanning, ipsiversive asymmetries on d 1 were observed over a wider range of DA lesions than in turning, namely up to a mean residual DA level of 78%. Similar to turning, animals had recovered to symmetry on d 7 unless mean residual DA was 17% or less. Thus, a critical level of about 20% residual DA in the lesioned neostriatum seems to be necessary, up to which some compensatory mechanisms may be efficient to provide functional recovery. A similar level has been shown in Parkinson's disease, indicating that this aspect of the lesion model may be especially relevant for investigating mechanisms of deficits and recovery similar to the human disease.

Other, unpublished behavioral data from rats with unilateral 6-OHDA lesions are presented in Figs. 6 and 7. These animals had almost total unilateral lesions of neostriatal DA, that is, the lesion depleted DA by either 95–99% ($n = 4$; Fig. 6) or even more than 99% ($n = 3$; Fig. 7) of the contralateral neostriatum. The analysis of spontaneous behavior on d 4 after the lesion yielded that animals with DA depletions of 95–99% or >99% showed similar levels of ipsiversive turning behavior and almost no contraversive turns. The ipsiversive turns were performed predominantly within a narrow diameter (<30 cm). In contrast to the lack of contraversive turning, contralateral scanning behavior could be observed, but there was less scanning contralateral than ipsilateral to the side of the lesion. It should be pointed out that this ipsilateral asymmetry in scanning is not simply an artifact of the ipsiversive asymmetry in turning. The turns were observed in a small-diameter class, that is, the animals were curved to the side of the lesion during this behavior. For ipsilateral scanning, however, the animal has to walk along the wall with the ipsilateral side of the body, a behavior that is incompatible with narrow turns. Thus, turning and scanning reflect two separate measures.

The ip injection of the DA receptor agonist apomorphine (0.5 mg/kg; d 11 after lesion) induced intensive contraversive turning, that could be observed already during the first 5 min

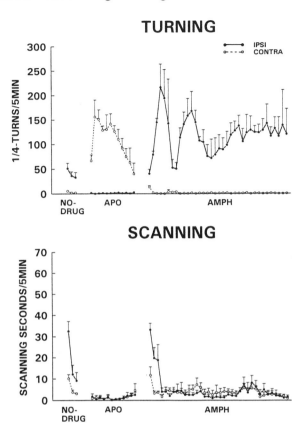

Fig. 6. Turning behavior (number of quarter turns within a diameter of <30 cm; mean + SEM; upper half) and thigmotactic scanning (distance in meters; lower half) in four animals with nigrostriatal DA depletions of 95–99%. Behavior was tested either in the undrugged state (NO-DRUG; **left**) after an ip injection of 0.5 mg/kg apomorphine (APO; **middle**), or 5.0 mg/kg amphetamine (AMPH; **right**). Results are expressed in time blocks of 5 min. Full lines correspond to ipsiversive turning or ipsilateral scanning; dotted lines correspond to contraversive turning or contralateral scanning.

after injection and reached its maximum at 6–10 min (95–99% DA depletions) or 21–25 min after injection (>99% DA depletions). Almost no ipsilateral turns could be observed during the 60 min of testing in animals with 95–99% depletions, whereas animals with the highest depletions (>99%) showed some ipsiversive turns during the second half of testing when

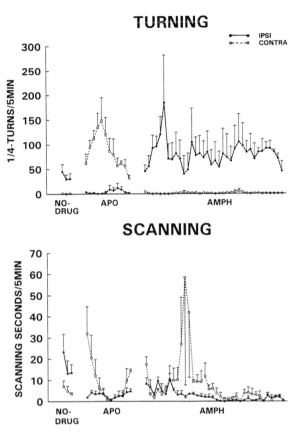

Fig. 7. Turning behavior (number of quarter turns within a diameter of <30 cm; mean + SEM; upper half) and thigmotactic scanning
(distance in meters; lower half) in four animals with nigrostriatal DA
depletions of more than 99%. Behavior was tested either in the
undrugged state (NO-DRUG; **left**) after an ip injection of 0.5 mg/kg
apomorphine (APO; **middle**), or 5.0 mg/kg amphetamine (AMPH;
right). Results are expressed in time blocks of 5 min. Full lines correspond to ipsiversive turning or ipsilateral scanning; dotted lines correspond to contraversive turning or contralateral scanning.

the rate of contraversive turns declined. The analysis of scanning under apomorphine indicated differences between the
95–99% and the >99% lesions: Whereas the animals with 95–99%
lesions showed almost no scanning (and therefore no asymmetry), those animals with lesions of >99% showed contraversive
scanning at the beginning (min 1–15) and at the end of testing

(min 51–60). Thus, not only turning, but also scanning behavior in 6-OHDA lesioned animals can be reversed by DA receptor stimulation. However, this reversal in scanning was observed when the levels of apomorphine-induced contraversive turning were relatively low, that is, at the beginning and toward the end of drug action.

Furthermore, these animals were tested after the ip injection of the indirect DA agonist amphetamine (d 18 after lesion). Here, a dose of 5.0 mg/kg was administered, since this dose is often used to select animals with severe DA lesions for the investigation of brain grafts (e.g., Björklund et al., 1980; Schmidt et al., 1982). This presynaptic DA challenge led to strong ipsiversive turning behavior, which outlasted the 3 h of observation, especially in the animals with 95–99% depletions. Furthermore, the time-course was clearly biphasic during the first 80 min in these animals. There were almost no contraversive turns after amphetamine. The levels of scanning were low and, in the case of the 95–99% group, an asymmetry of scanning could only be observed at the beginning of the test. Thereafter, scanning declined while the level of turning raised, probably reflecting the onset of drug action. The pattern in animals with >99% depletions was different, since there was a contraversive asymmetry at the beginning, which might reflect conditioning of the previous contraversive scanning asymmetry after apomorphine (*see also* Measurement of Conditioned Drug Effects in Rats with Unilateral 6-OHDA Lesions). Furthermore, there was an "outburst" of continuous contraversive scanning in one animal between 45 and 70 min after injection.

These data indicate the usefulness of measuring both turning and scanning in animals with unilateral brain lesions. For further details regarding the application of the two measures in the 6-OHDA model, the interested reader is referred to our previous publications (Steiner et al., 1988; Mattioli et al., 1992; Fornaguera et al., 1993, 1994, 1995). These experiments showed that turning and scanning may differ in sensitivity as indicators for the degree of brain damage and for the occurrence of functional recovery. Among others, these data suggest that thigmotactic scanning may not only provide an additional, but probably more sensitive indicator for functional asymmetries than turning.

Measurement of Conditioned Drug Effects in Rats with Unilateral 6-OHDA Lesions

In the previous example, it was shown that the DA receptor agonist apomorphine induces contraversive turning in animals with a severe unilateral 6-OHDA lesion. That is, the drug leads to a reversal of the otherwise ipsiversive deficit. This contraversive asymmetry under apomorphine is thought to reflect the stimulation of supersensitive DA receptors in the depleted neostriatum. However, in such animals, a single injection of apomorphine in a certain environment may also lead to a short phase of intense contraversive turning, when the animal is re-exposed to the same environment in the undrugged state. This "paradoxical" rotation is considered to be a conditioned drug effect: Usually a drug is thought to act as an unconditioned stimulus leading to an unconditioned response. However, association of a drug effect with a certain environment (conditioned stimulus) may lead to selective behavioral changes (conditioned reaction), when the animal is re-exposed to this environment in the undrugged state (for review, *see* Stewart and Eikelboom, 1987). In the case of the above-mentioned unilateral 6-OHDA model, such conditioned drug effects have been reported by several laboratories and especially after drugs, such as apomorphine and amphetamine (e.g., Silverman and Ho, 1981; Carey, 1986; Burunat et al., 1987).

In the following example, this behavior is presented in a minute-by-minute analysis in two animals with unilateral 6-OHDA lesions. Two weeks after the lesion, which induced a neostriatal DA depletion of more than 98%, the animals were tested for spontaneous behavior using the VIAS as described above. Three hours later, they received a first sc injection of 0.5 mg/kg apomorphine and were tested again. The drug procedure was repeated three times every 48 h. After an additional period of 7 wk without any treatment or test, the animals received an injection of vehicle instead of apomorphine and were tested again.

The analysis of narrow quarter turns (<30 cm in diameter) showed the following (Fig. 8): In the drug-free test (PRE) preceding the conditioning phase, the two animals showed only ipsiversive turns that approximated zero within the 4 min of testing. As indicated in the previous example, this is the typical ipsiversive asymmetry of animals with severe unilateral DA

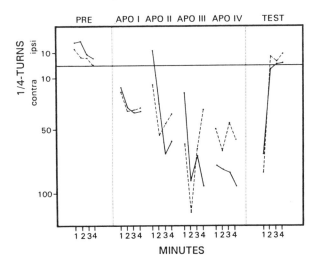

Fig. 8. Turning behavior (quarter turns per minute; within a diameter of <30 cm) of two rats with a unilateral 6-OHDA lesion of the nigrostriatal DA system. Behavior is expressed as either ipsiversive (upward) or contraversive (downward) to the side of the lesion. PRE: spontaneous behavior; APO I-IV: behavioral tests after an sc injection of 0.5 mg/kg apomorphine (every 48 h); TEST: spontaneous behavior (after a vehicle injection) performed 7 wk after the last apomorphine treatment.

lesions. Under apomorphine, these animals showed a strong contraversive asymmetry in turning, which increased during the time of 4 min. At the same time, almost no scanning was observed (results not shown). Furthermore, the contraversive asymmetry in turning started off at a higher level from test to test (APO I-IV). Finally, and most importantly, in the drug-free test performed 7 wk later (TEST), these animals again showed a strong contraversive asymmetry. When looking at the sum of these 4 min of testing, the animals displayed about 5–15 ipsiversive quarter turns compared to about 70 contraversive quarter turns. However, the minute-by-minute analysis shows that these contraversive turns were performed almost exclusively during the first minute, which confirms the data reported by Silverman and Ho (1981). During this first minute, the level of conditioned turning was similar to that observed with apomorphine testing. In the 3 min thereafter, the animals resumed their ipsiversive asymmetry, as otherwise observed in the undrugged state.

This example shows that the video-image analysis, especially when using a high temporal resolution, is qualified for a
refined analysis of conditioned drug effects in the 6-OHDA
model, where it may, for example, be used to analyze the important time-courses of these conditioned effects.

Measurement of Behavioral Asymmetries
After Unilateral Vibrissae Removal

The analysis of behavioral asymmetries in turning and
scanning is not restricted to the field of manipulations in
the brain. This is exemplified in the following, where a noninvasive peripheral manipulation was performed at the level
of the vibrissae, which provide an important sensory organ in
the rat (for review, *see* Huston et al., 1990). Thus, unilateral
removal of this input by means of clipping them on one side of
the face (hemivibrissotomy) leads to asymmetries in open field
behavior, because the animals now preferentially scan the walls
of the environment with the side of intact vibrissae (the contralateral side; Steiner et al., 1986; Milani et al., 1990). Furthermore, animals can recover from this asymmetry within about a
week, even though the vibrissae, which otherwise would
regrow, are clipped daily. The states of behavioral asymmetry
on the one hand and that of recovery on the other are related to
different states in the brain, comprising at least parts of the basal
ganglia and monoaminergic metabolism (for review, *see* Huston
et al., 1990). This close relationship between behavioral and
neural effects of unilateral vibrissae removal suggests that hemivibrissotomy may serve as a noninvasive approach to study
brain and behavior, for instance, in the context of recovery
(Huston et al., 1988).

Here, we give an example of how hemivibrissotomy can
affect open field behavior (Fig. 9). The data presented are reanalyzed video records taken from a previous study (Schwarting
et al., 1990). In short, animals were tested for spontaneous
behavior in the open field either 4 h or 10 d after unilateral
vibrissae removal. The duration of testing was 5 min. The group
that was tested shortly (4 h, $n = 6$) after hemivibrissotomy
showed an asymmetry in scanning; that is, they scanned more
with the side of intact vibrissae. In contrast, the group tested
after 10 d of vibrissae removal ($n = 6$) was balanced. Further-

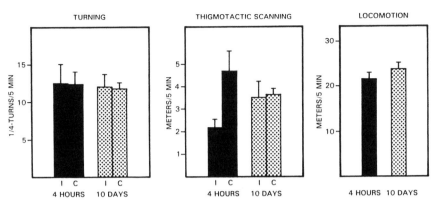

Fig. 9. Turning **(Left)**, thigmotactic scanning **(Middle)**, and loco-motion **(Right)** in animals with unilateral vibrissae removal (means + SEM). Behavior was tested during 5 min either 4 h (full bars) or 10 d (stippled bars) after hemivibrissotomy (n = 6 each). Thigmotactic scanning was measured as distance traveled and is expressed in m/5 min either ipsilateral (I) or contralateral (C) to the side of vibrissae removal. Turning was measured as quarter turns (within a diameter of <30 cm) ipsiversive (I) and contraversive (C) to the side of vibrissae removal. Locomotion is measured in meters.

more, when comparing the two groups, it can be seen that recovery to symmetry was probably the result of a contralateral reduction in scanning together with an ipsilateral increase. In contrast, the measure of turning behavior did not indicate functional asymmetries, since the levels of turning were symmetrical and similar after 4 h and 10 d. Finally, the levels of locomotion did not differ between the two groups.

These data illustrate that the measure of thigmotactic scanning can be a useful indicator for functional asymmetries not only in animals with unilateral brain lesions, but also in animals with unilateral peripheral manipulations, such as of the vibrissae system. Here, an asymmetry for the intact side is observed acutely after vibrissae removal, an asymmetry from which the animals can recover. These different behavioral states can then be used as a basis to study their relationships to possible changes in the brain, such as in the trigemino-thalamo-cortical system or in the basal ganglia (Huston et al., 1990; Adams et al., 1992). Finally, it should be pointed out that the analysis of scanning behavior may also be useful to indicate

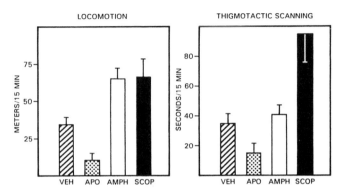

Fig. 10. Locomotion **(Left)** and thigmotactic scanning **(Right)** during a 15 min test performed 10 min after an ip injection of either
saline (VEH; striped bars), 1.5 mg/kg apomorphine (APO; stippled
bars), 1.5 mg/kg amphetamine (AMPH; open bars), or 1.5 mg/kg
scopolamine (SCOP; full bars; n = 5 each). Locomotion is expressed
in m/15 min; scanning is expressed in s/15 min.

functional asymmetries in entirely intact animals (Schwarting
et al., 1991b).

Measurement of Drug Effects in Intact Animals

The final example shows that automated video-image
analysis can not only be utilized for the analysis of behavioral
asymmetries, but also for that of open field behavior in general
(*see also* Schwarting and Huston, 1992). Here, the measures of
locomotion and scanning may be especially useful. To demonstrate such an application, intact and naive male Wistar rats
were injected intraperitoneally with one of the following drugs:
the DA receptor agonist apomorphine, the indirect DA agonist
amphetamine, the muscarinic receptor antagonist scopolamine,
or vehicle (saline). Each drug was administered in five animals
in a dosage of 1.5 mg/kg; an individual animal received only
one treatment. Ten minutes after injection, the animals were
tested in the open field as described above.

Compared to vehicle-injected controls, the level of
locomotion was reduced under apomorphine and increased
to a similar level after amphetamine or scopolamine (Fig. 10).
The time spent scanning under apomorphine was reduced
compared to that of the controls. Similar to locomotion,

VEH APO

AMPH SCOP

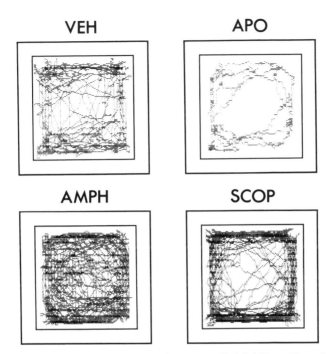

Fig. 11. Paths of locomotion in the open field (60 × 60 cm) printed out by the VIAS. The data are from individual animals during a 15-min test performed 10 min after an ip injection of either saline (VEH), 1.5 mg/kg apomorphine (APO), 1.5 mg/kg amphetamine (AMPH), or 1.5 mg/kg scopolamine (SCOP).

scanning was increased after scopolamine, but not after amphetamine.

The "paths of locomotion," taken from one representative animal of each group (Fig. 11) illustrate these quantitative data: control animals mainly walked along the walls of the open field, whereas apomorphine-treated animals often showed an uneven spatial pattern. Under amphetamine, animals locomoted along the walls, but also very often through the center of the open field, whereas scopolamine-treated animals only rarely entered the center.

These data, both quantitatively and qualitatively, confirm those reported by others (e.g., Geyer et al., 1986; Sanberg et al., 1987; Paulus and Geyer, 1991). They show that the VIAS can also be used for experiments with intact animals, where possible behavioral asymmetries are not necessarily of interest.

Here, the emphasis may lie on the measures of locomotion and scanning, which might serve as indicators of exploration, habituation, or emotionality (Royce, 1977; Sanberg et al., 1987; Treit and Fundytus, 1989).

Conclusions

We described a video-image analyzing system and its application for the measurement of turning, thigmotactic scanning, and locomotion in rats. The method is based on an open field situation, which is widely employed in behavioral studies of rodents. This approach generally has the practical advantage that no behavioral training is necessary. The size of the open field, which should be rectangular for the present way to measure scanning behavior, may be varied between a minimum of 20 × 20 up to about 100 × 100 cm. This video-image analysis can be used under different lighting conditions, ranging from dark (that is, dim red light) to daylight. The system detects the animal via the contrast to its environment, which is most easily achieved by using albino animals and a black open field. However, when working with different gray levels of video input, situations with low contrasts between subject and environment can also be mastered. Since the data are stored on disk (and/or video record), they may be reanalyzed at will. This is especially useful for the analysis of time; thus, for example, a given measurement of 20-min duration may be expressed as a 20-min segment, in blocks of 5 min, minute-by-minute, and so forth.

What are the general advantages and drawbacks of video systems compared to conventional approaches, that is, mechanical devices, infrared systems, or observational methods? With respect to the measurement of turning, video-image analysis has the advantage over other existing methods (the so-called rotometers) that no harnesses (or marks) are necessary. Thus, interference between experimenter and subject can be minimized, and possible artificial stimulation induced by a harness or mark is prevented. Recording and analyzing behavior by means of video and computer are fast and objective, have a high resolution—both with respect to space and time—and do not fatigue. Furthermore, several different behaviors can be detected simultaneously, the kind of analysis is flexible, and there are fast and alternative ways of data management (for example, in combination with spread sheets). On the other hand, when using video-image analysis, especially with unmarked animals, care must be taken to avoid artifacts. Thus, reflections from urine or feces may be detected and may interfere with the detection of the animal, that can be minimized, for example, by choosing the appropriate floor, the correct illumination, and/or image-filtering techniques. Video-image may reach its limits, especially when total darkness is required; however, most commercial video cameras can master very low levels of illumination and, when removing the internal infrared filter, can be used under infrared conditions. Finally, one might object that video-image analysis is costly; however, the prices for the necessary equipment (camera, digitizer, computer) have dropped considerably in the last few years. Thus, a system can be set up for <$5000. This includes a conventional computer, which can also be used for other purposes in the lab.

The given examples of behavioral studies indicate the applicability and usefulness of video-image analysis in experimental work, comprising the study of intact animals and the effects of peripheral or central manipulations. Video-image analysis is useful in analyzing open field behavior in intact animals, since it can, for example, differentiate the effects of various drugs. Most importantly, a system such as the VIAS can measure behavioral asymmetries (in turning and scanning) in open field situations. Thus, it shows functional asymmetries after unilateral vibrissae removal, and determines asymmetries both in spontaneous and drug-induced behavior in animals with unilateral brain lesions. Certainly, various other applications can be thought of. In contrast to other comparable methods (especially the rotometers), this system can measure turning and thigmotactic scanning. The possibility of measuring both turning and scanning may expand the study of asymmetries substantially, because the two measures not only complement each other, but can provide new information: For example, one measure might indicate functional asymmetries that might not be signaled by the other. Accordingly, our previous work has shown that thigmotactic scanning can serve as a sensitive indicator of asymmetry both after unilateral vibrissae removal (e.g., Steiner et al., 1986; Milani et al., 1990) and after unilateral lesions of the nigrostriatal DA system (Steiner et al., 1988; Mattioli et al., 1992; Fornaguera et al., 1993, 1994). Furthermore, the measure of thigmotactic scanning may provide an interesting behavioral variable for those working on functional aspects of the trigemino-thalamo-cortical representation of the vibrissae, since this measure is also sensitive, for example, to lesions of the barrel cortex (Adams et al., 1992).

Taken together, video-image analysis is a useful tool for the detailed analysis of turning behavior, thigmotactic scanning, and locomotion, which can provide the experimenter with refined and accurate behavioral measures. These may be used for the general analysis of open field behavior in rats, namely the analysis of scanning and locomotion, for example, in spontaneous behavior or after drug treatment. However, the main area of application certainly is that of behavioral asymmetries, which are usually performed either in intact animals, or after peripheral or central manipulations. The latter, especially the study of unilateral brain lesions, presumably is the one most frequently used. Here, video-image analysis can serve to define the kind and degree of different behavioral asymmetries, and their time-courses and recovery, as well as allowing testing of the effects of interventions, such as drugs, neurotrophic factors, or brain transplants.

Video and computer techniques are currently developing and improving enormously; thus, it can be anticipated that their applicability for behavioral measurement will expand substantially in the future, leading to an era when automated video-image analysis will be a standard technique in behavioral labs.

Acknowledgments

The development of the VIAS system and our experimental studies have been supported by grants from the Deutsche Forschungsgemeinschaft. R. K. W. Schwarting is a Heisenberg-fellow of the Deutsche Forschungsgemeinschaft.

Index of Instrumentation Companies

Poly-Track Video Tracking System
San Diego Instruments, Inc.
7758 Arjons Drive
San Diego, CA 92126-4391, USA
Tel. +619 530–2600, FAX +619 530–2646

Video Activity System
  TSE, Technical & Scientific Equipment GmbH
  Ludwigstr. 10
  6380 Bad Homburg v.d.H., Germany
  Tel. +49 6172 26421, FAX +49 6172 26428
Video Tracking Activity System
  (Activity Monitor, Video path analyzer)
  Coulburn Instruments
  7462 Penn Drive
  Allentown, PA 18106, USA
  Tel. +215–395–3771, FAX +215–391–1333
EthoVision
  The Observer, Video Tracking
    and Motion Analysis System
  Noldus Information Technology bv
  Business & Technology Center
  Costerweg 5, 6702 AA Wageningen
  The Netherlands
  Tel. +31–8370–97677, FAX +31–8370–24496
SMART,
  Spontaneous Motor Activity Recording & Tracking
  Letica S/A, Scientific Instruments
  Cromo 37, 08907 Hospitalet (Barcelona)
  Spain
  Tel. +34–9–33366062, FAX +34–9–33354210

References

Adams, F., Schön, H., Schwarting, R. K. W., and Huston, J. P. (1992) Behavioral and neurochemical indices of barrel cortex—basal ganglia interaction. *Brain Res.* **597**, 114–123.

Adams, F. S., Schwarting, R. K. W., and Huston, J. P. (1994) Behavioral and neurochemical asymmetries following unilateral trephination of the rat skull: Is this control operation always appropriate? *Physiol. Behav.* **55**, 947–952.

Adani N., Kiryati, N., and Golani, I. (1991) The description of rat drug-induced behavior: kinematics versus response categories. *Neurosci. Biobehav. Rev.* **15**, 455–460.

Barber, D. L., Blackburn, T. P., and Greenwood, D. T. (1973) An automatic apparatus for recording rotational behaviour in rats with brain lesions. *Physiol. Behav.* **11**, 117–120.

Björklund, A., Dunnett, S. B., Stenevi, U., Lewis, M. E., and Iversen, S. D. (1980) Reinnervation of the denervated striatum by substantia nigra transplants: functional consequences as revealed by pharmacological and sensorimotor testing. *Brain Res.* **199**, 307–333.

Bonatz, A. E., Steiner, H., and Huston, J. P. (1987) Video-image analysis of behavior by microcomputer: categorization of turning and locomotion after 6-OHDA injection into the substantia nigra. *J. Neurosci. Methods* **22,** 13–26.

Brann, M. R., Finnerty, M., Lenox, R. H., and Ehrlich, Y. H. (1982) A device for the automated monitoring of stereotypic behavior. *Prog. Neuro-Psychopharmacol. & Biol. Psychiat.* **6,** 351–354.

Burunat, E., Diaz-Palarea, M. D., Castro, R., and Rodriguez, M. (1987) Undrugged rotational response in nigrostriatal system lesioned rats is related to the previous early response to apomorphine when repeatedly administered. *Life Sci.* **41,** 309–313.

Carey, R. J. (1986) Conditioned rotational behavior in rats with unilateral 6-hydroxydopamine lesions of the substantia nigra. *Brain Res.* **365,** 379–382.

Carey, R. J. (1989) Stimulant drugs as conditioned and unconditioned stimuli in a classical conditioning paradigm. *Drug Dev. Res.* **16,** 305–315.

Crawley, J. N., Szara, S., Pryor, G. T., Creveling, C. R., and Bernard, B. K. (1982) Development and evaluation of a computer-automated color TV tracking system for automatic recording of the social and exploratory behavior of small animals. *J. Neurosci. Methods* **5,** 235–247.

Dusenbery, D. B. (1985) Using a microcomputer and video camera to simultaneously track 25 animals. *Comput. Biol. Med.* **15,** 169–175.

Etemadzadeh, E., Koskinen, L., and Kaakola, S. (1989) Computerized rotometer apparatus for recording circling behavior. *Methods Findings Exp. Clin. Pharmacol.* **11,** 399–407.

Fornaguera, J., Schwarting, R. K. W., Boix, F., and Huston, J. P. (1993) Behavioral indices of moderate nigrostriatal 6–hydroxydopamine lesion: a preclinical Parkinson's model. *Synapse* **13,** 179–185.

Fornaguera, J., Carey, R. J., Huston, J. P., and Schwarting, R. K. W. (1994) Behavioral asymmetries and recovery in rats with different degrees of unilateral striatal dopamine depletion. *Brain Res.* **664,** 178–188.

Fornaguera, J., Carey, R. J., Huston, J. P., and Schwarting, R. K. W. (1995) Stimulation of D1- or D2-receptors in drug-naive rats with different degrees of unilateral nigro-striatal dopamine lesions. *Psychopharmacology,* in press.

Geyer, M. A., Russo, P. V., and Masten, V. L. (1986) Multivariate assessment of locomotor behavior: pharmacological and behavioral analyses. *Pharmacol. Biochem. Behav.* **25,** 277–288.

Greenstein, S. and Glick, S. D. (1975) Improved automated apparatus for recording rotation (circling behavior) in rats or mice. *Pharmacol. Biochem. Behav.* **3,** 507–510.

Grossen, N. E. and Kelley, M. J. (1972) Species-specific behavior and acquisition of avoidance behavior in rats. *J. Comp. Physiol. Psychol.* **81,** 307–310.

Guilleux, H. and Peterfalvi, M. (1974) Le comportement de rotation après lésion unilatérale du striatum analysé à l'aide d'un rotomètre. *J. Pharmacol. (Paris)* **5,** 63–74.

Hodge, G. K. and Butcher, L. L. (1979) Role of the pars compacta of the substantia nigra in circling behavior. *Pharmacol. Biochem. Behav.* **10,** 695–709.

Höglund, A. U., Hägglund, J.-E., and Meyerson, B. J. (1983) A video interface for behavioural recordings with applications. *Physiol. Behav.* **30,** 489–492.

Hopper, D. L., Kernan, W. J., and Wright, J. R. (1990) Computer pattern recognition: an automated method for evaluating motor activity and testing neurotoxicity. *Neurotoxicol. Teratol.* **12,** 419–428.

Huston, J. P., Steiner, H., Schwarting, R. K. W., and Morgan, S. (1988) Parallels in behavioral and neural plasticity induced by unilateral vibrissae removal and unilateral lesion of the substantia nigra, in *Post-Lesion Neural Plasticity* (Flohr, H., ed.), Springer-Verlag, Berlin, pp. 537–551.

Huston, J. P., Steiner, H., Weiler, H.-T., Morgan, S., and Schwarting, R. K. W. (1990) The basal ganglia-orofacial system: studies on neurobehavioral plasticity and sensory-motor tuning. *Neurosci. Biobehav. Rev.* **14,** 433–446.

Jerussi, T. P. (1982) A simple, inexpensive rotometer for automatically recording the dynamics of circling behavior. *Pharmacol. Biochem. Behav.* **16,** 353–357.

Kernan, W. J., Mullenix, P. J., and Hopper, D. L. (1987) Pattern recognition of rat behavior. *Pharmacol. Biochem. Behav.* **27,** 559–564.

Kernan, W. J., Jr., Mullenix, P. J., and Hopper, D. L. (1989) Time structure analysis of behavioral acts using a computer pattern recognition system. *Pharmacol. Biochem. Behav.* **34,** 863–869.

Livesey, P. J. and Leppard, K. (1981) A TV monitored system for recording open field activity in the rat. *Behav. Res. Methods Instr. Comp.* **13,** 331–333.

Ljungberg, T. and Ungerstedt, U. (1976) Sensory inattention produced by 6–hydroxydopamine-induced degeneration of ascending dopamine neurons in the brain. *Exp. Neurol.* **53,** 585–600.

Lubinski, K. S., Dickson, K. L., and Cairns, J., Jr. (1977) Microprocessor-based interface converts video signals for object tracking. *Computer Design* **December,** 81–87.

Marshall, J. F., Richardson, I. S., and Teitelbaum, P. (1974) Nigrostriatal bundle damage and the lateral hypothalamic syndrome. *J. Comp. Physiol. Psychol.* **88,** 808–830.

Martin, B. R., Prescott, W. R., and Zhu, M. (1992) Quantitation of rodent catalepsy by a computer-imaging technique. *Pharmacol. Biochem. Behav.* **43,** 381–386.

Mattioli, R., Schwarting, R. K. W., and Huston, J. P. (1992) Recovery from unilateral 6–hydroxydopamine lesion of substantia nigra promoted by the neurotachykinin substance P_{1-11}. *Neuroscience* **48,** 595–605.

Milani, H., Schwarting, R. K. W., Kumpf, S., Steiner, H., and Huston, J. P. (1990) Interaction between recovery from behavioral asymmetries induced by hemivibrissotomy in the rat and the effects of apomorphine and amphetamine. *Behav. Neurosci.* **104,** 470–476.

Miller, R. and Beninger, R. J. (1991) On the interpretation of asymmetries of posture and locomotion produced with dopamine agonists in animals with unilateral depletion of striatal dopamine. *Prog. Neurobiol.* **36,** 229–256.

Morgan, S., Nomikos, G., and Huston, J. P. (1991) Changes in the nigrostriatal projection associated with recovery from lesion-induced behavioral asymmetry. *Behav. Brain Res.* **46,** 157–165.

Morgan, S., Nomikos, G., and Huston, J. P. (1993) Behavioral analysis of asymmetries induced by unilateral 6-OHDA injections into the substantia nigra. *Behav. Neural. Biol.* **60,** 241–250.

Naylor, R. J. and Olley, J. E. (1972) Modification of the behavioural changes induced by amphetamine in the rat by lesions in the caudate nucleus, the caudate-putamen and globus pallidus. *Neuropharmacology* **11,** 91–99.

Nilsson, G. E., Rosén, P., and Johansson, D. (1993) Anoxic depression of spontaneous locomotor activity in crucian carp quantified by a computerized imaging technique. *J. Exp. Biol.* **180,** 153–162.

Paulus, M. P. and Geyer, M. A. (1991) A temporal and spatial scaling hypothesis for the behavioral effects of psychostimulants. *Psychopharmacology* **104,** 6–16.

Pisa, M. and Szechtman, H. (1986) Lateralized and compulsive exteroceptive orientation in rats treated with apomorphine. *Neurosci. Lett.* **64,** 41–46.

Ploeger, G. E., Spruijt, B. M., and Cools, A. R. (1992) Effects of haloperidol on the acquisition of a spatial learning task. *Physiol. Behav.* **52,** 979–983.

Pycock, C. J. (1980) Turning behavior in animals. *Neuroscience* **5,** 461–470.

Pycock, C. J. and Kilpatrick, I. C. (1989) Motor asymmetries and drug effects. Behavioural analyses of receptor activation, in *Neuromethods, Vol. 13, Psychopharmacology* (Boulton, A. A., Baker, G. B., and Greenshaw, A. J., eds.), Humana, Clifton, NJ, pp. 1–93.

Renner, M. J., Pierre, P. J., and Schilcher, P. J. (1990) Contrast-based digital tracking versus human observers in studies of animal locomotion. *Bull. Psychon. Soc.* **28,** 77–79.

Royce, J. R. (1977) On the construct validity of open field measures. *Psychol. Bull.* **84,** 1098–1106.

Sanberg, P. R., Zoloty, S. A., Willis, R., Ticarich, C. D., Rhoads, K., Nagy, P. R., Mitchell, S. G., Laforest, A. R., Jenks, J. A., Harkabus, L. J., Gurson, D. B., Finnefrock, J. A., and Bednarik, E. J. (1987) Digiscan activity: automated measurement of thigmotactic and stereotypic behavior in rats. *Pharmacol. Biochem. Behav.* **27,** 569–572.

Schmidt, R. H., Ingvar, M., Lindvall, O., Stenevi, U., and Björklund, A. (1982) Functional activity of substantia nigra grafts reinnervating the striatum: neurotransmitter metabolism and [^{14}C]2–deoxy-D-glucose autoradiography. *J. Neurochem.* **38,** 737–748.

Schwarting, R. K. W. and Huston, J. P. (1987) Dopamine and serotonin metabolism in brain sites ipsi- and contralateral to direction of conditioned turning in rats. *J. Neurochem.* **48,** 1473–1479.

Schwarting, R. K. W. and Huston, J. P. (1992) Behavioral concomitants of regional changes in the brain's biogenic amines after apomorphine and amphetamine. *Pharmacol. Biochem. Behav.* **41,** 675–682.

Schwarting, R. K. W., Steiner, H., and Huston, J. P. (1990) Effects of hemivibrissotomy in the rat: time-dependent asymmetries in turning and biogenic amines induced by apomorphine. *Pharmacol. Biochem. Behav.* **35,** 989–994.

Schwarting, R. K. W., Bonatz, A. E., Carey, R. J., and Huston, J. P. (1991a) Relationships between indices of behavioral asymmetries and neurochemical changes following mesencephalic 6-hydroxydopamine injections. *Brain. Res.* **554,** 46–55.

Schwarting, R. K. W., Steiner, H., and Huston, J. P. (1991b) Asymmetries in thigmotactic scanning: evidence for a role of dopaminergic mechanisms. *Psychopharmacology* **103**, 19–27.

Schwarting, R. K. W., Goldenberg, R., Steiner, H., Fornaguera, J., and Huston, J. P. (1993) A video-image analyzing system for open field behavior in the rat focusing on behavioral asymmetries. *J. Neurosci. Methods* **49**, 199–210.

Schwarz, R. D., Stein, J. W., and Bernard, P. (1978) Rotometer for recording rotation in chemically or electrically stimulated rats. *Physiol. Behav.* **20**, 351–354.

Silverman, P. B. and Ho, B. T. (1981) Persistent behavioral effect of apomorphine in 6–hydroxydopamine-lesioned rats. *Nature* **294**, 475–477.

Simon, P., Dupuis, R., and Costentin, J. (1994) Thigmotaxis as an index of anxiety in mice. Influence of dopaminergic drugs. *Behav. Brain Res.* **61**, 59–64.

Spooner, R. I. W., Thomson, A., Hall, J., Morris, R. G. M., and Salter, S. H. (1994) The Atlantis platform: a new design and further developments of Buresova's on-demand platform for the water maze. *Learning & Memory* **1**, 203–211.

Spruijt, B. M. and Gispen, W. H. (1983) Prolonged animal observation by use of digitized videodisplays. *Pharmacol. Biochem. Behav.* **19**, 765–769.

Spruijt, B., Pitsikas, N., Algeri, S., and Gispen, W. H. (1990) Org2766 improves performance of rats with unilateral lesions in the fimbria fornix in a spatial learning task. *Brain Res.* **527**, 192–197.

Spruijt, B. M., Hol, T., and Rousseau, J. (1992) Approach, avoidance, and contact behavior of individually recognized animals automatically quantified with an imaging technique. *Physiol. Behav.* **51**, 747–752.

Steiner, H., Huston, J. P., and Morgan, S. (1986) Apomorphine reverses direction of asymmetry in facial scanning after 10 days of unilateral vibrissae removal in rat: vibrissotomy-induced denervation supersensitivity? *Behav. Brain Res.* **22**, 283–287.

Steiner, H., Bonatz, A. E., Huston, J. P., and Schwarting, R. K. W. (1988) Lateralized wall-facing versus turning as measures of behavioral asymmetries and recovery of function after injection of 6-hydroxydopamine into the substantia nigra. *Exp. Neurol.* **99**, 556–566.

Stewart, J. and Eikelboom, R. (1987) Conditioned drug effects, in *Handbook of Psychopharmacology, Vol. 19, New Directions in Behavioral Pharmacology* (Iversen, L., Iversen, S., and Snyder, S., eds.), Plenum, New York, pp. 1–57.

Sullivan, R. M., Parker, B. A., and Szechtman, H. (1993) Role of the corous callosum in expression of behavioral asymmetries induced by a unilateral dopamine lesion of the substantia nigra in the rat. *Brain Res.* **609**, 347–350.

Szechtman, H. (1983) Peripheral sensory input directs apomorphine-induced circling in rats. *Brain Res.* **264**, 332–335.

Tanger, H. J., Vanwersch, R. A. P., and Wolthuis, O. L. (1978) Automated TV-based system for open field studies: effects of methamphetamine. *Pharmacol. Biochem. Behav.* **9**, 555–557.

Tanger, H. J., Vanwersch, R. A. P., and Wolthuis, O. L. (1984) Automated quantitative analysis of coordinated locomotor behaviour in rats. *J. Neurosci. Methods* **10**, 237–245.

Torello, M. W., Czekajewski, J., Potter, E. A., Kober, K. J., and Fung, Y. K. (1983) An automated method for measurement of circling behavior in the mouse. *Pharmacol. Biochem. Behav.* **19**, 13–17.

Treit, D. and Fundytus, M. (1989) Thigmotaxis as a test for anxiolytic activity in rats. *Pharmacol. Biochem. Behav.* **31**, 959–962.

Ungerstedt, U. and Arbuthnott, G. W. (1970) Quantitative recording of rotational behavior in rats after 6–hydroxydopamine lesions of the nigrostriatal dopamine system. *Brain Res.* **24**, 485–493.

Vorhees, C. V., Acuff-Smith, K. D., Minck, D. R., and Butcher, R. E. (1992) A method for measuring locomotor behavior in rodents: contrast-sensitive computer-controlled video tracking activity assessment in rats. *Neurotoxicol. Teratol.* **14**, 43–49.

Walker, J. L., Resig, P., Guarnieri, S., Sisken, B. F., and Evans, J. M. (1994) Improved footprint analysis using video recording to assess functional recovery following injury to the rat sciatic nerve. *Rest. Neurol. Neurosci.* **6**, 189–193.

Walsh, M. J. and Silbergeld, E. K. (1979) Rat rotation monitoring for pharmacology research. *Pharmacol. Biochem. Behav.* **10**, 433–436.

Wise, R. A. and Holmes, L. J. (1986) Circling from unilateral VTA morphine: direction is controlled by environmental stimuli. *Brain Res. Bull.* **16**, 267–269.

Wolfer, P. D. and Lipp, H.-P. (1992) A new computer program for detailed off-line analysis of swimming navigation in the Morris water maze. *J. Neurosci. Methods* **41**, 65–74.

Ye, S. and Bell, W. J. (1991) A simple video position-digitizer for studying animal movement patterns. *J. Neurosci. Methods* **37**, 215–225.

Young, M. S., Li, Y. C., and Lin, M. T. (1993) A modularized infrared light matrix system with high resolution for measuring animal behaviors. *Physiol. Behav.* **53**, 545–551.

Ziegler, M. and Szechtman, H. (1988) Difference in the behavioral profile of circling under amphetamine and apomorphine in rats with unilateral lesions of the substantia nigra. *Behav. Neurosci.* **102**, 276–288.

CHAPTER 6

Automated Methods of Measuring Gnawing Behaviors

Donald E. Moss and Edward Castañeda

Introduction

Stereotyped behaviors have been defined as the repetitive performance of invariant sequences (fixed patterns) of purposeless movements (e.g., Randrup and Munkvad, 1967; Fog, 1972; Fray et al., 1980; Sanberg et al., 1983). Stereotypies, including gnawing, are well organized programs for normal species' typical behaviors that are activated (or disinhibited) by various drugs (e.g., Redgrave et al., 1982). Stereotyped behaviors occur without learning (Ungerstedt, 1979), and they occur in virtually identical form in the same species under the same conditions (Randrup and Munkvad, 1967).

Gnawing is a specialization in rodents (Byrd, 1981) shown by fossil evidence to exist since Pliocene times (Khozatskii and Vyalov, 1976). It is a typical behavior, by definition, in both of the orders of Rodentia and Lagomorpha. There is further specialization within the order Rodentia insofar as masticatory patterns are different in rats, hamsters, and guinea pigs (Byrd, 1981) and with a clearly defined pattern of postnatal development (Alleva et al., 1979). Detailed studies of mandibular movements have also shown that gnawing is different from chewing (Byrd, 1981), perhaps because incisors are used to bite and tear

From: *Motor Activity and Movement Disorders*
P. R. Sanberg, K. P. Ossenkopp, M. Kavaliers, Eds. Humana Press Inc., Totowa, NJ

in gnawing, whereas molars and premolars are used to crush in chewing.

Stereotyped gnawing (also described as compulsive biting) has long been known to be induced by apomorphine and amphetamine in rabbits, guinea pigs, and rats (Harnack, 1874; Amsler, 1923; Hauschild, 1939; Quinton and Halliwell, 1963; Ernst and Smelik, 1966; Ernst, 1967; Janssen et al., 1967) and, furthermore, this behavior appears to involve the striatum and dopaminergic functions (cf. Andén and Johnels, 1967; Ernst, 1967; Randrup and Munkvad, 1967).

Because stereotyped behaviors can be so closely tied to functional neurophysiological mechanisms, they are of significant interest in basic and clinical research related to the central nervous system. The significance of testing neuroleptics for "antiamphetamine and antiapomorphine" efficacy was recognized early (Janssen et al., 1967), and procedures were developed for measuring stereotyped gnawing and other behaviors (e.g., Janssen et al., 1960, 1967; Randrup and Munkvad, 1967). Thus, stereotyped gnawing has practical and theoretical significance, and measurement of stereotyped gnawing has been the purpose of many subsequent rating scales, tests of gnawing on wood and other materials, and recently, more fully automated methods.

Rationale for Selective Measurement of Gnawing Behavior

Gnawing has commonly been combined with other oral stereotypies, primarily sniffing and licking, as the highest intensity rating in scales that generally assume a behavioral continuum (cf. Janssen et al., 1967; Scheel-Kruger, 1971; Costall and Naylor, 1972; Rotrosen et al., 1972; Creese and Iversen, 1973; Peachey et al., 1976; Braestrup, 1977; Ljungberg, 1978; Pechnick et al., 1979; Sanberg et al., 1983). There have been numerous later modifications that combine stereotypies at different levels of a behavioral continuum (e.g., MacLennan and Maier, 1983). However, there is evidence that stereotypies do not occur on the ordered behavioral continuum that is often assumed by rating scales (cf. Peachey et al., 1976; Ljungberg and Ungerstedt, 1977b; Fray et al., 1980; Rebec and Segal, 1980).

There is, however, clear evidence that gnawing, licking, sniffing, and especially locomotor stereotypies are controlled by different neuroanatomical substrates (cf. Costall and Naylor, 1974; Jackson et al., 1975; Kelly et al., 1975; Andén and Johnels, 1977; Ungerstedt, 1979; Morelli et al., 1980; Redgrave et al., 1980; Imperato and Di Chiara, 1981; Gordon and Beck, 1984; Avdelidis and Spyraki, 1986; Havemann et al., 1986). It has even been argued that stereotypies represent fragments of behavior that are produced by a hierarchy of subsystems, and "microdescriptive" behavioral techniques, rating parts of behaviors, are required (Teitelbaum et al., 1990). At the very least, however, to obtain a clear understanding of functional neurophysiology of stereotypies, each should be measured separately, and gnawing must be viewed as only one of many behaviors in a neurological battery.

Measurement of Gnawing as a Separate Behavior

The definition and measurement of gnawing are improved by separating it from other stereotyped behaviors, and there is growing interest in measuring gnawing as one behavior among independent categories. The definition of gnawing as a separate behavior in these rating scales forms the basis for automated methods. Gripping or biting the wire cage floor with incisors was rated as gnawing by Risch et al. (1980), Hedley and Wallach (1983), Lewis et al. (1985), and Havemann et al. (1986). Biting the cage floor with a scratching noise (Peterson and Morin, 1984) and "gnawing" parts of the cage (Ridley et al., 1979) have also been used as criteria. Others rating gnawing as a separate behavior did not define the behavior precisely (e.g., Scheel-Kruger et al., 1977; Fray et al., 1980; Molloy et al., 1986).

Measurement of gnawing as a separate behavior is compatible with semiautomated rating scales wherein an observer uses a keyboard or event recorder to record data (e.g., Morelli et al., 1980; Imperato and Di Chiara, 1981; Gordon and Beck, 1984; Lewis et al., 1985; Koek et al., 1989). Microcomputers can be used for both data acquisition and analysis (e.g., Depaulis, 1983). These types of semiautomated rating scales facilitate separate assessment of drug effects on components of stereotypy (e.g., frequency, latency, intensity, and total time engaged in gnawing), and can reveal more detailed information about neu-roanatomical substrates and the hierarchy of behaviors in the animal's repertoire. It is important to note that the method of scoring behaviors is important. The use of "all-or-none" methods or scoring by frequency or intensity in rating scales can produce different conclusions (Dickinson et al., 1983). These same issues are important in evaluating automated methods.

In addition to rating scales that are, by nature, subjective, there have also been nonautomated objective methods of measuring gnawing on objects. One of these classic methods, used mostly with mice, involves covering the floor of the cage with corrugated paper with the corrugations upward, and then measuring gnawing by counting the bite marks or estimating the area of the floor containing marks (e.g., Ther and Schramm, 1962; Janssen et al., 1967; Pedersen, 1967; Scheel-Kruger, 1970).

A second nonautomated method, similar to measuring gnawing on corrugated paper, is to measure gnawing of wood, cork, or other materials by simply measuring the weight before and after gnawing. This method has been used widely in experiments that do not involve drug-induced stereotypies (cf. Kemble, 1973; Sclafani et al., 1973; Barnett and Smart, 1975; Martin, 1976; Tierney and Annan, 1978; Collier et al., 1981; Pollard and Howard, 1991) in addition to experiments using drug-induced stereotypies (cf. Ljungberg and Ungerstedt, 1976; Morelli et al., 1980; Imperato and Di Chiara, 1981; Klitenick and Wirtshafter, 1989; Winn, 1991). In some cases, time spent engaged in gnawing the wood blocks was also recorded by observers (e.g., Morelli et al., 1980; Imperato and Di Chiara, 1981). However, unless these methods are supplemented by observers, they have the disadvantage that they do not reveal information about latency, frequency, intensity, or total time engaged in the behavior.

Gnawing on wood, cork, or other materials in these tests is influenced by several variables. For example, there is a preference for certain types of wood (Cooper and Trowill, 1974) and for fresh blocks (Laties et al., 1969). Furthermore, it is increased by stress (e.g., Fray et al., 1982) and decreased by some anxiolytics (Pollard and Howard, 1991). The occurrence of spontaneous gnawing and its sensitivity to environmental and drug effects, and variability in the baseline behavior (e.g., Pollard and Howard, 1991) must be considered when wood or cork objects are incorporated into automated methods.

Automated Methods

Automated methods of measuring gnawing and other behaviors have significant advantages. Nonautomated observation methods typically time-sample behavior and are subjective (cf. Fray et al., 1980; Sanberg et al., 1983). Automation makes it possible to monitor behaviors continuously with consistent criteria in several animals and across several experiments.

Unfortunately, early attempts to automate behavioral recordings did not attempt to measure gnawing separately (e.g., Ljungberg, 1978; Sanberg et al., 1983), and nondescriptive records of motor activity were too crude to yield reliable and comparable results (Ljungberg, 1978). In view of these problems, several methods have been developed to record gnawing (both spontaneous and drug-induced) automatically as a separate behavior within the complex repertoire of laboratory animals.

The automated methods measure gnawing as a separate behavior because automation requires selection of a specific aspect of the behavior, usually quite limited, to be interfaced with some transducer. The requirement to select some limited aspect of the behavior is mutually exclusive with general ratings of combined behaviors. It would be difficult, for example, to find a single behavior common to sniffing, licking, and gnawing that is also reliable and easy to detect with automated procedures.

This chapter will review several automated methods for studying gnawing. Each of these methods uses a different operational definition of gnawing that will produce different types of data. The method selected will depend on the specific purpose of the experiment and level of behavioral analysis desired. The methods will be presented in ascending order from the most detailed to the most global analysis.

Mandibulogram Method

A method for recording an electromandibulogram from rats as a measure of stereotyped behavior in rats has been devel-

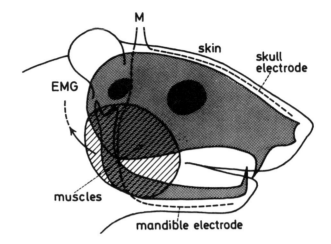

Fig. 1. Mandibulogram electrode placement under the skin and/ or muscles is shown with broken lines in this schematic drawing from Kubacki (1978). The bony parts are shadowed in gray. (Reprinted from Kubacki, A. [1978] Mandibulogram as a measure of stereotyped behavior in the rat. *Psychopharmacology* **59,** 209,210, with kind permission from Springer-Verlag, Heidelberg, Germany.)

oped by Kubacki (1978). Along with a myogram of the temporal and masseter muscles, this method allows a detailed analysis of oral behaviors. This type of recording method can differentiate among gnawing, chewing, and biting (e.g., Byrd, 1981), and should be useful in differentiating between spontaneous and drug- induced movements (Kubacki, 1978).

As shown in Fig. 1, the recording electrode is placed along the lower jaw, and a reference electrode is placed close to the tip of the animal's nose. This requires only minute cutaneous incisions. The electrodes are then connected to an implantation socket. The capacitance changes caused by changes in the distance between the lower jaw and the skull are then recorded.

The simple implantation of the electrodes and socket prepare the animal for continuous automatic monitoring of movements of the mandible. This is a chronic preparation, and the animals may be used for several experiments.

A method for measuring mandibular movement similar to this has also been developed by Ellison et al. (1987). In this method, however, mandibular movement is measured by a computer analysis of a video recording of two dots of UV-sensitive dye, one on the upper and one on the lower jaws of the rat. Using this technique, detailed movements of the jaws can also be continuously monitored. The video method of Ellison et al. (1987) has been used to measure spontaneous tremorous mouth movements that are an animal model of tardive dyskinesia. It would not be suitable for measuring gnawing because the dots would be obscured, and the movement of the animal must be restricted to maintain the video monitoring. The mandibulogram method of Kubacki (1978) is suitable for gnawing as well as a wide range of other behaviors, such as spontaneous tremorous oral movements and chewing, that might be of interest.

Bite Plate Methods

One behavioral characteristic of stereotyped gnawing involves biting with incisors. Kolasiewicz and Wolfarth (1976) have developed a method suitable for rabbits and rats that automatically records the number of bites. In this device, a pair of plastic plates protrudes upward at an angle through the floor into one corner of the test chamber. These have been designed to elicit gnawing and the angle of the plates can be adjusted to enhance the behavior. When the animal bites these plastic plates, a switch closes a circuit to a counter. When the force of the bite is released, the plates are pushed back to a parallel position by a spring. Several hundred counts were recorded per hour after rabbits were injected with apomorphine. These automatically recorded data showed a statistically significant difference between the effects of atropine and methacholine on apomorphine-induced gnawing that was not detected by a rating scale (Kolasiewicz and Wolfarth, 1976). Figures 2 and 3 show the design of the device and the relevant dimensions.

The bite plate method of Kolasiewicz and Wolfarth (1976) has been modified by Redgrave et al. (1982) to optimize further biting behavior and to monitor the behavior continuously by computer. The bite plate design used by Redgrave et al. (1982) is shown in Fig. 4. As in the earlier design, the plates protrude upward through the floor near a corner of the cage. This design incorporates three adjustments that can be used to enhance gnawing.

Depending on the particular program used to monitor the bite plate device, specific features of the behavior, such as num-

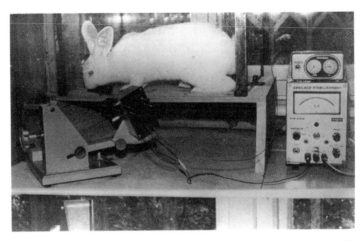

Fig. 2. View of the gnawing counter of Kolasiewicz and Wolfarth (1976) installed in a test cage with a rabbit. (Reprinted from Kolasiewicz, W. and Wolfarth, S. [1976] An objective and sensitive method for quantitative measurement of stereotyped gnawing. *Pharm. Biochem. Behav.* **4,** 201,202, with kind permission from Pergamon Press Ltd., Headington Hill Hall, Oxford 0X3 0BW, UK.)

ber of bites, total time plates closed, and average duration/ gnaw, can be recorded separately. Increasing doses of apomorphine caused an increase in total number of bites and the length of time engaged in biting, but the average duration/gnaw was not changed. The data show 2000–5000 bites induced by apomorphine (2–8 mg/kg) in rats over a 2-h test period.

The design was optimized for rats when the distance from the top of the plates to the floor of the box was 40 mm; the distance between plates was 2 mm; and "when a metal disk (diameter 25 mm) weighing 40 g placed on the edge of the flexible plate just effects a closure" (Redgrave et al., 1982, p. 874). This device was successful in recording virtually all gnawing that occurred. The gnawing that was not recorded was either of insufficient strength to close the plates or it was (rarely) directed toward the animal's own body.

An important feature of both bite plate methods is that the apparatus is designed so that no other objects in the cage are available for biting or gnawing. This increases the probability that gnawing will be directed toward the plates and recorded. Of course, other stereotyped behaviors and locomotion are not

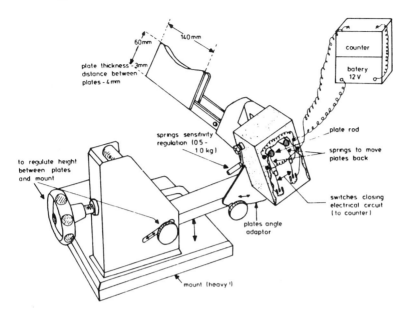

Fig. 3. Schematic drawing of the plate-bite gnawing counter from Kolasiewicz and Wolfarth (1976) showing relevant dimensions. (Reprinted from Kolasiewicz, W. and Wolfarth, S. [1976] An objective and sensitive method for quantitative measurement of stereotyped gnawing. *Pharm. Biochem. Behav.* **4**, 201,202, with kind permission from Pergamon Press Ltd., Headington Hill Hall, Oxford 0X3 0BW, UK.)

recorded automatically, and assessment of competing behaviors would require observation.

Biting and Pulling Methods

The biting and pulling methods measure a different aspect of gnawing behaviors than the mandibulogram and bite plate methods. Because naturally occurring gnawing involves biting with incisors and pulling the object apart (e.g., Cooper and Trowill, 1974), it is not surprising that stereotyped gnawing involves biting and pulling the object. This sequence of behavior is repeated in an invariant manner. Ernst (1967) gave the highest rating to jerking convulsively while the rats were clinging with their teeth around the wire of the cage bottom.

Moss et al. (1980) developed an automated method for measuring stereotyped gnawing by measuring the pulling of a

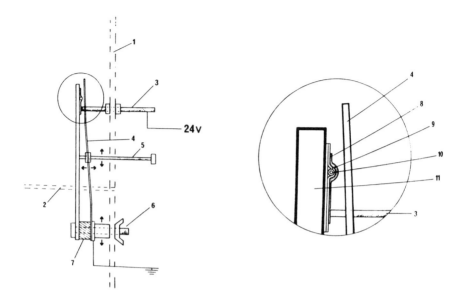

Fig. 4. Illustration of the plate-bite device from Redgrave et al. (1982) for the detection of stereotyped gnawing: (1) wall of box (3 mm Perspex); (2) floor of box; (3) threaded brass rod that provides the live connection to the counter; (4) flexible spring steel plate (0.25 mm) that is connected to ground; (5) nylon screw that can be used to adjust the plate positioning and force required to close the plates; (6) nylon screw that determines the distance between the top of the plates and the floor of the box; (7) insulated spacer; (8) aluminum foil for electrical contact; (9) double-sided insulating tape; (10) polyethylene tubing; and (11) inflexible aluminum plate (1 mm). (Reprinted from Redgrave, P., Dean, P., and Lewis, G. [1982] A quantitative analysis of stereotyped gnawing induced by apomorphine. *Pharm. Biochem. Behav.* **17,** 873–876, with kind permission from Pergamon Press Ltd., Headington Hill Hall, Oxford 0X3 0BW, UK.)

wire mesh placed on the floor of the observation chamber. The rats readily selected this object for drug-induced gnawing and vigorously pulled the wire mesh upward. To measure the gnawing behavior, the wire mesh was simply connected to a microswitch located below the floor, so that each time the wire mesh was pulled upward, the microswitch was activated. A diagram of this specific design is shown in Fig. 5.

It is important to note that in addition to the main wire extending down through the floor to the microswitch, there

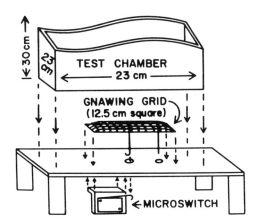

Fig. 5. Schematic drawing of bite-pull gnawing device from Moss et al. (1980). This is an exploded drawing, and proper positioning of parts is indicated by dashed lines and arrows. (Reprinted from Moss, D. E., McMaster, S. B., Castañeda, E., and Johnson, R. L. [1980] An automated method for studying stereotyped gnawing. *Psychopharmacology* **69,** 267–269, with kind permission from Springer Verlag, Heidelberg, Germany.)

must also be an additional wire extending through the floor to prevent rotation of the wire mesh while the rat is pulling on it. The exact dimensions of the wire mesh and test chamber are important. If the square piece of wire mesh is too large, the rats can place their entire bodies on the grid when they are lifting; this does not activate the microswitch. Second, if there is too much space between the walls of the test chamber and the edge of the wire mesh, gnawing is sometimes not elicited. With the dimensions shown in Fig. 5, gnawing is reliably elicited and recorded because the rats place themselves at a corner of the wire mesh, standing on the plastic floor, and pulling vigorously. Last, the microswitch was calibrated to operate with a lifting force of 40 g and a movement of 5 mm at any corner of the wire mesh.

The data collected with this apparatus show that 2000 counts or more can be expected in a 2-h test period from female rats treated with 30 mg/kg methylphenidate (Moss et al., 1980). In addition, this method of quantifying gnawing is exquisitely sensitive to suppression of methylphenidate-induced gnawing by other drugs. For example, gnawing behavior in rats was correlated –0.97 with CNS cholinesterase inhibition produced by physostigmine (Moss et al., 1985).

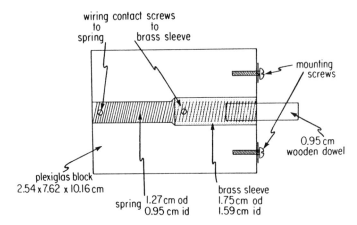

Fig. 6. Schematic drawing of gnawing counting device from Morin (1978). When the birch dowel is moved sufficiently to bring the spring into contact with the brass sleeve, a circuit is completed and a recorder is activated. (Reprinted from Morin, L. P. [1978] Rhythmicity of hamster gnawing: ease of measurement and similarity to running activity. *Physiol. Behav.* **21,** 317–320, with kind permission from Pergamon Press Ltd., Headington Hill Hall, Oxford 0X3 0BW, UK.)

The Moss et al. (1980) grid-pulling method has the advantage of sampling the whole behavior (biting and pulling) directed toward wire cage bottoms, which has been the criterion for gnawing in innumerable earlier rating scales. In addition, only drug-induced gnawing is recorded; there has been no spontaneous gnawing of the wire grid observed or recorded, and there have been no extraneous or "accidental" counts produced by any other activity (e.g., running over, sniffing, or licking the grid).

Other methods have also been developed to measure gnawing by pulling wood objects. These methods were developed to measure spontaneous (not drug-induced) gnawing.

Morin (1978) designed a very simple gnaw-bar apparatus in which a birch dowel is mounted so that it protrudes 2.5 cm into the wall of the test cage. The base of the dowel is fitted into a spring enclosed in a brass sleeve with a clearance of 1.6 mm around the spring (Fig. 6). Any movement in the *X-Y* plane sufficient to bring the spring into contact with the brass sleeve closes a circuit that operates a counter, an event recorder, or other device.

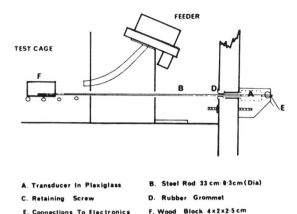

A. Transducer In Plexiglass B. Steel Rod 33 cm 0·3cm(Dia)

C. Retaining Screw D. Rubber Grommet

E. Connections To Electronics F. Wood Block 4x2x2·5 cm

Fig. 7. Schematic drawing of gnawing counter apparatus from Tierney and Annan (1978) showing the position of the wood block and sensor in the test cage. (Reprinted from Tierney, I. R. and Annan, A. [1978] A wood-gnawing sensor for recording collateral behavior emitted by rats on differential reinforcement of low rate schedules. *Behav. Res. Methods Inst.* **10,** 685–688, with kind permission from the Psychonomic Society, Inc., 1710 Fortview Road, Austin, TX.)

Tierney and Annan (1978) provided a beech wood block (4 × 2.5 × 2 cm) mounted on a steel rod that protrudes through the wall of the test cage. A ceramic-type phonograph cartridge was mounted against the base of the steel rod as a movement sensor transducer (Fig. 7), and the output was connected to an audio amplifier and a threshold control and pulse former circuit. The device was then calibrated by using a pendulum procedure (Altman and Hull, 1973) to determine the average momentum required to operate the sensor-output circuit.

The gnaw-bar and automated wood block procedures of Morin (1978) and Tierney and Annan (1978) are much more likely to elicit gnawing than the wire mesh procedure of Moss et al. (1980) and are much more suited to spontaneous gnawing. One potential disadvantage of these procedures if applied to drug-induced stereotyped gnawing is that the vigorous locomotor activity, rearing, and sniffing or licking might cause displacement of the arms protruding into the test cage from the walls and cause counts that are unrelated to gnawing. Of course, all of these methods will require an observer to measure any other behaviors.

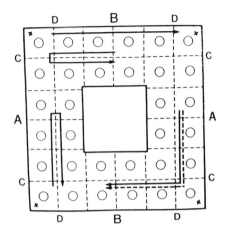

Fig. 8. Test box designed by Ljungberg and Ungerstedt (1978). Activity is monitored by number of interruptions of the ten photobeams shown by thin dotted lines. Total locomotion and forward locomotion are shown by solid and dotted arrows, respectively. Location of vertical photobeams is shown by "X" in the corners. Corner counts and corner time are defined as the number and accumulated time of interruptions. (Reprinted from Ljungberg, T. and Ungerstedt, U. [1978] A method for simultaneous recording of eight behavioral parameters related to monoamine neurotransmission. *Pharm. Biochem. Behav.* **8,** 483–489, with kind permission from Pergamon Press Ltd., Headington Hill Hall, Oxford 0X3 0BW, UK.)

Measurement of Gnawing with Other Behaviors

One of the disadvantages of all of the earlier automated methods for measuring gnawing is that no other behaviors are recorded. This is an important limitation if other behaviors are of interest or if they compete with gnawing. As an exception, however, Ljungberg and Ungerstedt (1978) have described a comprehensive testing device that automatically records eight behaviors in a modified open field (69 × 69 cm with a wall height of 25 cm). The center of the arena is blocked with a 25-cm cube. The open area of the floor contains 32 holes (2.5 cm diameter) arranged as shown in Fig. 8.

Locomotion in the open area of the apparatus is monitored with 10 photobeams crossing the area and a vertical photobeam in each corner (Fig. 8). Six photobeams are also aligned 0.5 cm under the rows of holes to detect nose pokes.

Gnawing behavior directed toward the edges of the holes produces a unique vibration that is measured with a miniature loudspeaker (5 cm diameter, 8 Ω) attached to the underside of the floor as a microphone. The output is amplified, filtered, and converted to TTL pulses through a Schmitt-Trigger, and then calibrated so that gnawing produces counting.

The eight behaviors that are automatically recorded are:

1. Total activity;
2. Forward locomotion;
3. Total locomotion;
4. Hole-poke counts;
5. Total hole time;
6. Corner counts;
7. Total time in corner; and
8. Gnawing.

Automatic recordings were highly reliable as compared to observer ratings. Hole pokes and gnawing were reliably separated insofar as it was uncommon for a rat to extend its nose 0.5 cm into a hole while gnawing. Similarly, there was a low background of false gnawing counts caused by vibrations from other behaviors.

The method for automatically, simultaneously, and continuously monitoring eight behaviors developed by Ljungberg and Ungerstedt (1978) is comprehensive. It has the advantage of recording gnawing within a context of other behaviors. Even with this comprehensive method, however, the authors caution that the automatic method should be supplemented with observers to avoid errors.

Summary and Conclusions

Gnawing is a common behavior in rodents and lagomorphs. Methods of automatically measuring spontaneous gnawing may have quite different requirements from those methods designed to measure stereotyped (drug-induced) gnawing. In addition, as reviewed above, there are widely different operational definitions of gnawing behavior both in rating scales and the automated methods. These definitions range from microanalyses of mandibulograms, sounds made by scraping during gnawing, to biting and pulling behaviors. Considerable attention must be given to the exact behavior being measured and the degree to which a specific method serves the particular purpose of an experiment. The experimental situation greatly influences gnawing (Ljungberg and Ungerstedt, 1977a) and, therefore, the overall environment in which gnawing is measured deserves detailed consideration. For example, the availability and type of objects for gnawing greatly affect the amount of spontaneous as well as drug-induced gnawing. Psychostimulants, such as apomorphine and amphetamine, induce a wide range of stereotyped behaviors, some of which may compete strongly with the measurement of gnawing. Animals may give different responses to drugs both because of different neuroanatomical substrates and because of competing behaviors. The availability and type of other objects in the test situation will strongly affect competing behaviors.

The main advantages of automated methods are that they are objective and are economical for continuous monitoring of behavior over long periods of time. However, because of unexpected effects of environmental design or competing behaviors, automatic methods should always be supplemented with observation in order to avoid errors in recording or interpreting the data (Ljungberg and Ungerstedt, 1978; Havemann et al., 1986).

Acknowledgments

The technical advice on the biology of Rodentia and Lagomorpha from Arthur H. Harris, Ph.D., Department of Biological Sciences, U. T. El Paso, is greatly appreciated. The preparation of this manuscript was supported, in part, by NIMH (MH47167).

References

Alleva, E., Castellano, C., and Oliverio, A. (1979) Ontogeny of behavioral development, arousal and stereotypies in two strains of mice. *Exp. Aging Res.* **5**, 335–350.
Altman, F. and Hull, L. D. (1973) Piezoelectric pecking key. *J. Exp. Anal. Behav.* **19**, 289–291.
Amsler, C. (1923) Beiträge zur Pharmakologie des Gehirns. *Arch. Exp. Pathol. Pharmakol.* **97**, 1–14.

Andén, N.-E. and Johnels, B. (1977) Effects of local applications of apomorphine to the corpus striatum and to the nucleus accumbens on reserpine-induced rigidity in rats. *Brain Res.* **133**, 386–389.

Avdelidis, D. and Spyraki, C. (1986) Dopamine dependent behaviors in rats with bilateral ibotenic acid-induced lesions of the globus pallidus. *Brain Res. Bull.* **16**, 25–32.

Barnett, S. A. and Smart, J. L. (1975) The movements of wild and domestic house mice in an artificial environment. *Behav. Biol.* **15**, 85–93.

Braestrup, C. (1977) Changes in drug-induced stereotyped behavior after 6-OHDA lesions in noradrenaline neurons. *Psychopharmacology* **51**, 199–204.

Byrd, K. E. (1981) Mandibular movement and muscle activity during mastication in the guinea pig *(Cavia procellus)*. *J. Morphol.* **170**, 147–169.

Collier, A. C., Cohn, M. U., Hothersall, D., and Berson, B. S. (1981) Effects of motivational variables and contextual stimuli on schedule-induced behavior. *Physiol. Behav.* **27**, 1005–1013.

Cooper, P. H. and Trowill, J. A. (1974) Wood gnawing preferences in rats. *Physiol. Behav.* **13**, 845–847.

Costall, B. and Naylor, R. (1972) Modification of amphetamine effects by intracerebrally administered anticholinergic agents. *Life Sci.* **11**, 239–253.

Costall, B. and Naylor, R. (1974) The involvement of dopaminergic systems with stereotyped behavior patterns by methylphenidate. *J. Pharm. Pharmacol.* **26**, 30–33.

Creese, I. and Iversen, S. D. (1973) Blockage of amphetamine induced motor stimulation and stereotypy in the adult rat following neonatal treatment with 6-hydroxydopamine. *Brain Res.* **55**, 369–382.

Depaulis, A. (1983) A microcomputer method for behavioral data acquisition and subsequent analysis. *Pharmacol. Biochem. Behav.* **19**, 729–732.

Dickinson, S. L., Jackson, A., and Curzon, G. (1983) Effect of apomorphine on behaviour induced by 5-methoxy-N,N-dimethyl tryptamine: three different scoring methods give three different conclusions. *Psychopharmacology* **80**, 196,197.

Ellison, G., See, R., Levin, E., and Kinney, J. (1987) Tremorous mouth movements in rats administered chronic neuroleptics. *Psychopharmacology* **92**, 122–126.

Ernst, A. M. (1967) Mode of action of apomorphine and dexamphetamine on gnawing compulsion in rats. *Psychopharmacologia (Berl.)* **10**, 316–323.

Ernst, A. M. and Smelik, E. (1966) Site of action of dopamine and apomorphine on compulsive gnawing behavior in rats. *Experientia* **22**, 837–839.

Fog, R. (1972) On stereotypy and catalepsy: studies on the effects of amphetamines and neuroleptics in rats. *Acta Neurol. Scand.* **48(Suppl. 50)**, 1–61.

Fray, P. J., Koob, G. F., and Iversen, S. D. (1982) Tail-pinch-elicited behavior in rats: preference, plasticity, and learning. *Behav. Neural. Biol.* **36**, 108–125.

Fray, P. J., Sahakian, B. J., Robbins, T. W., Koob, G. F., and Iversen, S. D. (1980) An observational method for quantifying the behavioural effects of dopamine agonists: contrasting effects of *d*-amphetamine and apomorphine. *Psychopharmacology* **69**, 253–259.

Gordon, D. and Beck, C. H. M. (1984) Subacute apomorphine injections in rats: effects on components of behavioral stereotypy. *Behav. Neural. Biol.* **41**, 200–208.

Harnack, E. (1874) Über die Wirkung des Apomorphins am Säugetier und am Frosh. *Arch. Exp. Pathol. Pharmacol.* **2**, 254–306.

Hauschild, F. (1939) Zur Pharmakologie des 1-phenyl-2-methylamino-propans (Pervitin). *Naunyn-Schmiedeberg's Arch. Exp. Pathol. Pharmak.* **191**, 465–468.

Havemann, U., Magnus, B., Möller, H. G., and Kuschinsky, K. (1986) Individual and morphological differences in the behavioural response to apomorphine in rats. *Psychopharmacology* **90**, 40–48.

Hedley, L. R. and Wallach, M. B. (1983) Potentiation of apomorphine-induced gnawing in mice. *Prog. Neuro-Psychopharmacology Biol. Psychia.* **7**, 47–56.

Imperato, A. and Di Chiara, G. (1981) Behavioural effects of GABA-agonists and antagonists infused in the mesencephalic reticular formation—deep layers of superior colliculus. *Brain Res.* **224**, 185–194.

Jackson, D. M., Andén, N.-E., and Dahlström, A. (1975) A functional effect of dopamine in the nucleus accumbens and in some other dopamine-rich parts of the rat brain. *Psychopharmacologia (Berl.)* **45**, 139–149.

Janssen, P. A. J., Niemegeers, C. J. E., and Jagenean, A. H. (1960) Apomorphine-antagonism in rats. *Arzneim-Forsch (Drug Res.)* **10**, 1003–1005.

Janssen, P. A. J., Niemegeers, C. J. E., Schellekens, K. H. L., and Lenaerts, F. M. (1967) Is it possible to predict the clinical effects of neuroleptic drugs (major tranquilizers) from animal data? *Arzneim-Forsch (Drug Res.)* **17**, 841–854.

Kelly, P. H., Seviour, P., and Iversen, S. D. (1975) Amphetamine and apomorphine responses in the rat following 6-OHDA lesions of the nucleus accumbens septi and corpus striatum. *Brain Res.* **94**, 507–522.

Kemble, E. D. (1973) Wood-gnawing in rats following cortical, olfactory bulb or limbic lesions. *Physiol. Behav.* **11**, 735–738.

Khozatskii, L. I. and Vyalov, O. S. (1976) Bite marks of Pliocene rodents on Harts horns. *Vestn. Leningr. Univ. Biol.* **1**, 64–69.

Klitenick, M. A. and Wirtshafter, D. (1989) Elicitation of feeding, drinking, and gnawing following microinjections of muscimol into the median raphe nucleus of rats. *Behav. Neural. Biol.* **51**, 436–441.

Koek, W., Colpaert, F. C., Woods, J. H., and Kamenka, J.-M. (1989) The phencyclidine (PCP) analog N-[1-(2-benzo(β)thiophenyl)cyclohexyl]piperidine shares cocaine-like but not other characteristic behavioral effects with PCP, ketamine and MK-801. *J. Pharmacol. Exp. Thera.* **250**, 1019–1027.

Kolasiewicz, W. and Wolfarth, S. (1976) An objective and sensitive method for quantitative measurement of stereotyped gnawing. *Pharmacol. Biochem. Behav.* **4**, 201,202.

Kubacki, A. (1978) Mandibulogram as a measure of stereotyped behavior in the rat. *Psychopharmacology* **59**, 209,210.

Laties, V. G., Weiss, B., and Weiss, A. B. (1969) Further observations on overt "mediating" behavior and the discrimination of time. *J. Exp. Anal. Behav.* **12**, 43–57.

Lewis, M. H., Baumeister, A. A., McCorkle, D. L., and Mailman, R. B. (1985) A computer-supported method for analyzing behavioral observations: studies with stereotypy. *Psychopharmacology* **85**, 204–209.

Ljungberg, T. (1978) Reliability of two activity boxes commonly used to assess drug induced behavioural changes. *Pharmacol. Biochem. Behav.* **8**, 191–195.

Ljungberg, T. and Ungerstedt, U. (1976) Reinstatement of eating by dopamine agonists in aphagic dopamine denervated rats. *Physiol. Behav.* **16**, 277–283.

Ljungberg, T. and Ungerstedt, U. (1977a) Apomorphine-induced locomotion and gnawing: evidence that the experimental design greatly influences gnawing while locomotion remains unchanged. *Eur. J. Pharmacol.* **46**, 147–151.

Ljungberg, T. and Ungerstedt, U. (1977b) Different behavioral patterns induced by apomorphine: evidence that the method of administration determines the behavioural response to the drug. *Eur. J. Pharmacol.* **46**, 41–50.

Ljungberg, T. and Ungerstedt, U. (1978) A method for simultaneous recording of eight behavioral parameters related to monoamine neurotransmission. *Pharmacol. Biochem. Behav.* **8**, 483–489.

MacLennan, A. J. and Maier, S. F. (1983) Coping and the stress-induced potentiation of stimulant stereotypy in the rat. *Science* **219**, 1091–1093.

Martin, J. R. (1976) Motivated behaviors elicited from hypothalamus, midbrain, and pons of the guinea pig *(Cavia porcellus). J. Comp. Physiol. Psych.* **90**, 1011–1034.

Molloy, A. G., Aronstam, R. S., and Buccafusco, J. J. (1986) Selective antagonism by clonidine of the stereotyped and non stereotyped motor activity elicited by atropine. *Pharmacol. Biochem. Behav.* **25**, 985–988.

Morelli, M., Porceddu, M. L., and Di Chiara, G. (1980) Lesions of substantia nigra by kainic acid: effects on apomorphine-induced stereotyped behaviour. *Brain Res.* **191**, 67–78.

Morin, L. P. (1978) Rhythmicity of hamster gnawing: ease of measurement and similarity to running activity. *Physiol. Behav.* **21**, 317–320.

Moss, D. E., McMaster, S. B., Castañeda, E., and Johnson, R. L. (1980) An automated method for studying stereotyped gnawing. *Psychopharmacology* **69**, 267–269.

Moss, D. E., Rodriguez, L. A., and McMaster, S. B. (1985) Comparative behavioral effects of CNS cholinesterase inhibitors. *Pharmacol. Biochem. Behav.* **22**, 479–482.

Peachey, J. E., Rogers, B., Brien, J. F., Maclean, A., and Rogers, D. (1976) Measurement of acute and chronic behavioural effects of methamphetamine in the mouse. *Psychopharmacology* **48**, 271–275.

Pechnick, R., Janowsky, D. S., and Judd, L. (1979) Differential effects of methylphenidate and d-amphetamine on stereotyped behavior in the rat. *Psychopharmacology* **65**, 311–315.

Pedersen, V. (1967) Potentiation of apomorphine effect (compulsive gnawing behavior) in mice. *Acta Pharmacol. Toxicol.* **25(Suppl. 4)**, 63–68.

Peterson, M. E. and Morin, L. P. (1984) Behavioral effects of *d*-amphetamine and apomorphine in the hamster. *Pharmacol. Biochem. Behav.* **20**, 855–858.

Pollard, G. T. and Howard, J. L. (1991) Cork gnawing in the rat as a screening method for buspirone-like anxiolytics. *Drug Dev. Res.* **22**, 179–187.

Quinton, R. M. and Halliwell, G. (1963) Effects of α-methyl-DOPA and DOPA on the amphetamine excitatory response in reserpinized rats. *Nature* **200,** 178,179.

Randrup, A. and Munkvad, I. (1967) Behavioural stereotypies induced by pharmacological agents. *Pharmakopsychatrie Neuro-Psychopharmakologie* **1,** 18–26.

Rebec, G. V. and Segal, D. S. (1980) Apparent tolerance to some aspects of amphetamine stereotypy with long-term treatment. *Pharmacol. Biochem. Behav.* **13,** 793–797.

Redgrave, P., Dean, P., Donohoe, T. P., and Pope, S. G. (1980) Superior colliculus lesions selectively attenuate apomorphine-induced oral stereotypy: a possible role for the nigrotectal pathway. *Brain Res.* **196,** 541–546.

Redgrave, P., Dean, P., and Lewis, G. (1982) A quantitative analysis of stereotyped gnawing induced by apomorphine. *Pharmacol. Biochem. Behav.* **17,** 873–876.

Ridley, R. M., Scraggs, P. R., and Baker, H. F. (1979) Modification of the behavioural effects of amphetamine by a GABA agonist in a primate species. *Psychopharmacology* **64,** 197–200.

Risch, C., Kripke, D., and Janowsky, D. (1980) Flurazepam effects on methylphenidate-induced stereotyped behavior. *Psychopharmacology* **70,** 79–82.

Rotrosen, J., Angrist, B. M., Wallach, M. B., and Gershon, S. (1972) Absence of serotonergic influence on apomorphine-induced stereotypy. *Eur. J. Pharmacol.* **20,** 133–135.

Sanberg, P. R., Moran, T. H., Kubos, K. L., and Coyle, J. T. (1983) Automated measurement of stereotypic behavior in rats. *Behav. Neurosci.* **97,** 830–832.

Scheel-Kruger, J. (1971) Comparative studies of various amphetamine analogues demonstrating different interactions with the metabolism of the catecholamines in the brain. *Eur. J. Pharmacol.* **14,** 47–59.

Scheel-Kruger, J. (1970) Central effects of anticholinergic drugs measured by the apomorphine gnawing test in mice. *Acta Pharmacol. Toxicol.* **28,** 1–16.

Scheel-Kruger, J., Golembiowska, K., and Mogilnicka, E. (1977) Evidence for increased apomorphine-sensitivity dopaminergic effects after acute treatment with morphine. *Psychopharmacology* **53,** 55–63.

Sclafani, A., Berner, C. N., and Maul, G. (1973) Differential effects of hypothalamic transections on the wood gnawing behavior of rats. *Physiol. Behav.* **10,** 451–454.

Teitelbaum, P., Pellis, S. M., and DeVietti, T. L. (1990) Disintegration into stereotypy induced by drugs or brain damage: a microdescriptive behavioral analysis in *Neurobiology of Stereotyped Behavior* (Cooper, S. J. and Dourish, C. T., eds.), Oxford University Press, Oxford, pp. 169–199.

Ther, L. and Schramm, H. (1962) Apomorphin-Synergismus (Zwangsnagen bei Mäusen) als Test zur Differenzierung Psychotroper Substanzen. *Arch. Int. Pharmacoldyn.* **138,** 302–310.

Tierney, I. R. and Annan, A. (1978) A wood-gnawing sensor for recording collateral behavior emitted by rats on differential reinforcement of low rate schedules. *Behav. Res. Methods Instru.* **10,** 685–688.

Ungerstedt, U. (1979) Central dopamine mechanisms and unconditioned behavior, in *The Neurobiology of Dopamine* (Horn, A. S., Korf, J., and Westerink, B. H. C., eds), Academic, New York, pp. 577–596.

Winn, P. (1991) Cholinergic stimulation of substantia nigra: effects on feeding, drinking and sexual behavior in the male rat. *Psychopharmacology* **104,** 208–214.

II. Preclinical Research

CHAPTER 7

The Catalepsy Test

Is a Standardized Method Possible?

Paul R. Sanberg, Rodrigo Martinez, R. Douglas Shytle, and David W. Cahill

Introduction

Catalepsy is most often operationally defined as the inability of an animal to correct an externally imposed posture. For example, a normal animal will correct itself quickly when placed into an awkward position, whereas a cataleptic animal will remain in the imposed posture for a prolonged period of time. Although similar states of behavioral arrest have been given different names, such as "freeze behavior" or "death feint," "catalepsy" has been used most extensively to describe the "active" immobility response produced by invasive experimental treatments, such as exposure to certain drugs. The word "active" is used to distinguish between this form of immobility from simple akinesia, since there must be some skeletal muscle activity in order to maintain a particular body posture for a period of time (Klemm, 1989).

Catalepsy is one of the behavioral tools most used by neuroscientists to study the behavioral mechanisms of neurochemical systems. It is of great use to the researcher because of its similarity to symptoms of such human disorders as Parkinsonism, catatonic schizophrenia, and abnormal behavior result-

From: *Motor Activity and Movement Disorders*
P. R. Sanberg, K. P. Ossenkopp, M. Kavaliers, Eds. Humana Press Inc., Totowa, NJ

ing from brain damage to the basal ganglia (e.g., Duvoisin, 1976; Garver, 1984; Sanberg and Coyle, 1984).

The most common method of quantifying catalepsy is the "standard bar test," first described by Kuschinsky and Hornykiewicz (1972). It consists of placing the animal's fore-paws on a horizontal bar and then recording the time taken to return to the ground. This time is regarded as an index of the intensity of catalepsy. Although small differences in the methods employed do not usually hinder its detection, they can influence the intensity of the cataleptic effect.

Over the years, various other tests have been devised that use wire grids, parallel bars, platforms, or pegs to situate the animals in unusual positions. Unfortunately, each new technique developed has brought with it new variables. Table 1 shows that there are even variations within the so-called standard bar test (Kuschinsky and Hornykiewicz, 1972). All of the parametrical discrepancies combine to produce data that are very difficult to compare across laboratories. To correct this problem, the bar test should be standardized or automated so that the same test can be performed by most researchers under similar conditions.

This chapter is an update of *The Catalepsy Test: Its Ups and Downs* (Sanberg et al., 1988), which suggested that the catalepsy test be standardized. In addition to identifying the most obvious parameters that need to be controlled in a standardized test, the authors seek to suggest the use of automated systems to eliminate much of the human subjectivity that could confound results.

Experimental Subject

Catalepsy tests are performed with many different animals, but by far the most popular subject is the rat. There are a few variables that must be considered when performing the test, such as the rat's strain, sex, weight, time of testing, and the way the rat is handled.

One of the first variables investigators must define and control is the strain of rat used in the catalepsy test. It has been shown that interstrain differences in catalepsy sensitivity to treatment with haloperidol do exist (Kinon and Kane, 1989).

Additionally, Shipley et al. (1981) showed that a hypertensive strain of rats was more cataleptic than other strains. In yet another experiment, it was proposed that pharmacological manipulations reveal genetic predisposition to catalepsy (Kolpalov et al., 1981). Catalepsy has been shown to be affected not only by the rat's strain, but also by its sex. Hannigan and Pilati (1991) found that male rats remained on the bar significantly longer than similarly treated females. In view of this evidence, researchers should try to keep rat strain and sex consistent throughout their studies.

The rat's weight is another factor that can contribute to variance in catalepsy scores. Sanberg et al. (1988) cited an example in which rats weighing 150 g were placed on a 12-cm-high bar, whereas 340 g rats were positioned on a bar only 6 cm high. Had there not been differences in other variables, such as timing criteria, maximal test time, method of scoring, and the number of test scores, it is likely that the heavier rats would have received higher catalepsy scores. It appears that the height and thickness of the bar used in the test may need to be adjusted according to the weight of the animal. Further experimentation should be done to determine the ideal bar height and diameter corresponding to each rat's weight.

The available evidence indicates that the behavioral responses to dopaminergic drugs vary greatly depending on the circadian rhythm of the animal (Campbell et al., 1982; Sanberg et al., 1984a; Russell et al., 1987a,b). For example, Campbell et al. (1982) demonstrated striking circadian changes in the behavioral sensitivity to haloperidol by measuring cataleptic responses in rats tested in a controlled lighting cycle (lights on 7:00 AM–7:00 PM). Under these conditions, catalepsy was maximal at about 4:00 PM and minimal at about 4:00 AM, virtually the opposite of the circadian rhythm of spontaneous behavioral activity in drug-free rats. Therefore, it is very important for researchers to describe the lighting conditions under which the animals are housed and tested for catalepsy.

Repeated handling of the rat can confound results. Catalepsy is known to be induced by neck and, to a lesser extent, tail-pinch (Beecham and Handley, 1974; Amir et al., 1981; Ornstein and Amir, 1981). The constant handling may potentiate catalepsy by triggering the animal's natural defense mech-

Table 1
Methodological Variations Across Studies Using the Standard Bar Test

Reference	Animal specifications		Bar dimensions		Behavioral criteria	Maximum test duration, s	Scoring
	Strain	Weight, g	Diam., cm	Height, cm			
Al-Khatib et al., 1989	Wistar		0.4	12.0	Time until at least one paw touched the plate surface to climb up bar or descent down of rat		Absolute time
Barnes et al., 1990	Sprague-Dawley	180–200		8.0	Time elapsed before rat changed position	300	1 point for every 6 s on the bar
Elazar and Paz, 1990	Wistar and Charles River	250–350		7.5	Time until rat descended with both limbs on floor		Absolute time
Emerich et al., 1991	Sprague-Dawley	250–350	0.6	9.0	Time for animal to remove itself from the bar	900	Absolute
Ferré et al., 1990	Sprague-Dawley	350–450	1.2	10.0	Time from placement until both forepaws touched the floor	1790	Logarithmic transformation of time
Kinon and Kane, 1989	Fisher and Brown Norway	180–200 140–170	1.0	11.0	Time elapsed before the rat returned both front paws to the table top	180	Absolute time

Study	Strain				Measure		
Meyer, 1990	Long-Evans	300–400	0.5	8.0	Latency until animal placed both forepaws on table		Absolute time
Parashos et al., 1989	Sprague-Dawley	180–200	0.5	12.0	Time elapsed before rat changed position	0 1 2 3	0–14 s 15–29 s 30–59 s 60+ s
Pertwee and Wickens, 1991	Sprague-Dawley	260–370	0.5	9.0	Length of time it remained in that position	1800	Absolute time
Rowlett et al., 1991	Sprague-Dawley		0.1	3.0 6.0	Time for pups to remove both forepaws from the bar	300	Absolute time
Schmidt and Bubser, 1989	Sprague-Dawley			9.0	Timespan from placement of paws until the first movement of one paw	180	Absolute time

[a]This table is an updated version from Sanberg et al. (1988).

anisms. In fact, morphine catalepsy itself may be an "artificially" magnified defense mechanism; the animal may simply be "playing dead" (De Ryck and Teitelbaum, 1984). Meyer (1990) showed that "continuous pressure applied to the dorsal midline at the nape of the neck...resulted in a form of horizontal bar grasp and vertical ding catalepsy" and that the durations of dorsal pressure immobility (DPI) were very similar to the effects of low doses of haloperidol. Meyer also mentioned that "all rodents elicit the species-typical immobility response" when they are grasped by the skin at the nape of the neck and then lifted up with no other part of their bodies touching anything. The duration of this DPI was about 60 s in untreated rats. The response is a transport reflex similar to one adopted by a young pup being picked up by its mother or by a prey being transported by a predator. In order to avoid the unwanted effects of handling, it seems best to very gently grab the animal from under the forelimbs and to drag it toward the bar, allowing the hindpaws to remain in contact with a surface top.

Because of the various factors that can influence the cataleptic behavior of each rat, all relevant information about the animals should be recorded. In additon, similar handling methods should be employed by investigators across laboratories.

Apparatus

Although the bar test is by far the most popular method for measuring catalepsy, many other tests that incorporate "platforms," "boxes," "walls," "single pegs," and so forth are based on the same principle, but are more difficult to standardize. Still others make use of an apparatus, such as the "grid," "four-peg," and "parallel bar" tests. Alternatively, catalepsy can be measured without the use of any apparatus whatsoever. Dunstan et al. (1981) placed rats into a "buddha" position (by crossing the limbs) and measured the latency until the posture changed. However, because most of these tests are used relatively infrequently and do not lend themselves to standardization easily, we will restrict our discussion to the bar test.

The diameter and height of the bar used are extremely important variables that are often ignored or overlooked by researchers (*see* Table 1). Sanberg et al. (1984) conducted a study in which the height and diameter of the bar in the "standard bar test" were manipulated in combination with four different behavioral criteria for cessation of timing. They demonstrated that when a thick bar is used, bar height and behavioral criteria adopted can markedly influence the catalepsy score. Conversely, a thin bar generally results in lower catalepsy scores, regardless of height. In short, the relative influence of bar diameter and shape may vary in either direction, depending on the bar height. This is probably because a thick high bar reduces the possibility of a good paw-hold when compared to a thick low bar, which can be easily rested on. However, once a critical height is reached, bar diameter and shape seem to have little effect. If a bar is too high, the animal is unlikely to rest on it for any appreciable length of time, regardless of other factors.

It is thus apparent that even relatively "small" procedural differences in the implementation of the bar test technique can result in important differences in results. Consequently, the much larger methodological variability concerning scoring criteria and dimensions of the apparatus from different laboratories may result in even greater inconsistencies in the measurement of cataleptic intensity. For this reason, researchers should agree on and report all apparatus parameters.

Testing Procedure

Although the occurrence of catalepsy is fairly obvious, various factors combine to confound its absolute intensity. Environmental distractions, differences in behavioral criteria, and variations in stopping and starting the timer all interact to produce data that vary from laboratory to laboratory.

The testing environment can contribute significantly to discrepancies in catalepsy test results. There are various factors to consider: lighting, the smell or sight of other animals, a novel environment, and noise. Kaufman and Rovee-Collier (1978) reported that the lighting condition during the test affected the animal's manifestation of multiple components of predation defense. Moderate changes in the intensity of the auditory and visual environment have also been shown to affect catalepsy scores (Ariyanayagam and Handley, 1975; Hinson et al., 1982).

To reduce sensory stimulation, Rowlett et al. (1991) tested the rats in wire cages wrapped with black cardboard. To eliminate auditory distractions, some researchers have performed the bar test in either "quiet" (Parashos et al., 1989) or soundproof (Al-Khatib et al., 1989; Ferre et al., 1990) rooms. No matter what is done to reduce distracting environmental stimulation, it is important to allow the animal ample time to habituate itself to a new testing environment.

Measurement of latency of descent should begin as soon as the hand is removed from the animal, and not as soon as the animal is placed on the bar, as some have suggested. Unfortunately, the exact moment when the timing should be stopped is not so easy to determine, because it depends on two disputable factors: the behavioral criteria adopted and the maximum allowable duration of the test.

Because of lack of standardization, the behavioral criteria chosen to judge when the animal has changed position is unfortunately subject to interpretation. This is demonstrated in Table 1, which quotes or closely paraphrases the authors' criteria. A few investigators chose to record the amount of time the rat maintained its imposed position. Others timed until one or both paws were removed from the bar, or until either one or both paws touched the table again. However, catalepsy scores have been shown to be significantly affected by very similar behavioral criteria (Sanberg et al., 1984b). Therefore, an agreement must be made on what action determines the end of the timing period. Otherwise, sensitive electronic measuring systems should be employed to eliminate any subjectivity or discrepancies.

The majority of researchers limit the amount of time the rat will remain immobile depending on their specific experimental design. However, certain discoveries can be made only after sufficiently long periods of observation. For example, Moss et al. (1981) found that catalepsy induced by reserpine alone would last for an average of 39 s, whereas catalepsy induced by a combination of reserpine and terahydrocannabinol lasted an average of 922 s. This dramatic effect may have gone unnoticed had one of the more commonly used short cutoff times been employed. The results of an experiment performed by Pertwee and Wickens (1991) suggested that more conclusive evidence for the relative potency of the two cannabinoids they tested may have been found had the animals been allowed to remain on the bar for a longer period of time. Although the ideal test would allow the animal to remain on the bar indefinitely, many investigators test the animal at specific intervals of time after injection and prefer to avoid overlap of the intervals.

Thus, even though catalepsy is a very obvious phenomenon, a review of the literature clearly demonstrates that its absolute intensity appears extremely sensitive to even small alterations in a variety of experimental and environmental parameters. Consequently, as many aspects of the testing procedure as possible should be controlled.

Repeated Testing and Scaling of Scores

Both the influence of repeated testing on catalepsy and the conversion of raw catalepsy scores to scaled values (e.g., Costall and Olley, 1971; Stanley and Glick, 1976) have been major points of controversy in catalepsy research. In 1976, Stanley and Glick demonstrated that if repeated catalepsy tests were given to the same animals, haloperidol-induced catalepsy was greater in subsequent tests. However, 2 yr later, Costall's group published a study (Costall et al., 1978) that showed no significant effects owing to repetition of tests or conversion of data to their six-point scale. It was, therefore, concluded that "workers who are unable to obtain stable catalepsy in rats may consider a change in animal strain or test procedure" (p. 763).

However, since then, evidence has demonstrated that increased catalepsy scores owing to experience alone are a real phenomenon, even in the absence of pharmacological manipulation (Brown and Handley, 1980; Sanberg et al., 1980; Ornstein and Amir, 1981; Klemm, 1983; Ferre et al., 1990). Costall et al. (1978) hypothesized that this experienced-induced increase in catalepsy is a result of "learning." Conversely, Sanberg et al. (1980) proposed that it was a form of tonic immobility that increases with repeated handling resulting from the short period of handling that preceded the positioning of the animal. However, both learning and tonic immobility may be interacting to produce the experience-induced increase in catalepsy scores. In any case, depending on experimental design, it is often impossible or undesirable to avoid repeated testing. Consequently, studies of cataleptic behavior that necessitate the use of repeated testing should include identical tests on drug-injected and control animals matched for experience.

Fuenmayor and Vogt (1979) suggested that inflated catalepsy scores owing to repeated testing could be avoided if attention was given to all movements. Using a grid test, they found that after repeated testing, the main factor differentiating between control rats and rats rendered cataleptic pharmacologically was head movements. If such findings are replicated using the bar test, confounding variables owing to repeated testing could be removed and more reliable scoring criteria established. However, no matter how experienced the observer is in picking up small movements, the data are still very subjective and are bound to differ across laboratories. Automated observation and recording systems, which will be discussed later, could aid in picking up these small, but significant head movements and thus avoid inflated scores.

The conversion of absolute latency measures to some type of scale is a common practice. Unfortunately, these scales tend to obscure rather than clarify results. In addition, the criteria for distinction between scaled points differ between investigators (see Table 1) making comparisons of data from different laboratories even more cumbersome. Although Ferre et al. (1990) used logarithmic transformation (ln) of results in order to normalize their data, this has not been adopted by many investigators.

To avoid the major discrepancies that hamper the comparison of results between laboratories, it is best to avoid both repeated testing and conversion of absolute scores to scales. If repeated testing must be done, then it must be accompanied by the testing of animals controlled for experience.

An Automated Catalepsy Test

Given the importance of the catalepsy test in elucidating the psychopharmacological actions of a wide range of neuroactive drugs, an accurate and reproducible measure of catalepsy would be a useful adjunct to such studies. The apparent influence of behavioral criteria on catalepsy scores suggests that the implementation of some type of automation in measurement for future studies would be desirable. This would reduce experimenter variability and would eliminate the problem of criteria differences between laboratories.

Moss et al. (1981) experimented with this, using a bar and platform with built-in electronic touch-sensors connected to a clock. The clock timed how long each rat stood at the bar. In addition to providing a very accurate measure of hypokinetic duration, the automated system allowed the simultaneous testing of four animals.

Sanberg et al. (1988) suggested another automated form of measurement, the Digiscan-16 Animal Activity Monitoring System (Omnitech Electronics, Columbus, OH). The bar test was performed within the system, which used infrared beams to record both the duration of the period of vertical movements and the latency to the first vertical movement. Comparison to a human observer's results showed a high correlation between the two. It thus appears that this automated measure may be a reliable test for catalepsy.

A third system that could eliminate the problem of human subjectivity was recently developed by Martin et al. (1992). It employs a computer-imaging technique designed to "acquire, store and analyze the data obtained simultaneously from either two rats or three mice...without the need for human assessment of movement." However, in performing the experiment, the investigators measured catalepsy not with the bar test, but using the ring test. The animal was placed on a ring attached to a stand, and the length of time the animal remained on the stand was recorded. As demonstrated by comparison of data, the electronic device was very sensitive to small movements, which the human observer could not readily perceive, but on adjusting the machine to make it comparable to the human visual threshold, results were found to be highly correlated. Although this device could solve the problem of discrepancies in behavioral criteria and timing, an experiment must be done in which this system records the bar test instead of the ring test, because, as pointed out by the authors, "these two tests may not be measuring the exact same behavior."

Despite the popularity and simplicity of the bar test, sometimes more accurate devices are necessary to detect small positional changes that the human eye cannot perceive. For example, Emerich et al. (1991) reported that "the Digiscan Animal Activity Monitors provide a more sensitive assessment of the interaction between nicotine and haloperidol than the bar test." Thus, it seems that the standardization of the bar test could include an automated system of observation to eliminate sources of subjectivity, such as the behavioral criteria adopted by researchers, and to allow more accurate assessments of cataleptic behavior.

Conclusions

Because catalepsy is such an important behavioral model to neuroscientists, it is extremely important for researchers to be able to compare results across laboratories. Obviously, the more standardized any test is, the more reliable and reproducible the conclusions. The authors therefore suggest that the bar test be adopted by all researchers to quantify cataleptic intensity. Moreover, owing to the high sensitivity of the bar test to a myriad of variables, all of its parameters should be standardized. In addition, it is critical that investigators include detailed descriptions of procedures in all future publications.

A consensus should be reached on what specific sex and strain of rat to consistently use in the test. The animal's weight must also be considered because it determines the optimum height and diameter of the bar. Also, the test should be performed with a minimum of environmental distractions to avoid arousing the animal and thus altering behavior.

In carrying out the test the researcher should be aware that handling, especially pressure behind the rat's neck, should be kept to a minimum and should be done very gently. The timing of descent latency should begin as soon as the experimenter's hand is removed. Furthermore, an agreement must be made regarding the judgment of when the animal has changed posture. A simple way of avoiding this subjectivity of behavioral criteria and of accurately recording very small positional changes is to perform the test in conjuction with one of the suggested automated systems. Additionally, repeated testing should be avoided owing to possible experiential learning as well as the effects of repeated handling. Finally, the easiest method to record, report, and compare catalepsy scores is to use the absolute time the animal remains on the bar.

If the aforementioned steps are followed by all researchers when performing the bar test for catalepsy, the results will be more reliable and will yield more readily comparable data.

Acknowledgments

This work was supported, in part, by grants from the Smokeless Tobacco Research Council and the National Institute of Neurological Disease and Stroke (RO1 NS. 32067-01A1) and from the USF Presidential and COM Research Funds.

References

Al-Khatib, I. M. H., Fujiwara M., and Ueki, S. (1989) Relative importance of the dopaminergic sytem in haloperidol-catalepsy and the anticataleptic effect of antidepressants and methamphetamine in rats. *Pharmacol. Biochem. Behav.* **33**, 93–97.

Amir, S., Brown, Z. W., Amit, Z., and Ornstein, K. (1981) Body pinch induces long lasting cataleptic-like immobility in mice: behavioral characterization and the effects of naloxone. *Life Sci.* **10**, 1189–1194.

Ariyanayagam, A. D. and Handley, S. L. (1975) Effect of sensory stimulation on the potency of cataleptogens. *Psychopharmacology* **41**, 165–167.

Barnes, D. E., Robinson, B., Csernansky, J. G., and Bellows, E. P. (1990) Sensitization versus tolerance to haloperidol-induced catalepsy: multiple determinants. *Pharmacol. Biochem. Behav.* **36**, 883–887.

Beecham, I. J. and Handley, S. L. (1974) Potentiation of catalepsy induced by narcotic agents during Haffner's test for analgesia. *Psychopharmacology* **40**, 157–164.

Brown, J. and Handley, S. L. (1980) The development of catalepsy in drug-free mice on repeated testing. *Neuropharmacology* **19**, 675–678.

Campbell, A., Herschel, M., Sommer, B., Madsen, J. R., Cohen, B. M., and Baldessarini, R. J. (1982) Circadian changes in the distribution and effects of haloperidol in the rat. *Neuropharmacology* **21**, 663–669.

Costall, B. and Olley, J. E. (1971) Cholinergic- and neuroleptic-induced catalepsy: modification by lesions in the caudate-putamen. *Neuropharmacology* **10**, 297–306.

Costall, B., Hui, S.-C. G., and Naylor, R. J. (1978) Correlation between multitest and single-test catalepsy assessment. *Neuropharmacology* **17**, 761–764.

De Ryck, M. and Teitelbaum, P. (1984) Morphine catalepsy as an adaptive reflex state in rats. *Behav. Neurosci.* **98**, 243–261.

Dunstan, R., Broekkamp, C. L., and Lloyd, K. G. (1981) Involvement of caudate nucleus, amygdala or reticular formation in neuroleptic and narcotic catalepsy. *Pharmacol. Biochem. Behav.* **14**, 169–174.

Duvoisin, R. C. (1976) Parkinsonism: animal analogues of human movement disorder, in *The Basal Ganglia* (Yahr, M. D., ed.), Raven, New York, pp. 293–303.

Elazar, Z. and Paz, M. (1990) Catalepsy induced by carbachol microinjected into the pontine reticular formation of rats. *Neurosci. Lett.* **115,** 226–230.

Emerich, D. F., Zanol, M. D., Norman, A. B., McConville, B. J., and Sanberg, P. R. (1991) Nicotine potentiates haloperidol-induced catalepsy and loco-motor hypoactivity. *Pharmacol. Biochem. Behav.* **38,** 875–880.

Ferré, S., Guix, T., Prat, G., Jane, F., and Casas, M. (1990) Is experimental catalepsy properly measured? *Pharmacol. Biochem. Behav.* **35,** 753–757.

Fuenmayor, L. D. and Vogt, M. (1979) Production of catalepsy and deple-tion of brain monoamines by a butyrophenone derivative. *Br. J. Pharma-cology* **67,** 115–122.

Garver, D. L. (1984) Disease of the nervous system: psychiatric disorders, in *Clinical Chemistry: Theory Analysis and Correlations* (Kaplen, L. A. and Pesce, A. J., eds.), CV Mosby, St. Louis, pp. 864–881.

Hanningan, J. H. and Pilati, M. L. (1991) The effects of chronic postweaning amphetamine on rats exposed to alcohol in utero: weight gain and behav-ior. *Neurotoxicol. Teratol.* **13,** 649–656.

Hinson, R. E., Poulos, C. X., and Thomas, W. L. (1982) Learning in tolerence to haloperidol-induced catalepsy. *Prog. Neuropsychopharmacol. Biol. Psy-chiatry* **6,** 395–398.

Kaufman, L. W. and Rovee-Collier, C. K. (1978) Arousal induced changes in the amplitude of death feigning and periodicity. *Physiol. Behav.* **20,** 453–458.

Kinon, B. J. and Kane, J. M. (1989) Difference in catalepsy response in in-bred rats during chronic haloperidol treatment is not predictive of the intensity of behavioral hypersensitivity which subsequently develops. *Psychopharmacology* **98,** 465–471.

Klemm, W. R. (1983) Cholinergic-dopaminergic interactions in experimen-tal catalepsy. *Psychopharmacology* **81,** 139–142.

Klemm, W. R. (1989) Drug effects on active immobility responses: what they tell us about neurotransmitter systems and motor functions. *Prog. Neu-robiol.* **32,** 403–422.

Kolpalov, V., Barykina, N., and Chepkasov, I. (1981) Genetic predisposition to catatonic behavior and methylphenidate sensitivity in rats. *Behav. Proc.* **6,** 269–281.

Kuschinsky, K. and Hornykiewicz, O. (1972) Morphine catalepsy in the rat. Relation to striatal dopamine metabolism. *Eur. J. Pharmacol.* **19,** 119–122.

Martin, B. R., Prescott, W. R., and Zhu, M. (1992) Quantitation of rodent catalepsy by a computer-imaging technique. *Pharmacol. Biochem. Behav.* **43,** 381–386.

Meyer, M. E. (1990) Dorsal pressure potentiates the duration of tonic immo-bility and catalepsy in rats. *Physiol. Behav.* **47,** 531–533.

Moss, D. E., McMaster, S. B., and Rogers, J. (1981) Tetrahydrocannabinol potentiates reserpine-induced hypokinesia. *Pharmcol. Biochem. Behav.* **15,** 779–783.

Ornstein, K. and Amir, S. (1981) Pinch-induced catalepspy in mice. *J. Comp. Physiol. Psychol.* **95,** 827–835.

Parashos, S. A., Marin, C., and Chase, T. N. (1989) Synergy between a selec-tive D1 antagonist and a selective D2 antagonist in the induction of catalepsy. *Neurosci. Lett.* **105,** 169–173.

Pertwee, R. G. and Wickens, A. P. (1991) Enhancement by chlorodiazepoxide of catalepsy induced in rats by intravenous or intrapallidal injections of enantiomeric cannabinoids. *Neuropharmacology* **30(3)**, 237–244.

Rowlett, J. K., Pedigo, N. W., Jr., and Bardo, M. T. (1991) Catalepsy produced by striatal microinjections of the D1 dopamine receptor antagonist SCH 23390 in neonatal rats. *Pharmacol. Biochem. Behav.* **40**, 829–834.

Russell, K. H., Hagenmeyer-Houser, S. H., and Sanberg, P. R. (1987a) Haloperidol produces increased defecation in rats in habituated environments. *Bull. Psychon. Soc.* **25(1)**, 13–16.

Russell, K. H., Hagenmeyer-Houser, S. H., and Sanberg, P. R. (1987b) Haloperidol-induced emotional defecation: a possible model for neuroleptic anxiety syndrome. *Psychopharmacology* **91**, 45–49.

Sanberg, P. R. and Coyle, J. T. (1984) Scientific approaches to Huntington's disease. *CRC Crit. Rev. Clin. Neurobiol.* **1**, 1–44.

Sanberg, P. R., Paulks, I. J., and Fibiger, H. C. (1980) Experiential influences on catalepsy. *Psychopharmacology* **69**, 225–226.

Sanberg, P. R., Johnson, D. A., Moran, T. H., and Coyle, J. T. (1984a) Investigation of locomotion abnormalities in animal models of extrapyramidal disorders: a commentary. *Physiol. Psychol.* **12(1)**, 48–50.

Sanberg, P. R., Pevsner, J., and Coyle, J. T. (1984b) Parametric influences on catalepsy. *Psychopharmacology* **82**, 406–408.

Sanberg, P. R., Giordano, M., Bunsey, M. D., and Norman, A. B. (1988) The catalepsy test: its ups and downs. *Behav. Neuroscience* **102**, 748–759.

Schmidt, W. J. and Bubser, M. (1989) Anticataleptic effects of the *n*-methyl-*d*-aspartate antagonist MK-801 in rats. *Pharmacol. Biochem. Behav.* **32**, 621–623.

Shipley, J. E., Rowland, N., Antelman, S. M., Buggy, J., Edwards, D. J., and Shapiro, A. P. (1981) Increased amphetamine stereotypy and longer haloperidol catalepsy in spontaneously hypertensive rats. *Life Sci.* **28**, 745–753.

Stanley, M. E. and Glick, S. D. (1976) Interaction of drug effects with testing procedures in the measurement of catalepsy. *Neuropharmacology* **15**, 393,394.

CHAPTER 8

Computer-Assisted Video Assessment of Dyskinetic Movements

Gaylord Ellison and Alan Keys

Introduction

Although tardive dyskinesia (TD) is a severe side effect of neuroleptics, animal models of TD have not proven generally fruitful in developing therapeutic counterstrategies or a coherent theory of why chronic neuroleptics induce this and other disorders. This is partially because relatively few primates have been maintained on the continuous proper drug regimen of neuroleptics for sufficient periods to produce TD, and also because rodent models based on the simple observational technique of counting mouth movements have yielded highly controversial results.

Many of the problems with rodent models seem to resolve considerably when the actual form of the oral dyskinesia is measured using a computerized technique based on measuring a rat's upper and lower jaw movements by remote video-tracking techniques. The analysis of these oral movement records indicates that several distinctive syndromes gradually develop in rats administered chronic neuroleptics. A syndrome similar to human TD develops with continuous, typical neuroleptics; it is characterized by oral movements of the same fast-Fourier trans-

From: *Motor Activity and Movement Disorders*
P. R. Sanberg, K. P. Ossenkopp, M. Kavaliers, Eds. Humana Press Inc., Totowa, NJ

form (FFT) energy spectrum as those seen in human TD patients, and it worsens on drug withdrawal. When rats pretreated with continuous neuroleptics are then given probe injections of a D1 dopamine agonist, the resulting oral movements appear indistinguishable from those of humans with TD. Although the long-term administration of several atypical neuroleptics does not lead to the TD-like syndrome, other unexpected alterations in oral behaviors occur that may reflect a TD-like predisposition.

A second model is based on giving weekly, intermittent injections of neuroleptics. In monkeys, this regimen induces a "primed dystonia" syndrome, and in rodents, a syndrome of very large, rapid oral movements with a much higher FFT energy spectrum. However, the two different syndromes in rodents do not respond to cholinergics and anticholinergics as might be expected from the substantial, although often contradictory, literature from humans.

It appears that provided measurements of the exact form of oral movements are made, neuroleptic-induced dyskinesias in rodents appear to be the most promising models of the syndromes that develop in humans. This is supported by recordings from humans using these same techniques that indicate a substantial similarity between the form of oral movements in the two species. Further studies directed toward locating the site of the actual pattern generator that is altered in dyskinetic oral movement syndromes may clarify this area considerably.

Neuroleptic-Induced Movement Disorders in Humans and Animals

There is a dramatic need for animal models related to the long-term side effects of antipsychotic drugs, because although there are clearly convincing models of acute effects of neuroleptics, such as pseudo-Parkinsonism, catalepsy, and so forth, the models of long-term effects have become very controversial. This is puzzling, since many antipsychotic drugs are not highly complex, and their major modes of action seem relatively simple, acting principally on receptors located in brain structures that are also present in subhuman species. The side effects of neuroleptics are widely believed to be exerted through extrapyramidal motor regions, particularly the nigrostriatal dopamine (DA) pathway (rather than cortical DA pathways; Marsden and Jenner, 1980; Creese, 1983). The symptoms produced are so overt and distinctive, neuroleptic-induced motor side effects would seem to be a perfect field for animal models. It would be very surprising if a simple motor dyskinesia induced in humans by the chronic administration of a relatively pure DA receptor blocker did not occur in primates or rodents as well.

Animal models of neuroleptic-induced side effects are becoming increasingly necessary. One reason for this is the difficulties encountered in assessment of favorable drug effects with antipsychotic drugs in patients. Most hospitalized schizophrenic patients have already received substantial quantities of several different antipsychotic drugs, over prolonged time periods, perhaps with interspersed drug holidays, and their drug treatment has often also included simultaneous treatment with other pharmacological agents designed to control some of the unpleasant side effects of the antipsychotics. Thus, most studies of antipsychotic medications in humans presently include patients with extraordinarily varied drug histories, which increasingly include the use of drugs of abuse as well (Khalsa et al., 1991). Because many of these drugs of abuse can have persisting effects on DA terminals, cell bodies, and other brain neurochemical systems, these complex drug-treatment histories have made it particularly difficult to assess causal factors of the extrapyramidal side effects that often accompany neuroleptic administration in humans. Yet all antipsychotic drugs may be ultimately proven capable of producing motor disorders to some extent, and these side effects have become the limiting factor in the prescription of these highly efficacious compounds.

Neuroleptics induce three reversible extrapyramidal side effects (i.e., ones that generally disappear when the drug is withdrawn); pseudo-Parkinsonism (hypokinesia, muscular stiffness, a gaunt, emotionless face, tremors, simian gait, and so on), akathisia (inability to sit still accompanied by unpleasant feelings of internal restlessness), and, more rarely, acute dystonia (writhing muscular spasms of the torso and neck, facial grimacing, and ocular spasms). Chronic neuroleptic treatment can also induce TD, which appears only after prolonged drug therapy (Jeste and Wyatt, 1982). The orofacial dyskinesias that are a predominant feature of this syndrome sometimes persist long after discontinuation of the drugs or may even be permanent (Crane, 1973; Casey, 1985). Age is a critical indicator of the presence of orofacial dyskinesia (Kane and Smith, 1982; Guy et al., 1985). Consequently, retrospective clinical studies have not been convincing as to what induces or worsens TD. Treatment is also not well understood, since although various strategies have been developed for reducing the incidence or treating TD (such as reducing the dopaminergic excess, or enhancing cholinergic or GABAergic tone), each has generally fallen into disfavor after further study (Jeste et al., 1988).

One obvious counterstrategy is to develop new antipsychotic medications that might not produce severe motor side effects. However, historically there have been a variety of promising candidates that led to falsely raised hopes. Jeste and Wyatt comment,

> A number of new neuroleptics, which were initially claimed to have a low risk of inducing TD, were later found to be as likely to produce dyskinesia as the older drugs. Probably all neuroleptics, when given in equivalent amounts and for similar periods to similar patients, carry a comparable risk of causing TD (1982, p. 86).

The most effective antipsychotic in use that appears not to induce TD is the dibenzodiazepine, clozapine (Pi and Simpson, 1983; Lindstrom, 1988). In addition, neuroleptics of the substituted benzamide class, most notably sulpiride, have also been reported to show a low incidence of TD (Gerlach and Casey, 1984). It is possible that these and other newly developed antipsychotic drugs may eventually replace presently used neuroleptics that do cause TD, but it is also possible that as more experience with these novel compounds is obtained, the same or new delayed side effects will also be found.

Animal Models of Neuroleptic-Induced Side Effects

Primate Studies

Initial research into effects of chronic neuroleptics on nonhuman primates led some authors to conclude that a variety of TD-like symptomatic movements, including orobuccolingual dyskinesias, clearly develop in primates after prolonged neuroleptic treatment (Deneau and Crane, 1969; Gunne and

Barany, 1976; reviewed by Domino, 1985), and that these are excellent animal models of TD. However, it is important to distinguish between acute dystonic effects of neuroleptics and TD-like effects (*see* Rupniak et al., 1986). These two reactions probably reflect very different mechanisms, yet are often confused or intermixed. Acute dystonic reactions are Parkinsonian-like: They often occur shortly after the drug is administered, and they are diminished by anticholinergics. TD, conversely, is usually maximal when the drug is wearing off (Gardos et al., 1978), and (although this is controversial), sometimes responds well to cholinomimetics (Moore and Bowers, 1980). However, this confusion of the two side effects has led to controversial practices, because many pharmaceutical companies use the propensity of a drug to induce acute dystonic reactions in "primed" monkeys as an indication of whether a new drug will induce extrapyramidal dysfunction (Neale et al., 1984) even though it has never been demonstrated that these acute dystonic reactions are an accurate predictor of TD development. The highly fluctuating (once a week) neuroleptic regimen in primates clearly contributes to the progressive appearance of this syndrome, and it is noteworthy that humans are never persistently maintained on such an extremely fluctuating drug regimen. Also, this reaction does not fit the typical pattern of TD. Pharmacologically, it behaves in an opposite fashion to TD, because it lessens when drug levels decline and responds oppositely to anticholinergics. Yet, because it is not convenient to administer neuroleptics to monkeys in the same way as given to humans (several times daily for many months), this is rarely the practiced drug regimen. Domino and Kovacic (1983), in fact, concluded that of 121 monkeys that have been given neuroleptics in several published studies, most developed only the acute dystonic or parkinsonism syndrome, and only 10 developed TD (some of those of Gunne and Barany, 1976; Bedard et al., 1977; McKinney et al., 1980). Kovacic and Domino (1984), Johansson (1989), and Gunne et al. (1984) subsequently increased this sample, but the number of documented cases is clearly very low.

Rodent Studies

Many studies have purported to show that TD-like behaviors develop in rats after chronic neuroleptic administration, and these have been put forth as viable rodent models of TD. Initial studies used brief neuroleptic exposure and reported only an increased sensitivity to injections of DA agonists, but later studies utilized more prolonged periods of neuroleptic treatment (3–18 mo) and focused on spontaneous changes in oral activity. Some reports noted an increase in "spontaneous stereotypies" (Sahakian et al., 1976) and "spontaneous mouth movements" (Iversen et al., 1980) after repeated neuroleptic treatment. Waddington et al. (1983) gave a variety of oral neuroleptics as well as im injections of fluphenazine decanoate to groups of rats. After 6 mo of drug administration, spontaneous increases in nondirected oral movements could be observed in many of the drugged animals when the rats were placed in an observation cage. Such late-onset increases in oral movements have also been reported by several other laboratories using various neuroleptics (Gunne and Haggstrom, 1983; Johansson et al., 1986; Mithani et al., 1987).

However, several important characteristics of these observer-rated vacuous chewing movements are controversial. Some investigators have failed to replicate these findings (e.g., Rodriguez et al., 1986), and still others argue that these vacuous oral movements in rats are not truly models of TD. Clow et al. (1979, 1980) described spontaneous mouthing movements in rats after 12 mo of neuroleptics, but reported that these movements disappeared quite rapidly after drug withdrawal, a pattern unlike that of TD. Furthermore, Gunne et al. (1986) and Rupniak et al. (1983) note that whereas cholinergic agents often ameliorate TD in humans (and monkeys) and anticholinergics exacerbate them, the vacuous chewing movements in rats induced by chronic HAL (HAL), trifluoperazine, or sulpiride are increased by cholinergic agents and decreased by anticholinergics.

An even more severe problem is that several laboratories have reported that "vacuous oral movements" occur in rodents as soon as 24 h after an acute injection (Glassman and Glassman, 1980; Rosengarten et al., 1983; Rupniak et al., 1985). Thus, in some situations, they are clearly not tardive in appearance. This entire issue has been reviewed by Rupniak et al. (1986) and Stewart et al. (1988), who conclude that vacuous oral movements actually represent acute dystonic reactions in rats rather than TD-like movements. A related problem is that although these oral movements are often labeled as being dyskinetic or dystonic, within the article there is often the comment that they seemed normal in appearance. As stated by Stewart et al.,

> However, it must be stressed again that the effect of neuroleptic treatment on the rodent is to increase the incidence of part of the normal repertoire of behavior of the animal. The purposeless chewing movements induced by neuroleptic drugs are not abnormal (1988, p. 352).

The definition of "opening the mouth" in all these studies is very vague, and attempts to categorize more precisely "oral movements" (such as including only those oral movements accompanied by audible tooth grinding, and so forth) have not proven successful in reconciling the discrepancies noted above. Indeed, one might question the basic premise of all of these attempts at developing rodent models, since the basic definition of dystonias and dyskinesias in humans is that of movement that is abnormal in form, and not just any movement, although at an increased frequency. Simply counting the number of mouth movements shown by schizophrenic patients treated with chronic neuroleptics yields scores that correlate very poorly with standard dyskinesia scales, including the AIMS (Chien et al., 1980).

Computerized Assessment of Oral Activity

We therefore set out several years ago to develop a method whereby the incidence and form of the oral movements of rats could be recorded using a computer as a sophisticated observer of behavior. A procedure was developed based on a simplification of a video image to its minimal essentials. The oral region of rats can be observed using close-up television images while the animals are in round Plexiglass™ tubes (Sant and Ellison, 1984; Gunne et al., 1986). In addition, a precise quantification of oral movements can be obtained by direct computerized measurements of jaw movements using a fluorescent dye marking that can be simultaneously verified by human observational scoring (Ellison et al., 1987). Rats are habituated to being placed for brief periods in loosely fitting Plexiglass™ tubes inside a soundproof chamber. At one end of the tube is a hole through which the rat's head protrudes. The rat is not tightly confined in this tube, so that in the habituated rat, there is minimal or no struggling during 6 min of testing sessions.

On one side of the animal, a closed circuit TV camera with a close-up lens provides a picture of the animal's muzzle; this information is fed to a large TV screen. The human observer watches this image and records the presence of oral activity by pressing keys on a computer-linked keyboard. The testing chamber is illuminated only by a 6-W UV ("black-light") bulb placed in front of and below the rat's muzzle. Prior to being placed in the tube, two small spots are painted, one on the upper and one on the lower jaw of the rat, using a UV-sensitive dye. A second TV camera with a close-up lens is positioned in front of and beneath the rat's muzzle. This camera has a powerful UV filter in front of the lens; consequently, it detects only the two fluorescent dots (and not the background UV illumination). The output of this camera is adjusted via potentiometers, so that only the two spots are visible and the resulting digital signal is fed to a computer with a movement-detection circuit containing a counter that stores the number of TV horizontal sweeps occurring between the bottom of the top spot and the top of the bottom spot. This distance represents the amount of mouth opening, and it, together with the human observer's reports, is stored 60 times each second in computer memory and, at the end of the testing session, to disk. This system was designed to be maximally sensitive to mouth movements, because it is the distance between the two spots, rather than their absolute position, that is recorded, so that horizontal or vertical movements of the entire head would be relatively undetected.

The raw data are analyzed in several ways. One basic unit of analysis is a unidirectional individual opening or closing of the mouth, a "computer-scored movelet" or "CSM," defined to begin each time the dots come closer together or go farther apart, and to terminate when the direction of change in distance between the dots reverses. CSMs are presently sorted into different amplitude categories corresponding to the number of horizontal sweeps crossed by the movelet. Simultaneous viewing of the observer's behavioral reports and the computerized files indicates that isolated oral movements are typically of amplitude between four and nine horizontal sweeps, with rise times that are neither extremely slow nor extremely fast. Sudden, extremely large head movements can be identified by their large amplitudes, steep slopes, and nonrepetitive nature. These two types of behavior can be sorted at least partially using restrictions ("filters") on amplitude, distance between CSMs, degree of repetitiveness, and waveform. The scoring programs also calculate the slope (amplitude divided by duration) of each CSM, determining the average slope for each animal within each amplitude category. We have also written programs to analyze the energy spectrum of oral activity using FFTs, which has proven an incredibly useful way to detect alterations in the form of oral movements.

Effects of Prolonged Neuroleptics on Oral Movements as Assessed by Computerized Measurement

In an initial study (Ellison et al., 1987), we repeatedly recorded from rats given continuous HAL using slow-release silastic implants. The literature led us to expect to see heightened oral movements gradually developing in the HAL animals during drug administration, but we did not. Instead, the human observer files indicated only a very small increase in total duration of observed oral movements during 8 mo of continuous HAL and no increase in oral movement (OM) frequency, but the device was clearly working, because the computerized two-spot files collected at 6 and 8 mo were well correlated with the behavioral reports by the observer, with observer reports of oral movements often preceded, by about 100 ms (reflecting the reaction time of the observer) by CSMs of amplitudes between four and nine rasters. These CSMs all showed a typical form of opening and closing of the jaws, but each individual animal had a characteristic waveform.

However, there were also distinctive changes in the HAL animals, since although the computer-analyzed files indicated that CSMs of the largest amplitudes were chronically depressed in the HAL animals, they also indicated that CSMs of the smallest amplitude (two rasters) were increased in these animals, which was an effect that clearly did not occur with acute injections of HAL. The meaning of this finding became clearer when the implants were removed, because total oral activity in the chronic HAL rats now dramatically increased to above control levels, and the amplitude distribution also shifted dramatically, since what had been isolated, small oral movements in the HAL rats now developed into repetitive trains of larger mouth movements. We replicated these findings in a separate study (Levin et al., 1987). These results led us to rethink our model of TD in rats, sicne our computer analysis had told us two things: (1) large-amplitude oral movements were not dramatically increased in chronic, continuous HAL implant animals, but were rather chronically suppressed in many ways, and (2) tiny tremorous movements, which are virtually undetected by the human eye under conditions used by all other laboratories, appeared to be our best indicator of "late-stage" neuroleptic effects.

Since these results were strikingly different from what we had expected based on the literature, we speculated that perhaps our observers were not seeing dramatically increased large-amplitude oral movements because our silastic implants were not releasing sufficient amounts of drug. Consequently, we conducted a long-term study using quite sizable depot injections (See et al., 1988). Rats were chronically administered either HAL or fluphenazine decanoate via depot injections for 8 mo. These were administered via im injections in the upper thigh muscle given once every 3 wk at a dose of 21 mg/kg, giving an average daily dose of 1 mg/kg/d. Since decanoate neuroleptics can persist for long times after drug withdrawal, the animals were then prepared for final drug withdrawal by switching to giving these same drugs in the nondecanoate form in the drinking water for the next 2 mo. Then the rats were withdrawn from all drugs. Throughout the experiment, the animals were tested repeatedly in the computerized tube device that measured CSMs and, in the latter half of the experiment, were also scored by a human observer in the tube, as well as in an open cage, for observed oral movements.

In the tube, the animals in both neuroleptic-treated groups showed initial decreases in the number of CSMs and made sluggish CSMs; these effects were generally larger in the fluphenazine-treated animals. After 6 mo of chronic neuroleptics, the HAL-treated animals showed slightly increased oral movements, as reported by both the human observer and in CSMs of all amplitudes. This effect of increased large oral movements only became significant on drug withdrawal. Fluphenazine-treated animals showed a more persistent depression of both oral movements and CSMs of large amplitudes during and after drug treatment. However, the behavior most characteristic of both neuroleptic-treated groups was again the gradual development of increases in CSMs of the smallest amplitudes measurable.

Open-Cage Observations of Oral Movements: A Confound Related to Activity

The observational technique used to assess oral activity can affect profoundly the conclusions drawn, sometimes because of unsuspected artifacts. This argument has been discussed in detail elsewhere (Ellison, 1991), since in several experiments we directly compared the computerized "tube test" we had developed with results from the same animals in the "open-cage test" described above.

These are not simply two different measures of the same behavior, because neuroleptics induce a different pattern of results in the two tests. When rats were administered chronic neuroleptics (orally, with depot injections, or via slow-release implants) and repeatedly observed in both tests (See et al., 1988; See and Ellison, 1990a), observers of the tube-test reported no or very small increases in OMs in the chronic neuroleptic animals compared to controls. Yet when these same animals were tested in the cage test, much larger differences were observed, often with strikingly increased OMs in the drug-treated groups. This might mean cage testing is simply more sensitive to drug effects, yet other results indicated profound discrepancies between these tests. Rats given acute HAL immediately showed increased OMs (a "dystonia-like result") in the cage test, but not in the tube test. When the drug was withdrawn after chronic administration, there were immediate increases in OMs in the tube test (a "TD-like result"), whereas in the cage test OMs rapidly returned to control levels (See and Ellison, 1990a). The results in the cage test were being profoundly influenced by other behaviors in which the rat was engaging. Rats given a large HAL injection and observed in both tests over the next 4 d showed disparate results in the two tests. Just after the HAL injection, the rats in the cage test were extremely inactive (not rearing and sniffing, and so on) but showed many oral movements, whereas in the tube test, they did not show a similar elevation in OMs. As the HAL wore off, activity in the cage increased and OMs correspondingly decreased.

It was concluded that there is an artifact related to activity levels in the cage test (Levy et al., 1987). Oral activity of normal rats in an open environment is dramatically inhibited during bouts of exploratory behavior, when the animals are rearing and sniffing. If all that is measured is number of OMs, misleading results can be obtained, since rats that have been injected with neuroleptics show inflated OM scores simply because they are less exploratory, apparently owing to an intrinsic incompatibility between the expression of gross skeletal activity and that of spontaneous, "vacuous" oral activity. This effect can be seen in several different tests. We have consistently observed that when rats (either neuroleptic-treated or drug naive) are placed into novel cages, exploration over the next 30 min progressively decreases, whereas oral movement scores correspondingly increase. Another situation where this effect may occur is in acute drug tests with tranquilizers, since we have confirmed the observations of others (Rupniak et al., 1985) that when tested in observation cages, rats can sometimes be observed to show elevated oral movements very soon after the beginning of neuroleptic administration (Rupniak et al., 1984) or even following acute injections (Rosengarten et al., 1983, 1986). Similarly, the elevated oral movements observed in the open-cage test following chronic neuroleptics returned to control levels soon after drug discontinuation. This may be because HAL withdrawal leads to hyperactivity (Owen et al., 1980), resulting in a suppression of the expression of oral movements because of the mechanism described above. When tested in the observation tubes, a different trend is often observed, especially in the initial stages of drug administration and testing. It is only after the animals have been receiving drugs and tested repeatedly in both conditions for months that the two measures begin to show some concordance (See et al., 1988). From these data it seems clear that simply counting normal-appearing oral activity will not develop into a noncontroversial model of a disorder that is fundamentally a dyskinesia. Future models must also demonstrate that the behavior measured is abnormal in form.

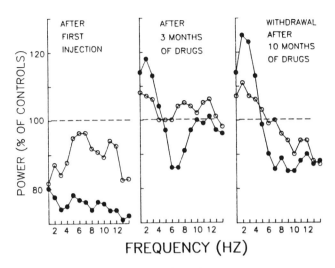

FREQUENCY (HZ)

Fig. 1. Fast-Fourier analyses of oral movement energy spectrum (expressed as percent of controls) at the various Hz at **(left)** 5 d after the first decaoate injection, **(center)** just after the fifth monthly injection, and **(right)** 10 d after drug withdrawal following 10 mo of drug administration. Open circles: HAL decanoate; closed circles: fluphenazine decanoate. After the first injection, there is decreased power at all parts of the spectrum, especially in the fluphenazine animals. This presumably reflects the general tranquilization induced. By 3 mo, a shift has occurred, with heightened energy at 1–3 Hz and decreased energy at around 6 Hz, again especially in the fluphenazine. This effect, which dramatically increases on drug withdrawal after 10 mo of neuroleptics, is the "TD-signature effect."

We next developed computer programs that would allow us to conduct FFT analyses of our data. We found that when the animals were first started on the drug, there was an initial period of sedation; this was reflected in the FFT analysis as decreased energy at all frequencies. With continued administration, to our surprise, the drugged animals began to show increased oral movements of 1–3 Hz, an effect that then increased substantially on drug withdrawal (Ellison and See, 1989); *see* Fig. 1. This increased energy was not the result of the very small CSMs, because it had a different time-course (appearing sooner in time), and different group characteristics (it was greater in the fluphenazine animals). Also, because the tiny CSMs are so small in amplitude, they did not possess sufficient energy to cause these changes. What this meant, then, was that the oral movements of appreciable and readily observable size were not necessarily increasing in frequency with chronic neuroleptic, but they were slowly changing in form, and that this effect showed the exact temporal characteristics expected of TD (late in appearance and exaggerated on drug withdrawal). This was a remarkable finding, because this is precisely the altered energy spectrum (increased energy only at 1–2 Hz) observed in humans with TD. Five different studies (Alpert et al., 1976; Fann et al., 1977; Lees, 1985; Rondot and Bathien, 1986; Wirshing et al., 1989) have recorded the oral or extremity movements of humans with TD using various devices, and all report this same place in the energy spectrum where the movements of TD show exaggerated energy. In one other investigation that reported on the altered energy spectra of TD (Tryon and Pologe, 1987), not enough data were presented to determine at exactly what frequency the alterations occurred, although it was clear that the results were not incompatible with the 1–2 Hz result. This very close agreement between the altered form of the movements recorded in this experiment in rats administered chronic neuroleptics and those of humans with TD would appear to indicate that the altered form of these computer-recorded oral movements is indeed a quite valid animal model (Ellison, 1991).

We recently extended this technique to recording from humans. We initially tried to apply our fluorescent spot technique directly to humans, but encountered limitations with this procedure. Placing the patients in a darkened room with UV lighting ("black-light") made many of the patients very uncomfortable, especially the paranoid schizophrenics. However, it should be noted that Gattaz and Buchel (1993) have directly applied our procedures to humans and do report increased energy at 1–2 Hz in the oral movements of humans with TD.

In our human studies, conducted in collaboration with William Wirsching of the VA hospital, Wadsworth, we have employed a color-based tracking system (Zeigler et al., 1991). A small spot (1.5-cm cube of foam with a bright neon red adhesive paper attached) is affixed to the upper lip and another one to the lower lip. Using a color-based tracker board based on construction details very similar to our previous board, the horizontal and vertical coordinates of the spots are placed into computer memory 30 times/s. Figure 2 shows the actual data recorded from a patient. The upper trace shows the mouth movements, with their characteristic 1–2 Hz repetitive form. In this record, a second camera was zoomed in on the hand, to which a colored spot was also affixed, and the two camera images were simultaneously recorded on videotape using a video mixer. In addition to showing the 1–2 Hz signature of TD movements, this figure further shows how these devices can be used to analyze correlations between various extremities. In the completely unrestrained patient, we can study whether the waxing and waning of the TD movements correlate with Parkinsonian-like movements of the hand or akathisia-like movements of the feet, for example. Patients can also be filmed in situations in which the movements are maximal, such as in their home environment, while watching television, and so on.

However, it should be noted that the apparent reasons for the increase in 1–2 Hz energy in the rats is often not identical to that in the humans. There are two ways in which an oral movement could have increased energy at 1–2 Hz. One is if the movement, no matter what its form, is repetitively repeated once or twice a second, and as can be seen in Fig. 2, humans with severe TD are frequently characterized by quite repetitive oral movements. A second way, however, is if the movement is aperiodically repeated, but adopts a smooth waveform similar to that of a single 1–2 Hz sine wave (rather than the more jerky spontaneous movements of the controls). This is also often the case with humans, and also is the predominant effect in the rats given chronic and continuous neuroleptics.

When the oral movements in drug-naive rats were studied following injections of one of three doses of either a D1 agonist (SKF38393) or a D2 agonist (LY171555) by Johansson et al. (1987), it was found that the SKF38393-induced increase and the LY171555-induced decrease in oral movements were owing to altered energy at all frequencies, indicating that the basic form of the oral movements had not been altered.

However, in a second study (Ellison et al., 1988), the oral movements in rats chronically administered continuous HAL or fluphenazine displayed oral movements with the characteristic form previously described (increased energy at 1–2 Hz), and this altered form was exaggerated by SKF38393 in a dose-

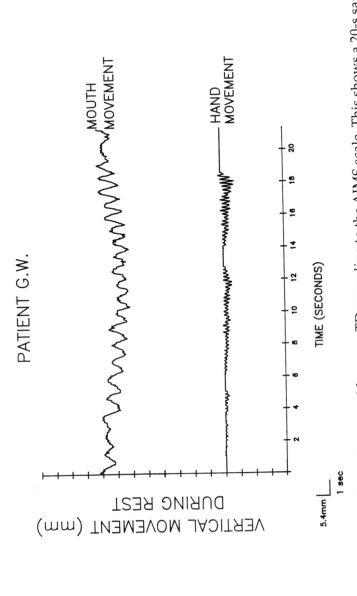

Fig. 2. Recording from a human with severe TD according to the AIMS scale. This shows a 20-s sample from a patient. Mouth movements and hand movements were simultaneously recorded. This shows the possibility of recording and crosscorrelating various extrapyramidal motor effects.

228

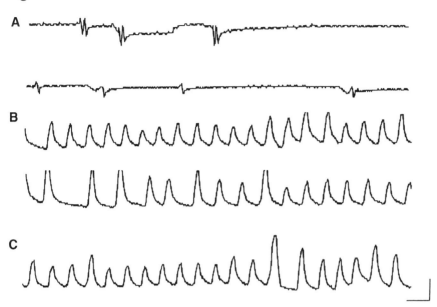

Fig. 3. Recording from rat given 9 mo of neuroleptics and then probe injections of **(A)** saline (top tracing), or **(B)** the D2 agonist LY171555 (second tracing), or **(C)** the D1 agonist SKF38393 (next three tracings), which are contiguous. The smooth and repeated oral movements following the D1 agonist are apparent, and are remarkably similar to those from the human in Fig. 4.

dependent manner, whereas LY171555 did not induce these distinctively formed oral movements. These data indicate that the altered oral movement pattern generator in this neuroleptic-induced dyskinesia can be driven by D1, but not D2 stimulation. Figure 3 shows tracings from one of these animals. The extent to which the animal pretreated with chronic HAL and then given a probe injection of the D1 agonist SKF38393 shows oral movements of a form and repetitive nature similar to those of the human with TD is striking.

In a further study designed to determine how drug regimen would affect these results, rats were given HAL for 8 mo, either via highly fluctuating weekly injections or more continuous administration (via their drinking water and osmotic minipumps). The two groups received approximately equivalent total doses of HAL. The results were quite striking, because they indicated that these two groups developed distinctively

Fig. 4. Two representative tracings of mouth movements in animals 10 d after discontinuation of 8 mo of chronic HAL. (A) Following continuous HAL (oral and minipump). The observer reported five oral movements during this segment. Smooth, repetitive movements with energy at 1–2 Hz are apparent. (B) Another animal following 8 mo of intermittent HAL at same total dosage. Repetitive, sharp oral movements occur, each of which has a similar waveform. Calibration: 0.5 s and 1.5 mm (i.e., five rasters).

different changes in oral movement form with time, as determined both by the computer system and by the human observer. Continuous administration again resulted in late-onset oral activity changes at 1–3 Hz and withdrawal increases in CSMs, a pattern indicating TD. However, intermittent treatment (weekly large injections) produced a pattern more analogous to a primed dystonia-like reaction: large-amplitude CSMs that had steep onset slopes and a peak energy at 4–7 Hz (Figs. 4 and 5). These results demonstrate the importance of drug regimen in determining the type of neuroleptic-induced dyskinesias that develop with prolonged neuroleptic treatment in rodents, and validates our highly unique measuring device, because we can now distinguish between two forms of neuroleptic-induced motor dyskinesias (TD and tardive dystonia), and can now study drugs that exacerbate or ameliorate these two different syndromes.

This experiment will go a long way toward solving a great deal of controversy in this field. The intermittent injection group appears to be showing the "primed dystonia" effect that has been demonstrated repeatedly in monkeys (Liebman and Neale, 1980; Kovacic et al., 1986).

We have also studied, using our rodent model, some promising atypical neuroleptics. These are drugs that have a low propensity for inducing TD (Tamminga and Gerlach, 1987). One such promising drug is clozapine, a nonselective antipsychotic

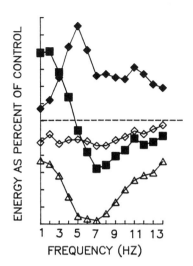

Fig. 5. Fast-Fourier analysis (compared to controls) of oral movements 10 d after drug withdrawal in rats that had received 8 mo of continuous HAL, intermittent HAL, clozapine, or raclopride. Abcissa: percent of control levels, going from 40 to 160% in steps of 10%, with the dashed line representing control levels. The distinctively different waveforms of oral movements in animals that had been pretreated with continuous vs intermittent HAL are clear. Although raclopride animals show no distinctive alteration in energy, the clozapine animals appear to have a form similar to continuous HAL, but decreased energy at all parts of the spectrum, reflecting the dramatically decreased number of oral movements that progressively developed in these animals. Note that the controls for the intermittent HAL animals received weekly saline injections, whereas the controls for all other groups received nondrugged, flavored drinking water and sham minipump implants. (■) HAL contin; (◆) HAL inject; (△) clozapine; and (◇) raclopride.

that can be differentiated from typical neuroleptics by its noncataleptogenic nature (Niemegeers and Janssen, 1979) and its inability to block DA agonist-induced stereotypies (Robertson and MacDonald, 1984). Clozapine has been shown to be a potent antipsychotic (Gerlach et al., 1974) that apparently does not induce TD (Lindstrom, 1988). A second direction in the search for atypical neuroleptics has been the development of more specific compounds for the dopamine D2 receptor, particularly the substituted benzamides, of which sulpiride is the most widely established in the treatment of schizophrenia (Tamminga and Gerlach, 1987). The substituted benzamides also appear to show less acute and tardive extrapyramidal side effects (EPS) (Gerlach and Casey, 1984). We chose to employ one of the most promising substituted benzamides, raclopride (Hall et al., 1988). In this experiment (See and Ellison, 1990b), rats were administered HAL, clozapine, raclopride, or no drug for either 28 d or 8 mo, and then withdrawn from drug treatment for 3 wk. Four weeks of neuroleptic administration produced no changes in CSMs in any drug-treated group. Long-term administration induced distinctively different patterns of oral activity in the three drug groups, both in number of CSMs and the form of these movements. The oral movements that developed in the HAL-treated rats fit our previously described syndrome of late-onset oral dyskinesias, which increased on drug withdrawal. The clozapine- and HAL-treated rats did not show the increased oral movements seen in the HAL animals, but each exhibited uniquely different CSM characteristics compared to controls. Figure 5 compares the form of oral movements, as determined by FFT analysis, of rats administered 8 mo of chronic HAL (either continuously or intermittently) with that of rats administered clozapine or raclopride for the same length of time. This figure graphically demonstrates the radically different energy spectrum of continuous versus intermittent HAL animals, but it further demonstrates a novel finding. The clozapine animals, which progressively developed a profound suppression of oral movements, correspondingly showed decreased energy at all parts of the spectrum. However, although these animals showed profound differences from HAL animals, it is also apparent that they eventually developed an altered form of oral activity as determined by FFT that in some ways showed similar characteristics to that in the HAL animals (i.e., peak energy at 1–3 Hz, and decreased activity at 6–8 Hz). Raclopride clearly did not induce this altered profile of oral movement form (although it is possible that the doses of raclopride given were somewhat lower than the other drugs on a clinical-effectiveness scale).

These results provide a strong rodent model for assessing different oral behaviors following chronic administration of typical and atypical neuroleptics, and imply that HAL, but not clozapine or raclopride, produces late-onset oral dyskinesias in rats that fit the pattern expected for TD.

Do the Two Different Oral Syndromes that Follow Continuous or Intermittent Haloperiodol Respond Differently to Cholinergics?

We recently conducted a lengthy study in which animals were pretreated for 8 mo with either continuous HAL, intermittent HAL, both regimens combined, or vehicle. The animals were then tested in the computerized oral movement device following injections of saline, one of two doses of physostigmine, or one of two doses of scopolamine. One result of this experiment was rather surprising: When both intermittent and continuous regimens were combined, the resulting behavior was not an admixture of the two syndromes, but rather, similar to the continuous HAL syndrome. This suggests that it is not so much the highly fluctuating HAL levels that give rise to the syndrome in the intermittent animals, but rather the fact that plasma levels periodically go to zero.

In this experiment, it was found that when rats are concurrently administered both intermittent and continuous HAL, the continuous syndrome predominates, both with respect to oral movement form and in the response to probe injections of physostigmine or scopolamine. Thus, the decline of serum HAL levels to near zero levels is necessary for the production of the intermittent OM syndrome, a finding that has implications for the etiology of TD as opposed to primed dystonia.

Although the different drug groups in this experiment did respond differently to cholinergics and anticholinergics, the results only partially fit a simple dystonia-TD model (Fig. 6). Although the baseline measures of the HAL-INJ regimen and the regimen's worsened response to the physostigmine challenge suggest that their syndrome, in some ways, resembles dystonia, other measures, such as the lack of an attenuating response owing to scopolamine challenge, suggest that the HAL-INJ regimen may not be a perfect model of human or primate dystonia. Similarly, the measures of the HAL-CONT and HAL-BOTH groups following the probe injections did not con-

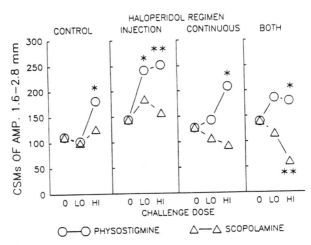

Fig. 6. Frequency of CSMs of amplitude 1.6–2.8 mm in the four groups chronically pretreated with control vehicle, intermittent HAL injections, continuous HAL, or both drug regimens. The data are presented following saline injections, systemic physostigmine, or systemic scopolamine injections. *Significantly different from controls, $p < .05$; **$p < .01$.

sistently indicate the presence of a pure TD-like syndrome. For example, the HAL-BOTH regimen's response included a scopolamine-induced decrease and physostigmine-induced increase of large observable OMs. Although this is opposite to the anticholinergic-induced exacerbation of TD symptoms seen in humans, a scopolamine-induced elevation of 5–7 Hz energy was observed with this regimen, and physostigmine challenge did decrease the activity in the 1–3 Hz energy spectrum, which was abnormally increased at baseline for this regimen.

However, the results were not at all clear as to whether the two different syndromes responded differently to the cholinergic vs the anticholinergic. The physostigmine made the syndrome of the intermittent animals worse, as would be predicted by the dystonia model, but the scopolamine did not benefit them, as would be predicted, and the continuous animals also did not respond as predicted, although there were some interesting alterations in energy at 5–7 vs 1–3 Hz observed. Clearly, further and more detailed analysis of cholinergic and anticholinergic modulation of neuroleptic-induced oral movements in rats is needed to determine the extent to which the rodent syndromes model aspects of human dystonia and TD.

These results indicate that there are several basic problems involved in conducting these experiments. One is that in rats, systemic cholinergics induce oral movements directly, whereas systemic anticholinergics make the rats hyperactive. Both of these effects, which do not seem to be such a major problem in humans and primates, dramatically complicate testing in the tube apparatus. Peripheral effects of these drugs, such as inducing salivation (which then leads to oral movements) or a dry mouth, are related problems. Several of these problems appear idiosyncratic to rodents.

Thus, this question is proving much more difficult to resolve than one would have anticipated. In order to overcome some of these problems of systemic cholinergics, we have begun studies of direct injections of drugs into the site in ventrolateral striatum where oral movements can be elicited in rats by intracerebral injections, thinking that perhaps the results would be clearer if more general effects, including peripheral ones, were eliminated.

In another study (Ellison et al., 1994), we studied effects of direct injections of pilocarpine or SKF38393 into the striatum in rats pretreated with continuous HAL, intermittent HAL, or vehicle (*see* Fig. 7). Both of these drugs exacerbated the preexisting behavioral syndrome (i.e., they worsened the 1–3 Hz effect in continuous HAL animals and worsened the higher-frequency effect in the intermittent animals). Thus, these drugs had different effects depending on the prior regimen of drug exposure, which suggests that the site of the altered pattern generator underlying the different oral movement syndromes was either in ventrolateral striatum or downstream from it. It seems likely that moving now toward the study of local drug effects will clarify considerably the pharmacology of TD-like and primed dystonic-like effects of chronic neuroleptic exposure to rodents.

Conclusions

A comprehensive rodent model of TD would greatly facilitate our understanding of this disorder and provide a means

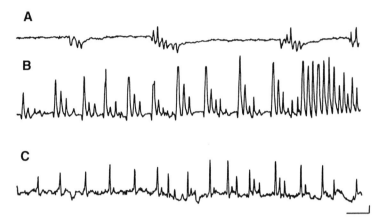

Fig. 7. Representative tracings of oral movements following pilo-carpine infusions into ventrolateral striatum in **(A)** an animal pre-treated with control (vehicle), showing the characteristic rhythymic (4–5 Hz) bursting pattern induced by pilocarpine, **(B)** an animal pre-treated with continuous HAL, showing the dramatic exaggeration of this pattern, where the bursts were more frequent, with larger oral movements, and gradually developed into a 4–5 Hz rhythmic pat-tern intermixed with a strong 1–2 Hz component, and **(C)** an animal pretreated with intermittent HAL, where the bursting pattern was much less apparent and the oral movements had a sharper form. Calibration: 1 s and CSM of amplitude 10 vertical sweeps.

of assessing treatment approaches, including novel neuroleptic development. The results described here indicate that simple observational techniques will probably not convincingly pro-duce models of neuroleptic-induced disorders, and that record-ing the actual form of the abnormal movements appears essential. When this is done, the results are much more con-vincing, but still not without problems, because simple notions of effects of how systemic pharmacological agents should alter these syndromes are only partially validated.

Several clear avenues of study remain for further exten-sion of this rodent model of TD. Further studies of the responsivity of these oral movement syndromes to various phar-macological probes are needed to assess what drugs may ame-liorate or exacerbate the symptomatic movements, including drugs that have been previously been found to modulate oral activity in humans and animals, such as cholinergic, dopaminergic, and GABAergic drugs. It is now of paramount importance to determine, using this model, where in the brain the altered pattern generator for these oral dyskinesias lies, since this will shed considerable light on the mechanisms underlying TD and guide crucial experiments.

References

Alpert, M., Diamond, F., and Friedhoff, A. J. (1976) Tremographic studies in tardive dyskinesia. *Psychopharmacol. Bull.* **12(2)**, 5.

Bedard, P., Delean, J., Lafleur, J., and Larochelle, L. (1977) Haloperidol-induced dyskinesias in the monkey. *J. Can. Sci.* **4**, 197–201.

Casey, D. E. (1985) Tardive dyskinesia, reversible and irreversible, in *Dyskinesia—Research and Treatment* (Casey, D. E., Chase, T. N., Christensen, A. V., and Gerlach, J., eds.), Springer, Heidelberg, pp. 88–97.

Chien, C., Jung, K., and Ross-Townsend, A. (1980) Methodological approach to the measurement of tardive dyskinesia: piezoelectric recording and concurrent validity test on given clinical rating scales, in *Tardive Dyskinesia—Research and Treatment* (Fann, W., Smith, R., Davis J., and Domino, E., eds.), Spectrum, New York, pp. 233–241.

Clow, A., Jenner, P., and Marsden, C. D. (1979) Changes in dopamine-mediated behaviour during one year's neuroleptic administration. *Eur. J. Pharmacol.* **57**, 365–375.

Clow, A., Theodorou, A., Jenner, P., and Marsden, C. D. (1980) Cerebral dopamine function in rats following withdrawal from one year of continuous neuroleptic administration. *Eur. J. Pharmacol.* **63**, 145–157.

Crane, G. E. (1973) Persistent dyskinesia. *Br. J. Psychiatry* **122**, 395–405.

Creese, I. (1983) Receptor interactions of neuroleptics, in *Neuroleptics, Neurochemical, Behavioral, and Clinical Perspectives* (Coyle, J. T. and Enna, S. J., eds.), Raven, New York, pp. 183–223.

Deneau, G. and Crane, G. (1969) Dyskinesia in rhesus monkeys tested with high doses of chlorpromazine, in *Psychotropic Drugs and Dysfunction of the Basal Ganglia* (Crane, G. and Gardner, R., eds.), US Public Health Service, Washington, DC, pp. 12–14.

Domino, E. and Kovacic, B. (1983) Monkey models of tardive dyskinesia. *Mod. Prob. Pharmacopsychiatry* **21**, 21–33.

Domino, E. F. (1985) Induction of tardive dyskinesia in Cebus apella and macaca speciosa monkeys, a review, in *Dyskinesia—Research and Treatment* (Casey, D. E., Chase, T. N., Christensen, A. V., and Gerlach, J., eds.), Springer, Heidelberg, pp. 217–223.

Ellison, G. (1991) Spontaneous orofacial movements in rodents induced by long-term neuroleptic administration: a second opinion. *Psychopharmacology* **104**, 404–408.

Ellison, G., Johansson, P., Levin, E., See, R., and Gunne, L. (1988) Chronic neuroleptics alter the effects of the D1 agonist SK&F 38393 and the D2 agonist LY171555 on oral movements in rats. *Psychopharmacology* **96**, 253–257.

Ellison, G., Liminga, U., and Keys, A. (1995) Oral dyskinesias in rats elicited by local injections of cholinergic or dopamine agonists into ventrolateral striatum. *Psychopharmacology*, in press.

Ellison, G. D. and See, R. E. (1989) Rats administered chronic neuroleptics develop oral movements which are similar in form to those in humans with tardive dyskinesia. *Psychopharmacology* **98**, 564–566.

Ellison, G. D., See, R. E., Levin, E. D., and Kinney, J. (1987) Tremorous mouth movements in rats administered chronic neuroleptics. *Psychopharmacology* **92**, 122–126.

Fann, W. E., Stafford, J., Malone, R., Frost, J., and Richman, B. (1977) Clinical research techniques in tardive dyskinesia. *Am. J. Psychiatry* **134**, 759.

Gardos, G., Cole, J. O., and Tarsy, D. (1978) Withdrawal syndromes associated with antipsychotic drugs. *Am. J. Psychiatry* **135(11)**, 1321–1324.

Gattaz, W. and Buchel, C. (1993) Assessment of tardive dyskinesia by means of digital image processing. *Psychopharmacology* **111**, 278–284.

Gerlach, J. and Casey, D. E. (1984) Sulpiride in tardive dyskinesia. *Acta Psychiatr. Scand.* **(Suppl. 311)**, 93–102.

Gerlach, J., Koppelhus, P., Helweg, E., and Monrad, A. V. (1974) Clozapine and haloperidol in single-blind cross-over trial: therapeutic and biochemical aspects in the treatment of schizophrenia. *Acta Psychiatr. Scand.* **50**, 410–424.

Glassman, R. and Glassman, H. (1980) Oral dyskinesias in brain-damaged rats withdrawn from a neuroleptic: implications for models of tardive dyskinesia. *Psychopharmacology* **69**, 19–25.

Gunne, L. M. and Barany, S. (1976) Haloperidol-induced tardive dyskinesia in monkeys. *Psychopharmacology* **50**, 237–240.

Gunne, L. M. and Haggstrom, J. E. (1983) Reduction of nigral glutamic acid decarboxylase in rats with neuroleptic-induced oral dyskinesia. *Psychopharmacology* **81**, 191–194.

Gunne, L. M., Haggstrom, J. E., and Syoquist, B. (1984) Association with persistent neuroleptic-induced dyskinesia of regional changes in brain GABA synthesis. *Nature* **309(24)**, 347–349.

Gunne, L. M., Andersson, U., Bondesson, U., and Johansson, P. (1986) Spontaneous chewing movements in rats during acute and chronic antipsychotic drug administration. *Pharmacol. Biochem. Behav.* **25**, 897–901.

Guy, W., Ban, T. A., and Wilson, W. H. (1985) An international survey of tardive dyskinesia. *Prog. Neuro–Psychopharmacol. Biol. Psychiatr.* **9**, 401–405.

Hall, H., Kohler, C., Gawell, L., Farde, L., and Sedvall, G. (1988) Raclopride, a new selective ligand for the dopamine-D2 receptors. *Prog. Neuro-Psychopharmacol. Biol. Psychiatry* **12**, 559–568.

Iversen, S. D., Howells, R. B., and Hughes, R. P. (1980) Behavioral consequences of long-term treatment with neuroleptic drugs. *Adv. Biochem. Psychopharmacol.* **24**, 305–313.

Jeste, D. V. and Wyatt, R. J. W. (1982) Therapeutic strategies against tardive dyskinesia: two decades of experience. *Arch. Gen. Psychiatry* **39**, 803–816.

Jeste, D. V., Lohr, J. B., Clark, K., and Wyatt, R. J. W. (1988) Pharmacological treatments of tardive dyskinesia in the 1980s. *J. Clin. Psychopharmacol.* **8**, 385–485.

Johansson, P. (1989) *Characterization and Application of Animal Models for Tardive Dyskinesia.* Acta Universitatis Upsaliensis, 222, 46 pp., Uppsala, Sweden.

Johansson, P., Casey, D. E., and Gunne, L. M. (1986) Dose-dependent increases in rat spontaneous chewing rates during long-term administration of HAL but not clozapine. *Psychopharmacol. Bull.* **22(3)**, 1017–1019.

Johansson, P., Levin, E., Ellison, G., and Gunne, L. (1987) Opposite effects of a D1 and a D2 agonist on oral movements in rats. *Eur. J. Pharmacol.* **134**, 83–88.

Kane, J. M. and Smith, J. M. (1982) Tardive dyskinesia: prevalence and risk factors, 1959–1979. *Arch. Gen. Psychiatry* **39**, 473–481.

Khalsa, H., Shaner, A., Anglin, M., and Wang, J. (1991) Prevalence of substance abuse in a psychiatric evaluation unit. *Drug Alcohol Depend.* **28**, 215–223.

Kovacic, B. and Domino, E. F. (1984) Fluphenazine-induced acute and tardive dyskinesia in monkeys. *Psychopharmacology* **84**, 310–314.

Kovacic, B., Ruffing, D., and Stanley, M. (1986) Effect of neuroleptics and of potential new antipsychotic agents (MJ13859-1 and MJ13980-1) on a monkey model of tardive dyskinesia. *J. Neural. Tran.* **65**, 39–49.

Lees, A. J. (1985) in *Tics and Related Disorders.* Churchill Livingstone, Edingburgh, pp. 191–234.

Levin, E. D., Galen, D., and Ellison, G. D. (1987) Chronic haloperidol effects on radial-arm maze performance and oral movements in rats. *Pharmacol. Biochem. Behav.* **26**, 1–6.

Levy, A. D., See, R. E., Levin, E. D., and Ellison, G. D. (1987) Neuroleptic-induced oral movements in rats: methodological issues. *Life Sci.* **41**, 1499–1506.

Liebman, J. and Neale, R. (1980) Neuroleptic-induced acute dyskinesias in squirrel monkeys: correlation with propensity to cause extrapyramidal side effects. *Psychopharmacology* **68**, 25–29.

Lindstrom, L. H. (1988) The effect of long-term treatment with clozapine in schizophrenia: a retrospective study in 96 patients treated with clozapine for up to 13 years. *Acta Psychiatr. Scand.* **77**, 524–529.

Marsden, C. D. and Jenner, P. (1980) The pathophysiology of extrapyramidal side effects of neuroleptic drugs. *Psychol. Med.* **10**, 55–72.

McKinney, W. T., Moran, E. C., Kraemer, G. W., and Prange, A. J. (1980) Long-term chlorpromazine in rhesus monkeys: production of dyskinesias and changes in social behavior. *Psychopharmacology* **72**, 35–39.

Mithani, S., Atmadja, S., Baimbridge, K. G., and Fibiger, H. C. (1987) Neuroleptic-induced oral dyskinesias: effects of progabide and lack of correlation with regional changes in glutamic acid decarboxylase and choline acetyltransferase activities. *Psychopharmacology* **93**, 94–100.

Moore, D. C. and Bowers, M. B. (1980) Identification of a subgroup of tardive dyskinesia patients by pharmacologic probes. *Am. J. Psychiatry* **137**, 1202–1205.

Neale, R., Gerhardt, S., and Liebman, J. M. (1984) Effects of dopamine agonists, catecholamine depletors, and cholinergic and GABAergic drugs on acute dyskinesias in squirrel monkeys. *Psychopharmacology* **82**, 20–26.

Niemegeers, C. J. E. and Janssen, P. A. J. (1979) A systematic review of the pharmacological activities of dopamine antagonists. *Life Sci.* **24**, 2201–2216.

Owen, F., Cross, A. J., Waddington, J. L., Poulter, M., Gamble, S. J., and Crow, T. J. (1980) Dopamine-mediated behaviour and 3H-spiperone binding to striatal membranes in rats after nine months haloperidol administration. *Life Sci.* **26(1)**, 55–59.

Pi, E. H. and Simpson, G. M. (1983) Atypical neuroleptics: clozapine and the benzamides in the prevention and treatment of tardive dyskinesia. *Mod. Probl. Pharmacopsychiatr.* **21**, 80–86.

Robertson, A. and MacDonald, C. (1984) Atypical neuroleptics clozapine and thioridazine enhance amphetamine-induced stereotypy. *Pharmacol. Biochem. Behav.* **21**, 97–101.

Rodriguez, L. A., Moss, D. E., Reyes, E., and Camarena, M. L. (1986) Perioral behaviors induced by cholinesterase inhibitors: a controversial animal model. *Pharmacol. Biochem. Behav.* **25**, 1217–1221.

Rondot, P. and Bathien, N. (1986) Movement disorders in patients with coexistent neuroleptic-induced tremor and tardive dyskinesia: EMG and pharmacologic study. *Adv. Neurol.* **45**, 361.

Rosengarten, H., Schweitzer, J. W., and Friedhoff, A. J. (1983) Induction of oral dyskinesias in naive rats by D1 stimulation. *Life Sci.* **33**, 2479–2482.

Rosengarten, H., Schweitzer, J. W., and Friedhoff, A. J. (1986) Selective dopamine D2 receptor reduction enhances a D1 mediated oral dyskinesia in rats. *Life Sci.* **39**, 29–35.

Rupniak, N. M. J., Jenner, P., and Marsden, C. D. (1983) Cholinergic manipulation of perioral behaviour induced by chronic neuroleptic administration to rats. *Psychopharmacology* **79**, 226–230.

Rupniak, N. M. J., Mann, S., Hall, M. D., Fleminger, S., Kilpatrick, G., Jenner, P., and Marsden, C. D. (1984) Differential effects of continuous administration for 1 year of haloperidol or sulpiride on striatal dopamine function in the rat. *Psychopharmacology* **84**, 503.

Rupniak, N. M. J., Jenner, P., and Marsden, C. D. (1985) Pharmacological characterisation of spontaneous or drug-associated purposeless chewing movements in rats. *Psychopharmacology* **85**, 71–79.

Rupniak, N. M. J., Jenner, P., and Marsden, C. D. (1986) Acute dystonia induced by neuroleptic drugs. *Psychopharmacology* **88**, 403–419.

Sahakian, B. J., Robbins, T. W., and Iversen, S. D. (1976) Fluphenthixol-induced hyperactivity by chronic dosing in rats. *Eur. J. Pharmacol.* **37**, 169–178.

Sant, W. W. and Ellison, G. (1984) Drug holidays alter onset of oral movements in rats following chronic haloperidol. *Biol. Psychiatr.* **19**, 95–99.

See, R. E. and Ellison, G. (1990a) Intermittent and continuous haloperidol regimens produce different types of oral dyskinesias in rats. *Psychopharmacology* **100(3)**, 404–412.

See, R. E. and Ellison, G. (1990b) Comparison of chronic administration of haloperidol and the atypical neuroleptics, clozapine and raclopride, in an animal model of tardive dyskinesia. *Eur. J. Pharmacol.* **181(3)**, 175–186.

See, R. E., Levin, E. D., and Ellison, G. D. (1988) Characteristics of oral movement in rats during and after chronic haloperidol and fluphenazine administration. *Psychopharmacology* **95**, 421–427.

Stewart, B. R., Rupniak, N. M. J., Jenner, P., and Marsden, C. D. (1988) Animal models of neuroleptic-induced acute dystonia, in *Advances in Neurology, vol. 50: Dystonia 2* (Fahn, S., Marsden, C. D., and Calne, D. B., eds), Raven, New York, pp. 343–359.

Tamminga, C. A. and Gerlach, J. (1987) New neuroleptics and experimental antipsychotics in schizophrenia, in *Psychopharmacology—The Third Generation of Progress* (Meltzer, H. Y., ed.), Raven, New York, pp. 1129–1140.

Tryon, W. W. and Pologe, B. (1987) Accelerometric assessment of tardive dyskinesia. *Am. J. Psychiatry* **144**, 1584–1587.

Waddington, J. L., Cross, A. J., Gamble, S. J., and Bourne R. C. (1983) Spontaneous orofacial dyskinesia and dopaminergic function in rats after 6 months of neuroleptic treatment. *Science* **220**, 530–532.

Wirshing, W. C., Cummings, J. L., Lathers, P., and Engel, J. (1989) The machine measured characteristics of tardive dyskinesia. *Schizophrenia Res.* **2(1–2)**, 240.

Zeigler, S., Keys, A., Ellison, G., and Wirsching, W. (1991) A rapid color-based videotracking system for behavioral studies: application to studies of tardive dyskinesia in humans. *Soc. Neurosci. Abst.* 1761.

CHAPTER 9

The Computer Pattern Recognition System for Study of Spontaneous Behavior of Rats

A Diagnostic Tool for Damage in the Central Nervous System?

Phyllis J. Mullenix

Introduction

For many years, investigators have been analyzing spontaneous behavior to develop measures that delineate malfunction of the central nervous system (CNS). Initial efforts focused on technology that improved behavioral identification and quantification. As the technology achieved a level of sophistication, attention turned increasingly to questions of data interpretation. In 1989, Norton aptly described the situation by stating that "we have a plethora of numbers and a paucity of principles." She emphasized a need for developing principles that help relate behavior to other parameters in the CNS. Such development is important because of criticisms that spontaneous activity measures, when used for neurotoxicity screening, generate data that provide little or no information about the origin of the problem (Maurissen and Mattsson, 1989). In other words, changes in spontaneous activity do not identify the brain structure involved.

From: *Motor Activity and Movement Disorders*
P. R. Sanberg, K. P. Ossenkopp, M. Kavaliers, Eds. Humana Press Inc., Totowa, NJ

If it is theoretically possible to link a brain lesion with a behavioral pattern, then activity data should reflect a change in pattern with a change in the site of brain damage. The hyperactivities resulting from lesions in the frontal cortex, hippocampus, septal area, substantia nigra, caudate putamen, and globus pallidus should be distinguishable in one experiment. Furthermore, the behavioral pattern must stay the same when lesions consistently involve a particular site or sites (common patterns). For example, lesions in the granule cell layer of the hippocampus, regardless of how induced, should produce the same array of changes in behavior.

No systematic investigation has demonstrated common patterns because of certain experimental stumbling blocks. The device measuring spontaneous activity needed redesign to detect behavioral changes other than just the extremes of hyperactivity and hypoactivity. When data consist simply of increased or decreased horizontal or vertical movements, all hyper- and hypoactivities look alike and make detection of different patterns virtually impossible (Mullenix, 1989). When data do not reveal changes per behavior, the equivalents of cognitive problems, IQ deficits, and learning disabilities go undetected because of their inherent behavior specificity. Poor inter- or intraobserver reliability, variation in results, and tediousness of data accumulation needed work before a search for lesion-specific behavioral patterns could even be attempted.

As an attempt to remove such stumbling blocks, a computer pattern recognition system was introduced for the study of spontaneous behavior in rats. Replacing photocell devices, this system automated behavioral act identification and quantified the initiations, total times, and time structure of multiple motor acts. Thus, it is a tool that analyzes the microstructure of behavior, advancing perspective as did the step from light to electron microscopy. Since its introduction in 1987, the computer system has generated enough data to begin a search for potential structure–function correlations. Outlined in this chapter are the details of this system and some of its applications as a screening procedure for neurotoxicity. Also presented is a retrospective look at behavioral changes detected by the system following various neurotoxic exposures. These exposures were conducted in different experiments and different years, but theoretically, they shared the same site of action in the brain. It is demonstrated here that these neurotoxic exposures indeed share the same array of effects on behavior, statistically an unlikely finding considering that the computer system performs well over 100 comparisons between treated and control animals in any one experiment. With mounting evidence of lesion-dependent behavioral patterns, the image of activity measures being a nonspecific "predictor" of CNS problems is quickly fading.

Computer Pattern Recognition Analysis of Spontaneous Behavior

The computer pattern recognition system eliminates variability inevitable with human observers. In fact, from behavior identification through data analysis, minimal human intervention is required. Because the system includes direct video recording of behavior, "between the beam" loss of information and the need for calibration procedures also are eliminated. Specific motor acts are identified, and quantification of the initiations, total times, and time structures of each are facilitated by large data storage capacity. Finally, a key benefit of this system is speed. The rate-limiting step in detecting CNS malfunction is no longer animal training sessions, recording observations, or data analysis. It is exposing the animals and waiting until they reach the age chosen for study.

Test Environment and Procedures

Tests are usually conducted during the diurnal period between 9:00 AM and 12:00 PM. Other times and red light conditions can be selected (Kernan et al., 1991), as long as the test conditions are consistent when comparing data from different experiments. The test chamber is a Plexiglas™ box made invisible to videocameras by its trapezoidal shape. This box is positioned on a permanent platform located in an isolated observation room where disturbances from external noise are minimized. White "noise" is not necessary in this situation. Accidental noises during tests would be experienced by both the control and experimental animals because they are always tested simultaneously. One meter in front and above the test chamber are videocameras linked to computers located outside the observation room. Clear horizontal and vertical views of albino rats are created by contrasts with black backgrounds built below and behind the test chamber and uniform lighting positioned throughout the test area. To conduct a test, two rats, one treated and one control matched for sex, age, and experience in the test environment, are placed simultaneously on opposite sides of a partition dividing the Plexiglas box. The clear partition has small holes, which allow the rats to see and smell each other, but not to interact while exploring this novel environment for 15 min. Further details of the test chamber, cameras, and observation room can be found elsewhere (Kernan et al., 1987).

Behavioral Identification

During the test, the cameras monitor behavior at 1 frame/s, a rate sufficient to detect over 90% of the occurrences of spontaneous behaviors in the rat. The original computer hardware and software that processed the video signals were described in detail by Kernan and coworkers in 1987. Since then, the computer configuration has been updated to allow even faster pattern analysis and behavioral classification of the data. The original MICRO VAX I and DEC VAX 11/750 computers have been replaced with a Barco monitor, a Gateway 2000, and a DEC 3100. The same discrete acts are identified by the new computer system, which include five major body positions (stand, sit, rear, walk, lying down) and eight modifiers (blank or no recognized activity, groom, head turn, turn, look, smell, sniff, wash face).

Behavioral Measures

There are three measures performed on the computer-identified behaviors: behavioral initiations (BI), behavioral total times (BTT), and behavioral time structures (BTS). These measures were found to provide independent information about CNS malfunction (Kernan et al., 1988; Mullenix and Kernan, 1989). They are performed on the discrete acts identified by the computer and each act's associated group, either attention, groom, or explore (Norton et al., 1976; Kernan et al., 1988). The attention group includes the acts of stand, blank, look, smell, and head turn. Groom includes sit, lying down, groom, and wash face, whereas explore includes rear, walk, turn, and sniff. The BI, BTT, and BTS measures are routinely computed for 18–22 pairs of animals, a pair consisting of one experimental rat and one matching control.

Calculation of BI

The number of frames in which a specific behavioral act begins is totaled for the 15-min observation period for each rat. The mean number of initiations for each act is determined for the control and experimental group of rats.

Calculation of BTT

The number of frames that a behavior continues, including the frame in which it is initiated, is totaled for the 15-min observation period. The mean total time for each act in control and experimental rats is then determined.

Calculation of BTS

This involves the calculation of the K function, a measure of the distribution of behavioral acts with respect to time (Kernan et al., 1988; Mullenix and Kernan, 1989). Of the three measures, BTS analysis provides the most stable and reproducible results (Kernan and Mullenix, 1991), and general experience has proven it to be the most sensitive measure. K functions are determined for discrete behavioral acts (e.g., sit or rear) or sequences of discrete acts (e.g., sit…rear). In addition, K functions are determined for combined acts (attention, groom, or explore), co-occurring combined acts (e.g., attention/exploratory), or sequences of combined acts (e.g., attention…explore) and co-occurring combined acts (e.g., attention/exploratory…grooming/attention). The K function formula for the time distribution of discrete acts is the following,

$$K_\alpha (t) = [\tau_\alpha/(n_\alpha)^2] \mid \sum_{i \neq j} \sum W_{ij}^{-1} I_t (U_{ij}^\alpha) \mid \qquad (1)$$

In this equation, n_α is the number of initiations of act α, τ_α is total observational time corrected for the extension of act α, W_{ij} is an edge correction term, and $I_t (U_{ij}^\alpha)$ is 1 (or 0) according to whether the pair (i, j) of initiations of act α occur (do not occur) within a time separation t. The function $K_\alpha(t)$, evaluated at eight time-points (2, 5, 10, 20, 30, 45, 100, and 200 s), is referred to as the time distribution of act α. For each of the time-points, a $\Delta K(t)$ (the difference between $K(t)$ for the control and the experimental groups) is calculated. At any one time-point, when K values increase (compared to controls) for a behavior, it means that that particular behavior (or sequence) is "clustering" in time (as seen in hypoactivity), whereas a decrease means it is "dispersing" in time (it has increased regularity of timing between initiations as seen in hyperactivity). Whenever a behavioral act is initiated < 10 times on average per animal, control or experimental, $K(t)$ values are not determined for that behavior and related sequences.

The bootstrap technique is used to estimate standard deviation at each time-point of the K function for a behavior. This technique uses Monte Carlo methods to generate an estimate of the variance of a statistic based only on the data. A random number generator is used to construct 1000 simulations of this calculation, each time generating a list of 20 pairs randomly selected from the original set of animal pairs. Obviously, one or more pairs may be dropped in any one of these simulations, whereas others are included in the calculation more than once. Standard statistical formulae are then used on the 1000 simulations to obtain an estimate of the standard deviations of $K(t)$ for the control and experimental groups separately or of $\Delta K(t)$.

Significance of Observed Changes

Considering BI or BTT measures alone, the statistical significance of any one change is assessed using the t-test, and a $p \leq 0.05$ is required for a change to be labeled as significant. Considering the BTS measure alone, significance of a change in any one behavioral time structure is determined by ad hoc criteria consisting of multiple tests that must be simultaneously satisfied. The first test evaluates for each time-point the following quantity:

$$[\mid K_{exp}(t) - K_{con}(t) \mid / SD\ (K_{exp}(t) - K_{con}(t))] = [\Delta K(t)/SD\ (\Delta K(t))] \quad (2)$$

In this equation, the subscripts "exp" and "con" refer to the exposed and control groups, whereas $\Delta K(t)$ represents the difference between the exposed and control groups. The SD $[\Delta K(t)]$ represents the estimated standard deviation in this measure derived from the bootstrap calculation. A given time distribution or sequence is not flagged as corresponding to a "real" change unless the value calculated above is ≥ 2.0 for three adja-

Table 1
Dose–Response Evaluations with Behavioral Response
Indicated by the RS Statistic

CNS agent	Data source	Dose or plasma level	RS statistic
d-Amphetamine sulfate	Mullenix et al., 1989	0.25 (mg/kg, sc)	0.131[a]
		0.50	0.575[b]
		1.00	0.583[b]
		2.00	0.838[b]
Triethyltin bromide	Kernan et al., 1991	3 (mg/kg, oral)	0.245[b]
		5	0.644[b]
Sodium fluoride	Mullenix et al., 1995	0.066 ± 0.02 (ppm F ± SD)	0.052
		0.107 ± 0.03	0.115[a]
		0.150 ± 0.03	0.359[b]

[a]$p < 0.01$.
[b]$p < 0.001$.

cent time-points with the same sign for $\Delta K(t)$. To assess the importance of the number of behaviors showing this "real" change in time structure for any one experiment, false-positive error rates were determined in untreated rats, and they were found to be 10% or less for this measure (Kernan et al., 1989).

Performing BI, BTT, and BTS measures on multiple behavioral acts and sequences generates well over 100 comparisons of experimental and control animals in one experiment. To distinguish significant behavioral effects from noise among so many comparisons, Kernan and Meeker (1992) developed the ad hoc RS statistic. This simple statistic, which encompasses all of the BI, BTT, and BTS data produced in an experiment, indicates whether behavior is changed overall and the confidence level associated with that change. Using this statistic, significance is set at the $p < 0.01$ level. As shown in Table 1, the single RS statistic is an effective indicator of behavioral response for dose–response evaluations.

Applications
of the Computer Pattern Recognition System

The computer pattern recognition system has detection capabilities that no other activity measures possess. However, there are limitations to its use. Before testing in the system, experimental animals should pass at least a crude observational battery that screens for ataxias, severe gait or motor coordination disturbances, or tremors. The computer system is not programmed to distinguish these neurotoxic effects, and results obtained if these dyskinesias exist would be unpredictable. Overall, the ideal animals to test in the computer system are those with no observable deficits or those given doses in a range not usually associated with toxicity. Changes in body weight or size in experimental animals do not hinder accurate behavioral identification by the computer system.

Stereotypies

Prior to development of the computer pattern recognition system, time-lapse photography provided the most extensive analysis of the microstructure of spontaneous behavior (Norton, 1968). Many features of its experimental design, therefore, were adopted as the starting point for the computer system. For example, the computer was programmed to identify most of the behaviors previously identified by a human observer from film, except for a few rapid movements of one portion of the body, such as head bobbing, scratching, and pawing. Neither time-lapse photography nor the computer pattern recognition system effectively quantifies sterotypies involving orofacial tremors, licking, or chewing. In contrast, stereotypies consisting of repetitive rearing, circling (turning), head turning, and sniffing (nose poking) are readily identified. The structure of stereotypy induced by amphetamine at 2 mg/kg has been defined with the computer pattern recognition system (Mullenix et al., 1989), and this stereotypy resulted in the highest RS statistic (greatest behavioral disruption) yet recorded in the computer system (see Table 1).

Hyperactivity and Hypoactivity

In 1973, Norton used time-lapse photography to demonstrate that an amphetamine hyperactivity consisted of increased frequencies of certain movements (e.g., walking and turning), shorter durations of most acts, and a randomization of behavioral initiations. There were indications that all hyperactivities had in common the same basic structure (Norton et al., 1976). It was not realized that one hyperactivity differed from another until a measure sensitive to behavioral time distribution was introduced. Using an index of dispersion and time-lapse photography, Norton (1977) concluded that hyperactivities induced by morphine (1.0–2.0 mg/kg) and amphetamine (0.25–1.0 mg/kg) were indeed very different. The morphine effect was generalized, whereas amphetamine affected some behavioral linkages more than others.

The amphetamine model of hyperactivity was reanalyzed using the computer pattern recognition system. The computer system detected amphetamine hyperactivity at the same dose levels (0.25–1.0 mg/kg) as those detected using time-lapse photography (Mullenix et al., 1989). The computer BI, BTT, and BTS measures confirmed that amphetamine consistently increased initiations and total times of certain movements (e.g., walking, turning or head turning), and when time distribution of behavior was affected, it almost always was more dispersed in time relative to controls. Thus, according to the computer system, amphetamine hyperactivity was structured about the same as that found using time-lapse photography. The computer system also had the sensitivity to distinguish hyperactivities. Using the BTS measure, the hyperactivity induced by prenatal exposure to phenytoin was readily distinguished from that induced by amphetamine (Mullenix, 1989).

The triethyltin model of hypoactivity also was reanalyzed using the new computer system. Kernan and coworkers (1991) compared results from the computer system with those reported by Bushnell and Evans (1986), who used a measure of horizontal and vertical home cage activity. During the 11 d postexposure to 3 or 5 mg/kg triethyltin, both techniques generated clear evidence of hypoactivity. The computer system, however, further demonstrated that triethyltin altered the time structure of many behaviors. After the second day postexposure, the change relative to controls was predominately behavioral clustering, the opposite response observed in amphetamine hyperactivity.

Cognitive Dysfunction

It is rare that measures of spontaneous behavior are used to study cognitive dysfunction. The lack of use stems from problems of measurement and not from the application being inappropriate. Early neurophysiological evidence indicated that the CNS, independent of sensory feedback, was the source of temporal regulation of behavior (Weiss, 1941). High-speed autonomous oscillators in the CNS were the proposed regulators of temporal structure of rhythmical motor sequences (Lashley, 1951). Spontaneous sucking rhythms were measured in infants to compare brain damage from different situations of perinatal distress (Wolff, 1968), and the frequency and time structure of spontaneous stereotyped behaviors were studied in autistic children (Sorosky et al., 1968) and mentally retarded individuals (Pohl, 1976; Baumeister, 1978). In retarded individuals, analysis of spontaneous behaviors (e.g., body rocking, head nodding, hand waving, and head banging) indicated that the frequency of such acts correlated negatively with measured intelligence (Baumeister, 1978). Studies of brain neurotransmitters indicated that depletion of brain dopamine mostly impaired sequential organization of behavior (Iversen, 1977). Rats with depleted brain norepinephrine appeared not to habituate, but continually explore their environment (Mason and Fibiger, 1979), indicating that the norepinephrine system modulates the brain's responsivity to external stimulation, specifically novel input (Aston-Jones and Bloom, 1981). Although mechanisms are a matter of some conjecture, there certainly is evidence to suggest that these neurotransmitter substrates and their control of motor acts comprise neural systems that ultimately regulate cognitive function (Tucker and Williamson, 1984).

The relationship between spontaneous behavior and cognitive function emphasizes the importance of quantifying the organization or microstructure of motor behavior. There are two features of the computer pattern recognition system that enable it to screen for cognitive malfunction. The system identifies a more complete repertoire of spontaneous behaviors triggered by novelty, and it measures temporal structure of discrete motor acts and sequences of motor acts, not just single-act initiations or durations. To demonstrate that these features in fact facilitate detection of cognitive deficits, the computer system was used to determine neurotoxicity in an animal model to study therapy for childhood leukemia.

Treatment of the CNS is a standard component of therapy for childhood acute lymphoblastic leukemia. However, this CNS therapy with well-documented efficacy is not without late-occurring neurotoxic consequences (Inati et al., 1983). Children given CNS therapy have significant reductions in overall IQ scores, along with impaired verbal and visual-spatial memory abilities, attention, organization, and motor output (Meadows et al., 1981; Moss et al., 1981; Brouwers et al., 1985; Waber et al., 1990). They are best characterized as learning disabled, not mentally retarded or hyperactive, and this impairment is sufficient to warrant supplemental special education or even education outside the regular classroom setting. Because CNS therapy is so crucial to event-free survival, it cannot be systematically manipulated to determine neurotoxic mechanisms. Thus, an animal model had to be developed to determine which CNS therapy agent or agents were responsible for the cognitive deficits in leukemia survivors.

The computer pattern recognition system readily detected behavioral disturbances induced by CNS therapies in rats (Mullenix et al., 1990, 1994). After study of individual agents and different combinations of CNS therapy agents, in both sexes and at two ages, the system demonstrated only rare occurrences of the usual components of hyperactivity, e.g., increased initiations or total times of walking, rearing, or turning. The few single-acts that were affected more were sit, groom, wash face, stand, and look, all stationary acts in the groom and attention clusters not easily monitored by most activity devices. The BTS measure revealed the behaviors and sequences with altered time structure, and whether the change was one of dispersion or clustering. Together, the BI, BTT, and BTS measures defined sex- and dose-dependent behavioral patterns that varied with the agents comprising the CNS therapy combination (Mullenix et al., 1994). With the components of the combination dictating the behavioral "signatures" in the animals, the overall realization was that neurotoxicity was primarily caused by combinations of CNS therapy agents and not by any one agent alone. These findings had greater impact because parallels could be found among results from clinical studies. As an example, some studies in humans indicated that neurotoxicity might be the result of combinations of agents rather than single-agent therapy (Bleyer, 1981, 1988), and that neurotoxicity associated with some CNS therapies was sex-dependent (Bleyer et al., 1990; Waber et al., 1990, 1992). The point most important to realize here, however, is that a measure of spontaneous behavior is successfully being used to detect cognitive dysfunction, specifically the equivalent of learning disabilities.

Structure and Function Correlation: Search for Common Patterns

Although sensitive to a range of neurotoxicities, the computer pattern recognition system is still in its infancy as a screening procedure that pinpoints the origin of a problem. Two hurdles complicate achievement of diagnostics where damage-specific behavioral patterns are identified. The first concerns identification of brain damage. The hyperactivity and cognitive dysfunction detected by the computer system may relate to microneuronal hypoplasia rather than qualitatively evident neuropathology, such as demyelination, glial reactions, or degenerative changes in neuronal perikarya (Altman, 1987). A case in point, neuropathologic changes have not been correlated with the learning disabilities associated with current leukemia treatment protocols, nor has neuropathology been confirmed in rats given CNS therapy agents (Yadin et al., 1983; Yanovski et al., 1989). Demonstrating brain damage may have to await tedious morphometric analysis of neuronal population size in specific brain regions or even newer techniques, such as in vivo proton spectroscopy (Mullenix et al., 1993). While awaiting such information, the brain regions affected might at least be predicted by knowing the timing of exposure. Detailed timetables for brain region development have been described that give an idea regarding the neuronal populations at risk for any particular age (Dekaban, 1968; Hicks and D'Amato, 1968; Altman, 1969, 1971, 1972, 1987; Rakic, 1974; Bayer, 1980; Bayer et al., 1982).

The second diagnostic hurdle concerns definition of common behavioral patterns. The computer pattern recognition system has been used to study spontaneous behavior in only two laboratories for fewer than 10 yr. Therefore, the pool of experiments that can be searched for recurring behavioral patterns is small, and is smaller still when experiments are matched for sex, age at exposure, and age at testing. The results from any one experiment span over 100 comparisons between experimental animals and their matched controls. Such high numbers of comparisons make it unlikely that any two different experiments will match in results unless a common pattern with biological relevance actually exists. Searching through so many comparisons, however, is an optical challenge. Attention is readily drawn to significant ($p \leq 0.05$ for BI and BTT; ad hoc criteria satisfied for BTS) behavioral changes because they are highlighted in the computer printout for each experiment. In contrast, when comparing results from different experiments, trends are also important to consider, but they are not highlighted in the printout and must be evaluated on a behavior-by-behavior basis. However, they are worth considering because BI and BTT results have behavior-dependent standard deviations that are quite high (Kernan and Mullenix, 1991), which generally assures that only extreme changes ever reach statistical significance. Standard deviations in BTS results are likewise behavior-dependent, but they are much smaller and better suited for low-level sensitivity (Kernan and Mullenix, 1991). Although smaller, they are still uncertain by a few percent, since they are an estimate taken from the bootstrap technique, which is not considered in the rigid structure of the ad hoc criteria that establish the significance of time structure changes for each behavior.

If there is merit to the simplistic notion that one brain lesion causes one pattern of behavioral disruption, common outcomes should be easy to find despite very different neurotoxic exposures. A search among data gathered to date using the computer pattern recognition system revealed one such pattern made evident here in a comparison of results from two different experiments. These experiments shared certain characteristics:

1. They both studied rats with interrupted development in the same brain region based on the timing of exposure;
2. They both had clear behavioral changes with significant ($p \leq 0.01$) RS statistics; and
3. They both included rats of the same strain, sex, and approximate age when tested.

Otherwise, the experiments appeared unrelated. One experiment was a study of neurotoxicity from CNS therapy for leukemia (Mullenix et al., 1990, 1994), and the other a study of neurotoxicity from exposures to fluoride (Mullenix et al., 1995).

Different Neurotoxic Exposures, but Same Brain Regions Affected

According to the time of exposure, development of the hippocampus theoretically was interrupted, causing the behavioral changes found in both the leukemia and fluoride experiments (Mullenix et al., 1994, 1995). Learning disabilities from leukemia therapy are certainly congruent with hippocampal damage that other studies have linked with hyperactivity and cognitive dysfunction (Altman, 1987). The hippocampus may in fact be the central processor that integrates inputs from the environment, memory, and motivational stimuli for creating behavioral decisions and modifying memory (DeLong, 1992).

In the leukemia experiment compared here, the exposure group, radiation, prednisolone, and methotrexate, called RPM, received three agents typically involved in CNS therapy for leukemia, i.e., steroids, methotrexate and cranial irradiation. The RPM exposure consisted of an ip injection of 18 mg/kg prednisolone on postnatal d 17 and 18, and also on postnatal d 18, an ip injection of 2 mg/kg methotrexate followed by 1000 cgy cranial radiation. Exposure in the fluoride study consisted of 125 ppm fluoride in the drinking water for 16 wk starting on postnatal d 21. Granule cells in the dentate gyrus of the hippocampus start developing postnatally and continue well into adulthood in the rat (Bayer and Altman, 1974; Bayer et al., 1982). Granule cells in the cerebellum also form postnatally, but their development is completed in the rat by postnatal d 21, whereas other cerebellar microneurons (stellate and basket cells) complete development by d 12 (Altman, 1969, 1972). Therefore, relative to the experiments considered here, the cerebellum might be impacted, but not to the extent expected in the hippocampus. Minimal effects would be expected in other brain regions, too (i.e., the cortex, thalamus, hypothalamus, septum, and pyramidal cells of the hippocampus), especially when development occurs prenatally (Bayer et al., 1973). Although the RPM and fluoride exposures were not administered exactly at the same ages, they both certainly occurred when the hippocampal granule cell layer was developing.

One Brain Lesion, One Behavioral Pattern

The RPM and fluoride animals were male pathogen-free Sprague-Dawley rats from Charles River, and the behavioral effects compared in the two experiments were those found when the rats were 4–5 mo old. The computer pattern recognition system determined RS statistics of 0.273 for the RPM rats and 0.311 for the fluoride rats, indicating that behavior changed significantly ($p \leq 0.001$) overall in both experiments. In the RPM experiment, there were 139 comparisons between RPM and control animals, whereas there were 126 comparisons between fluoride rats and their matching controls. Total comparisons differed because some behaviors occurred fewer than 10 times on average/control or experimental group, preventing calculation of K functions for those behaviors. Using only the behaviors (or sequences) for which K functions were determined in both experiments, comparison of RPM and fluoride results spanned a total of 126 exposed–control comparisons. Of these 126, neither the RPM nor the fluoride animals differed significantly from their matching controls in 86 of the same comparisons. In another 28 of these 126 comparisons, significant behavioral changes were found in either or both the fluoride and RPM experiments, and are listed in Table 2. This table shows that eight of the same BI, BTT, or BTS changes occurred in both experiments, and involved the behaviors sit and groom, the clusters groom/explore and groom/attention, and the sequences groom/explore...attention/groom and attention/groom...groom/explore. Figure 1 further illustrates the similarity in BTS changes found in the two experiments. Another 19 changes statistically significant in one experiment had matching trends in the second experiment (Table 2). BTS trends of dispersion or clustering are listed in Table 2 only if the value in Eq. (2) was ≥ 1 for 3 time-points with the same sign for $\Delta K(t)$. In general, results in the two experiments appeared the same in approx 90% of the original 126 comparisons (113/126). In the remaining 10%, there were statistically significant changes in one experiment, but for the exact same measures in the second experiment, there were no clear trends in any direction. More importantly, however, there was never a measure where the change was statistically significant in both experiments, but in opposite directions.

Table 2
Different Neurotoxicants, But One Pattern of Behavioral Change (±SE) in 4–5-Mo-Old Male Rats

Behavior	Control	125 ppm F for 16 wk	Control	RPM
Stand				
BTT	532.5 ± 20.2	599.0 ± 22.0[a]	526.6 ± 18.6	545.9 ± 24.0
Sit				
BI	57.7 ± 3.3	42.8 ± 3.5[b]	53.9 ± 3.7	40.9 ± 3.5[a]
BTT	245.6 ± 21.8	174.4 ± 23.6[a]	189.6 ± 22.8	121.9 ± 15.4[a]
BTS	—	Clustered[d]	—	Clustered
Groom				
BI	30.0 ± 5.5	14.3 ± 2.2[a]	30.5 ± 4.8	18.7 ± 2.7[b]
BTT	70.3 ± 16.1	35.8 ± 8.9	67.3 ± 13.2	32.8 ± 5.9[b]
Turn				
BI	97.2 ± 5.0	81.7 ± 5.4[a]	125.9 ± 6.4	137.1 ± 7.0
Groom/explore (cluster)				
BI	37.5 ± 3.0	23.7 ± 2.9[b]	31.2 ± 3.0	23.8 ± 3.4
BTT	49.5 ± 4.3	32.5 ± 4.6[b]	41.1 ± 4.0	30.8 ± 4.6
BTS	—	Clustered[d]	—	Clustered[d]
Groom/attention (cluster)				
BI	72.1 ± 5.2	46.1 ± 4.1[c]	60.3 ± 4.7	43.0 ± 3.8[a]
BTT	184.9 ± 19.8	131.4 ± 18.7	136.7 ± 16.5	86.4 ± 9.2[a]
BTS	—	Clustered[d]	—	Clustered[d]
Groom (cluster)				
BI	22.6 ± 3.8	11.7 ± 2.0[a]	20.5 ± 3.3	13.4 ± 2.1
BTT	42.4 ± 7.8	23.6 ± 4.5[a]	41.4 ± 8.6	22.7 ± 4.5

Attention (cluster)				
BTT	418.4 ± 21.4	499.3 ± 20.9^a	410.4 ± 17.3	440.7 ± 23.6
Attention/explore (cluster)				
BI	91.3 ± 4.7	92.1 ± 5.7	108.3 ± 4.9	126.7 ± 5.6^a
BTT	149.5 ± 8.9	159.9 ± 9.6	154.1 ± 7.9	183.8 ± 8.8^a
Blank . . . groom				
BTS	—	Clustered^d	—	Clustered
Groom . . . head turn				
BTS	—	Dispersed	—	Dispersed^d
Groom . . . turn				
BTS	—	Clustered	—	Clustered^d
Head turn . . . smell				
BTS	—	Clustered^d	—	Clustered
Groom . . . explore				
BTS	—	Clustered^d	—	Clustered
Groom . . . attention/groom				
BTS	—	Clustered^d	—	Clustered
Groom/explore . . . groom				
BTS	—	Clustered	—	Clustered^d
Groom/explore . . . attention/groom				
BTS	—	Clustered^d	—	Clustered^d
Attention/groom . . . groom				
BTS	—	Clustered^d	—	Clustered
Attention/groom . . . groom/explore				
BTS	—	Clustered^d	—	Clustered^d

^{a} $p < 0.05$, t-test.
^{b} $p < 0.01$, t-test.
^{c} $p < 0.001$, t-test.
^{d} Ad hoc criteria for significance satisfied (value in Eq. [2] is ≥ 2.0 for 3 adjacent time-points, all having the same sign for $\Delta K[f]$).

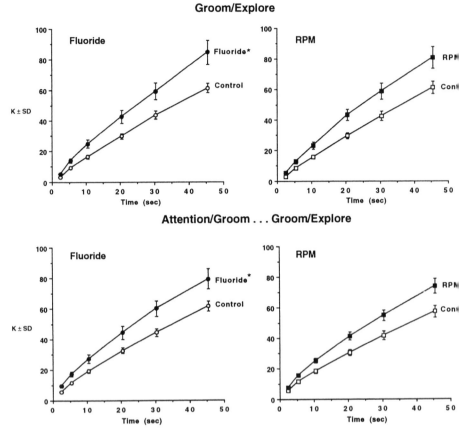

Fig. 1. Radiation, prednisolone, and methotrexate (RPM, ■), and fluoride (●) exposures significantly altered the groom/explore behavior and the sequence of attention/groom...groom/explore in 4–5 mo-old male rats compared to respective controls (□ and ○). These K functions illustrate similar clustering (increased K values) of behavioral time structures by the two different exposures. Error bars indicate ±SD.

The array of behavioral changes induced by RPM and fluoride exposures was similar to that induced by exposure on postnatal d 18 to 1000 cgy cranial irradiation (Mullenix et al., 1990). Irradiated animals matched those given RPM and fluoride exposures with respect to the strain and sex studied, the age at exposure, and the age at testing. In contrast to the RPM and fluoride experiments, however, the RS statistic for the

Table 3
A Familiar Pattern of Behavioral Changes (± SE)
After Cranial Irradiation of Male Rats

Behavior	Control	Cranial radiation
Sit		
BI	41.9 ± 2.8	31.8 ± 3.0[a]
BTT	116.2 ± 8.9	82.8 ± 9.4[a]
BTS	—	Clustered
Groom		
BI	24.5 ± 3.7	16.5 ± 4.3
BTT	49.4 ± 10.0	40.4 ± 19.0
BTS	—	Clustered[c]
Groom/explore (cluster)		
BI	23.5 ± 1.9	15.6 ± 1.9[b]
BTT	29.1 ± 2.6	21.1 ± 2.6[a]
BTS	—	Clustered
Groom/attention (cluster)		
BI	46.4 ± 3.6	35.3 ± 4.5
BTT	101.5 ± 10.3	78.8 ± 16.8
BTS	—	Clustered[c]
Groom (cluster)		
BI	12.9 ± 1.6	8.9 ± 2.3
BTT	19.9 ± 3.1	13.9 ± 4.0
Attention/explore (cluster)		
BI	109.5 ± 5.3	117.8 ± 7.0
BTT	162.8 ± 9.2	182.2 ± 11.5
Groom/explore . . . attention/groom		
BTS	—	Clustered
Attention/groom . . . groom/explore		
BTS	—	Clustered

[a]$p \leq 0.05$, t-test.
[b]$p \leq 0.01$, t-test.
[c]Ad hoc criteria for significance satisfied (value in Eq. [2] is ≥ 2.0 for 3 adjacent time-points, all having the same sign for $\Delta K[t]$).

radiation experiment (0.092) was not statistically significant. Yet, the fewer behavioral changes that did occur (Table 3) were in the same direction, and tended to involve the same behaviors and sequences as those seen with RPM and fluoride exposures (Table 2). Similarities in the RPM, fluoride, and radiation effects on behavioral time structures are illustrated in Fig. 2.

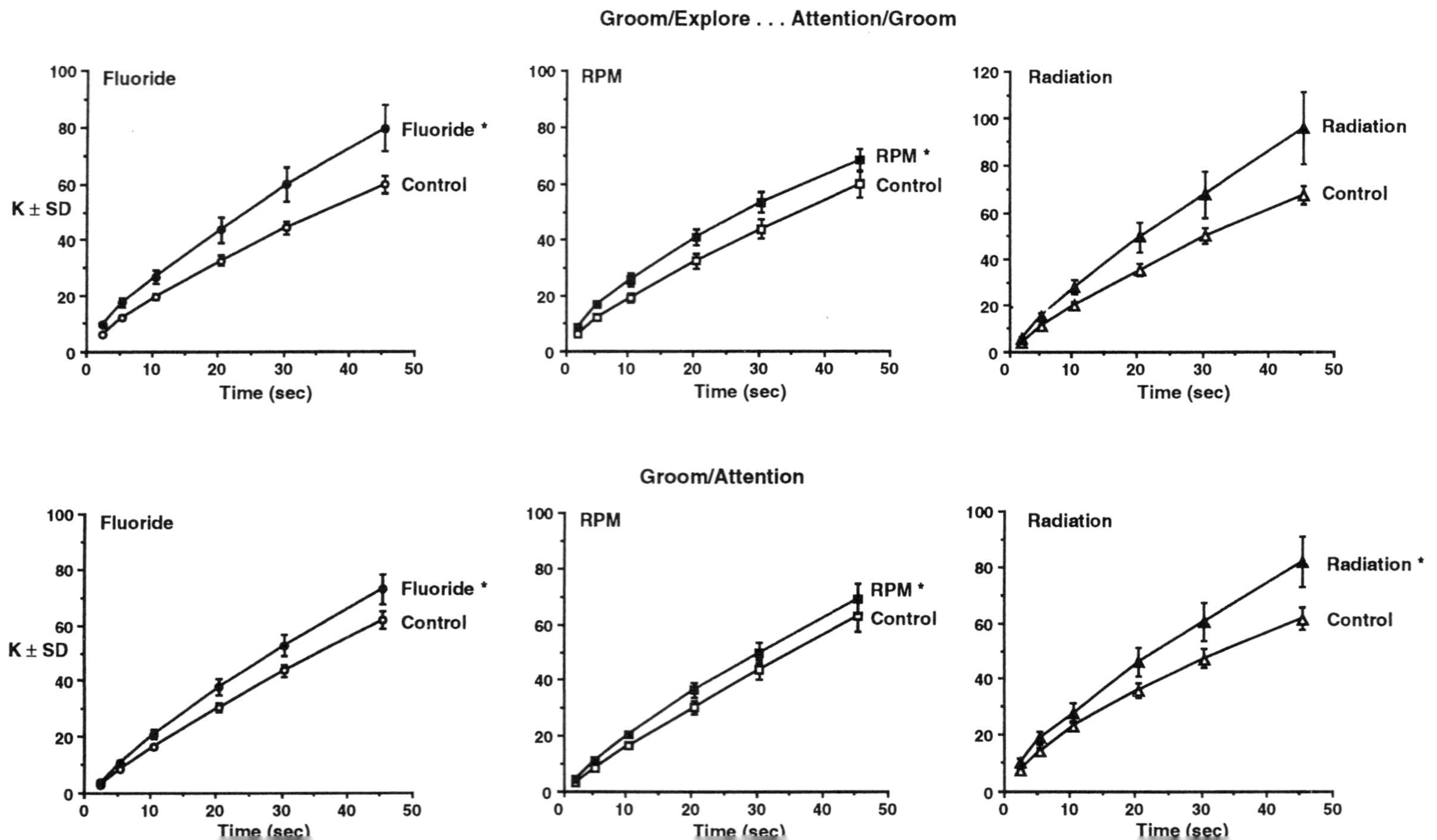

Groom/Explore . . . Attention/Groom
Fluoride
Fluoride *
Control
K ± SD
Time (sec)
RPM
RPM *
Control
Time (sec)
Radiation
Radiation
Control
Time (sec)
Groom/Attention
Fluoride
Fluoride *
Control
K ± SD
Time (sec)
RPM
RPM *
Control
Time (sec)
Radiation
Radiation *
Control
Time (sec)

Conclusions

The computer pattern recognition system is a sensitive, reliable, and rapid tool for the study of spontaneous behavior in rats. The detail it reveals about the microstructure of CNS output demonstrates that spontaneous behavior is neither chaotic nor unpredictable. As indicated in the evidence presented here, brain damage-specific behavioral patterns appear to exist. Using three very different neurotoxic exposures resulted in one pattern of behavioral disruption that was neither hyperactivity nor hypoactivity. Now when a measure of spontaneous behavior is used, cognitive dysfunction is no longer beyond the limits of detection, and the source of the problem, or the brain structure involved, is not a total mystery.

Demonstration of common patterns is important to neurotoxicity screening. It provides a means of classifying outcomes without using terms with nebulous meanings, i.e., hyperactivity, hypoactivity, and cognitive dysfunction. These terms have been referred to as "subtle" effects, which misrepresents their importance and underestimates the consequences. Each common pattern with defined characteristics would predict neurological consequences for a class of exposures. For example, the common pattern defined in Table 2 predicts learning disabilities for one class of exposures to RPM, fluoride, and radiation. When this particular common pattern is observed in trials of a new compound, learning disabilities should be considered as a potential risk. Note that this does not mean that all exposures to RPM, fluoride, and radiation will cause learning disabilities. Different exposures can mean different common patterns that predict other consequences.

Rat behaviors *per se* do not extrapolate to humans. Therefore, to use the common pattern described here as a predictor

---

Fig. 2. (*previous page*) Radiation, prednisolone, and methotrexate (RPM, ■), fluoride (●), and radiation (▲) exposures similarly altered the time structure of the groom/explore...attention/groom sequence and the behavior groom/attention in 4–5 mo-old male rats compared to respective controls (□, ○, and △). These *K* functions illustrate clustering (increased *K* values) induced by all three exposures. An asterisk indicates significant changes (ad hoc criteria satisfied) and error bars indicate ±SD.

for learning disabilities, a link to clinical evidence was necessary. The RPM exposure provided that link because the learning disabilities associated with agents used in leukemia therapy are well known. Cranial irradiation, although a frequent laboratory tool to study CNS development, is not used as a single agent to treat childhood leukemia. Thus, its link to learning disabilities is not as well established. The same can be said for fluoride. In fact, no clinical studies searching for learning disabilities have even considered exposures to fluoride. Fluoride-induced neurotoxicity has only recently been realized even in laboratory studies (Mullenix et al., 1995).

The common behavioral pattern defined here also has meaning for the prevention of neurotoxicity. It provides a template by which one can judge the effectiveness of new treatments designed to lessen neurotoxic side effects. Steroids and methotrexate alter behavioral and growth responses to cranial irradiation (Mullenix et al., 1990, 1994; Schunior et al., 1990, 1994). New ways to administer these drugs and prevent the RPM and radiation common pattern would be an important contribution to CNS therapy for leukemia.

The biological relevance of this common pattern is still unknown, although by virtue of the timing of exposure, it may relate to damage in the granule cell layer of the hippocampus. Irradiating infant rats has been shown to prevent acquisition of dentate granule cells in the hippocampus and in turn produce behavioral changes resembling those observed following surgical destruction of the hippocampus in adults (Bayer et al., 1973). Thus, one might expect the RPM and radiation exposures to interrupt development of this particular region of the brain. In contrast, until now there has been very little evidence to suggest that a fluoride exposure would interrupt development in any brain region. Now it is reasonable to suspect that fluoride is indeed a developmental neurotoxin. It produced the common pattern defined here, and it appears capable of producing a different pattern when the timing of exposure is changed. The pattern of behavioral changes produced by prenatal exposure to fluoride on gestational d 17–19 differed from that induced by postnatal fluoride exposures (Mullenix et al., 1995). On gestational d 17–19, a different brain region develops, the pyramidal cells of the hippocampus (Bayer, 1980), so different behavioral outcomes from prenatal and postnatal fluoride exposures would be the logical expectation. As lesion-specific behavioral patterns are established, activity data assume a biological relevance that previously was not thought possible.

References

Altman, J. (1969) Autoradiographic and histological studies of postnatal neurogenesis. III. Dating the time of production and onset of differentiation of cerebellar microneurons in rats. *J. Comp. Neurol.* **136,** 269–294.

Altman, J. (1971) Irradiation of cerebellum in infant rats with low level X-ray: histological and cytological effects during infancy and adulthood. *Exp. Neurol.* **30,** 492–509.

Altman, J. (1972) Postnatal development of the cerebellar cortex in the rat. I. The external germinal layer and the transitional molecular layer. *J. Comp. Neurol.* **145,** 353–398.

Altman, J. (1987) Morphological and behavioral markers of environmentally induced retardation of brain development: an animal model. *Environ. Health Perspect.* **74,** 153–168.

Aston-Jones, G. and Bloom, F. E. (1981) Norepinephrine-containing locus coeruleus neurons in behaving rats exhibit pronounced responses to nonnoxious environmental stimuli. *J. Neurosci.* **1,** 887–900.

Baumeister, A. A. (1978) Origins and control of stereotyped movements, in *Quality of Life in Severely and Profoundly Mentally Retarded People: Research Foundations for Improvement: Monograph no. 3* (Meyers, C. E., ed.), American Association on Mental Deficiency, Washington, DC, pp. 353–384.

Bayer, S. A. (1980) Development of the hippocampal region in the rat. I. Neurogenesis examined with ^{3}H-thymidine autoradiography. *J. Comp. Neurol.* **190,** 87–114.

Bayer, S. A. and Altman, J. (1974) Hippocampal development in the rat: cytogenesis and morphogenesis examined with autoradiography and low-level X-irradiation. *J. Comp. Neurol.* **158,** 55–80.

Bayer, S. A., Brunner, R. L., Hine, R., and Altman, J. (1973) Behavioural effects of interference with the postnatal acquisition of hippocampal granule cells. *Nat. New Biol.* **242,** 222–224.

Bayer, S. A., Yackel, J. W., and Puri, P. S. (1982) Neurons in the rat dentate gyrus granular layer substantially increase during juvenile and adult life. *Science* **216,** 890–892.

Bleyer, W. A. (1981) Neurologic sequelae of methotrexate and ionizing radiation: a new classification. *Cancer Treat. Rep.* **65 (Suppl. 1),** 89–98.

Bleyer, W. A. (1988) Central nervous system leukemia. *Pediatr. Clin. North Am.* **35,** 789–814.

Bleyer, W. A., Fallavollita, J., Robinson, L., Balsom, W., Meadows, A., Heyn, R. M., Sitarz, A., Ortega, J., Miller, D., Constine, L., Nesbit, M., Sather, H., and Hammond, D. (1990) Influence of age, sex, and concurrent intrathecal methotrexate therapy on intellectual function after cranial irradiation during childhood. *Pediatr. Hematol. Oncol.* **7,** 329–338.

Brouwers, P., Riccardi, R., Fedio, P., and Poplack, D. G. (1985) Long-term neuropsychologic sequelae of childhood leukemia: correlation with CT brain scan abnormalities. *J. Pediatr.* **106**, 723–728.

Bushnell, P. J. and Evans, H. L. (1986) Diurnal patterns in homecage behavior of rats after acute exposure to triethyltin. *Toxicol. Appl. Pharmacol.* **85**, 346–354.

Dekaban, A. S. (1968) Abnormalities in children exposed to X-radiation during various stages of gestation: tentative timetable of radiation injury to the human fetus. *J. Nucl. Med.* **9**, 471–477.

DeLong, G. R. (1992) Autism, amnesia, hippocampus, and learning. *Neurosci. Biobehav. Rev.* **16**, 63–70.

Hicks, S. P. and D'Amato, C. J. (1968) Cell migrations to the isocortex in the rat. *Anat. Rec.*, **160**, 619–634.

Inati, A., Sallan, S. E., Cassady, J. R., Hitchcock-Bryan, S., Clavell, L. A., Belli, J. A., and Sollee, N. (1983) Efficacy and morbidity of central nervous system "prophylaxis" in childhood acute lymphoblastic leukemia: eight years experience with cranial irradiation and intrathecal methotrexate. *Blood* **61**, 297–303.

Iversen, S. D. (1977) Brain dopamine systems and behavior, in *Handbook of Psychopharmacology, vol. 8. Drugs: Neurotransmitters and Behavior* (Iversen, L. L., Iversen, S. D., and Snyder, S. H., eds.), Plenum, New York, pp. 333–384.

Kernan, W. J. and Meeker, W. Q. (1992) A statistical test to assess changes in spontaneous behavior of rats observed with a computer pattern recognition system. *J. Biopharm. Statist.* **2**, 115–135.

Kernan, W. J. and Mullenix, P. J. (1991) Stability and reproducibility of time structure in spontaneous behavior in male rats. *Pharmacol. Biochem. Behav.* **39**, 747–754.

Kernan, W. J., Mullenix, P. J., and Hopper, D. L. (1987) Pattern recognition of rat behavior. *Pharmacol. Biochem. Behav.* **27**, 559–564.

Kernan, W. J., Mullenix, P. J., Kent, R., Hopper, D. L., and Cressie, N. A. C. (1988) Analysis of the time distribution and time sequence of behavioral acts. *Int. J. Neurosci.* **43**, 35–51.

Kernan, W. J., Mullenix, P. J., and Hopper, D. L. (1989) Time structure analysis of behavioral acts using a computer pattern recognition system. *Pharmacol. Biochem. Behav.* **34**, 863–869.

Kernan, W. J., Hopper, D. L., and Bowes, M. P. (1991) Computer pattern recognition: spontaneous motor activity studies of rats following acute exposure to triethyltin. *J. Am. Coll. Toxicol.* **10**, 705–718.

Lashley, K. (1951) The problem of serial order in behavior, in *The Neuropsychology of Lashley* (Beach, F. A., Hebb, D. O., Morgan, C. T., and Nissen, H. W., eds.), McGraw-Hill, New York, pp. 506–528.

Mason, S. T. and Fibiger, H. C. (1979) NE and selective attention. *Life Sci.* **25**, 1949–1956.

Maurissen, J. P. J. and Mattsson, J. L. (1989) Critical assessment of motor activity as a screen for neurotoxicity. *Toxicol. Indust. Health* **5**, 195–201.

Meadows, A. T., Gordon, J., Massari, D. J., Littman, P., Fergusson, J., and Moss, K. (1981) Decline in IQ scores and cognitive dysfunction in children with acute lymphocytic leukemia treated with cranial irradiation. *Lancet* **2**, 1015–1018.

Moss, H. A., Nannis, E. D., and Poplack, D. G. (1981) The effects of prophylactic treatment of the central nervous system on the intellectual functioning of children with acute lymphocytic leukemia. *Am. J. Med.* **71**, 47–52.

Mullenix, P. J. (1989) Evolution of motor activity tests into a screening reality. *Toxicol. Indust. Health* **5**, 203–216.

Mullenix, P. J. and Kernan, W. J. (1989) Extension of the analysis of the time structure of behavioral acts. *Int. J. Neurosci.* **44**, 251–262.

Mullenix, P. J., Kernan, W. J., Tassinari, M. S., and Schunior, A. (1989) Generation of dose–response data using activity measures. *J. Am. Coll. Toxicol.* **8**, 185–197.

Mullenix, P. J., Kernan, W. J., Tassinari, M. S., Schunior, A., Waber, D. P., Howes, A., and Tarbell, N. J. (1990) An animal model to study toxicity of central nervous system therapy for childhood acute lymphoblastic leukemia: effects on behavior. *Cancer Res.* **50**, 6461–6465.

Mullenix, P. J., Schunior, A., Cook, C., and Mulkern, R. V. (1993) In vivo ^{1}H MR spectroscopy of rats for the detection of neurotoxicity from leukemia therapy agents. *Soc. Magnetic Resonance Med. Abstr.* **241**, 1506.

Mullenix, P. J., Kernan, W. J., Schunior, A., Howes, A., Waber, D. P., Sallan, S. E., and Tarbell, N. J. (1994) Interactions of steroid, methotrexate, and radiation determine neurotoxicity in an animal model to study therapy for childhood leukemia. *Pediatr. Res.* **35**, 171–178.

Mullenix, P. J., DenBesten, P. K., Schunior, A., and Kernan, W. J. (1995) Neurotoxicity of sodium fluoride in rats. *Neurotoxicol. Teratol.* **17**, 169–177.

Norton, S. (1968) On the discontinuous nature of behavior. *J. Theoret. Biol.* **21**, 229–243.

Norton, S. (1973) Amphetamine as a model for hyperactivity in the rat. *Physiol. Behav.* **11**, 181–186.

Norton, S. (1977) The structure of behavior of rats during morphine-induced hyperactivity. *Commun. Psychopharm.* **1**, 333–341.

Norton, S. (1989) Correlation of cerebral cortical morphology with behavior. *Toxicol. Indust. Health* **5**, 247–254.

Norton, S., Mullenix, P. J., and Culver, B. (1976) Comparison of the structure of hyperactive behavior in rats after brain damage from X-irradiation, carbon monoxide and pallidal lesions. *Brain Res.* **116**, 49–67.

Pohl, P. (1976) Spontaneous fluctuation in rate of body rocking: a methodological note. *J. Ment. Defic. Res.* **20**, 61–65.

Rakic, P. (1974) Neurons in rhesus monkey visual cortex: systematic relation between time of origin and eventual disposition. *Science* **183**, 425–427.

Schunior, A., Zengel, A. E., Mullenix, P. J., Tarbell, N. J., Howes, A., and Tassinari, M. S. (1990) An animal model to study toxicity of central nervous system therapy for childhood acute lymphoblastic leukemia: effects on growth and craniofacial proportion. *Cancer Res.* **50**, 6455–6460.

Schunior, A., Mullenix, P. J., Zengel, A. E., Landy, H., Howes, A., and Tarbell, N. J. (1994) Radiation effects on growth are altered in rats by prednisolone and methotrexate. *Pediatr. Res.* **35**, 416–423.

Sorosky, A. D., Ornitz, E. M., Brown, M. B., and Ritvo, E. R. (1968) Systematic observations of autistic behavior. *Arch. Gen. Psychiatry* **18,** 439–449.

Tucker, D. M. and Williamson, P. A. (1984) Asymmetric neural control systems in human self-regulation. *Psychol. Rev.* **91,** 185–215.

Waber, D. P., Urion, D. K., Tarbell, N. J., Sollee, N. D., Dinklage, D., Gioia, G., Paccia, J., Sherman, B., Niemeyer, C., Gelber, R., and Sallan, S. E. (1990) Late effects of central nervous system treatment of childhood acute lymphoblastic leukemia are sex-dependent. *Dev. Med. Child Neurol.* **32,** 238–248.

Waber, D. P., Tarbell, N. J., Kahn, C. M., Gelber, R. D., and Sallan, S. E. (1992) The relationship of sex and treatment modality to neuropsychological outcome in childhood acute lymphoblastic leukemia. *J. Clin. Oncol.* **10,** 810–817.

Weiss, P. A. (1941) Autonomous versus reflexogenous activity of the central nervous system. *Proc. Am. Philos. Soc.* **84,** 53.

Wolff, P. H. (1968) The serial organization of sucking in the young infant. *Pediatrics* **42,** 943–956.

Yadin, E., Bruno, L., Micalizzi, M., Rorke, L., and D'Angio, G. (1983) An animal model to detect learning deficits following treatment of the immature brain. *Child's Brain* **10,** 273–280.

Yanovski, J. A., Packer, R. J., Levine, J. D., Davidson, T. L., Micalizzi, M., and D'Angio, G. (1989) An animal model to detect the neuropsychological toxicity of anticancer agents. *Med. Pediatr. Oncol.* **17,** 216–221.

CHAPTER 10

Circling Behavior in Rodents

Methodology, Biology, and Functional Implications

Jeffrey N. Carlson and Stanley D. Glick

Introduction

Circling is a readily measurable behavior that is exhibited by many organisms and that may indicate that a lateral preference is present for carrying out other behavioral functions. In this respect, it is similar to handedness, the best known human index of lateral preference. Most modern accounts of the determinants of lateral preference suggest that it occurs as a result of some functional lateralization of systems in the brain. Lateralization simply means that the two hemispheres are differentially proficient in controlling various behavioral activities. The significance of lateralization extends beyond the simple fact of brain asymmetry. Rather, it implies that certain features of behaviors that are controlled by asymmetric brain areas will vary according to the kind and degree of asymmetry that is present. In this chapter, we will assess the utility of circling behavior as an index of brain asymmetry and the relationship of this asymmetry to various aspects of behavioral function.

History

Functional lateralization of the human brain has been known since Broca's discovery that lesions that resulted in lan-

From: *Motor Activity and Movement Disorders*
P. R. Sanberg, K. P. Ossenkopp, M. Kavaliers, Eds. Humana Press Inc., Totowa, NJ

guage disorders were, in most cases, located in the left cerebral hemisphere (Kilshaw and Annett, 1983). Numerous neuroanatomical differences between the left and right hemispheres of the human brain have since been related to differences in the control of various aspects of behavior (Corballis, 1991; Hellige, 1993). Findings have accumulated, for example, to suggest that although the human left hemisphere is specialized for the processing of language, the right hemisphere has a dominant role for musical, visuospatial, and emotional processing (Springer and Deutsch, 1981). The direction and degree of hemispheric dominance differ among individuals, and may predict differences in behavioral function. Variation in human cerebral asymmetry has been shown to vary with differences in functional characteristics, such as handedness (Bryden, 1982; Annett, 1985), cognitive ability (O'Boyle and Hellige, 1989), emotional function (Davidson, 1992), and psychopathology (Flor Henry, 1986).

The issue of whether an equivalent functional lateralization exists in nonhuman species has not been as clearly resolved as it has been in humans. Until relatively recently, it had been assumed that cerebral lateralization of function is a uniquely human trait (e.g., Levy, 1977). Nonetheless, there are scientifically compelling reasons for the study of cross-species comparisons of brain asymmetry. To the extent that animal brain lateralization is present, it provides an opportunity to establish models of human behavioral function that allow a more precise determination of the relationship between chemical and morphological asymmetries and behavioral processes. Demonstrations that animals have asymmetric brains have been repeatedly made for morphological characteristics. Data indicate that morphologic brain asymmetries exist in species such as primates, rodents, birds, reptiles, amphibians, and fish (cf. Galaburda et al., 1985). Evidence that functional brain asymmetry occurs in species other than humans has also accumulated steadily over the last 20 years. Functional brain asymmetries have been reported in species that range from songbirds (Nottebohm, 1977; Arnold and Bottjer, 1985) to nonhuman primates (LeMay and Geschwind, 1975; LeMay, 1985). Lateral specialization for motor performance, the perception and production of vocalizations, visuospatial processes, and motivation and emotion have been reported in a range of species that includes nonhuman primates, songbirds, chickens, dolphins, and rodents (Hellige, 1993). Clearly, findings have suggested that cerebral lateralization is an evolutionary principle and not simply one indigenous to humans, in many species, an important manifestation of functional brain lateralization is circling or turning behavior.

The notion that turning behavior gives us clues about functional brain asymmetry has been with us for more than a century. Nothnagel in 1873 found that injections of chromic acid into the striatum caused a bending of the trunk to the side of the lesion (Wilson, 1914). According to Ferrier (1876), "irritation of the corpus striatum causes general muscular contraction on the opposite side of the body. The head and body strongly flexed to the opposite side, so that the head and tail become approximated." These results were seen in monkeys, cats, dogs, jackals, and rabbits. Wilson (1914) reported that small unilateral lesions of the striatum caused monkeys to show "a preference, in the taking of nuts or bananas, for the homolateral limb." Lashley (1921) described a rotation syndrome in rats after combined unilateral destruction of the caudate nucleus and the motor cortex above it. "Circus" movements after "partial and unsymmetrical injuries to the striatum" in rats were also observed by Herrick (1926). In a 1942 study by Mettler and Mettler (1942) unilateral striatal lesions in dogs caused circus movements and a forced posture toward the side of the lesion.

In the early 1970s, observations that D-amphetamine (D-A) would enhance side preferences (Glick, 1973; Glick and Jerussi, 1974) and induce rotation (or circling) in normal unlesioned rats (Jerussi and Glick, 1974, 1976) suggested that the rat brain might also exhibit an endogenous form of functional lateralization. At the time of these observations, it had become apparent that afferent neurons of the nigrostriatal dopamine (DA) system played an important role in the control of motor behavior. The discovery of L-DOPA's efficacy in treating the symptoms of Parkinsonism (Cotzias et al., 1967) led to a search for animal models that could be used to evaluate new and potentially more useful agents for this disorder. An important research strategy that emerged was to perform complete or partial lesions of the nigrostriatal pathways, and then to evaluate changes in the direction and intensity of whole-body motor behavior and their further alteration by various drugs (Ungerstedt and Arbuthnott, 1970; Ungerstedt, 1971a). Ungerstedt (1971b) first demonstrated that rats with unilateral lesions of the nigrostriatal pathways would rotate in response to a variety of dopaminergic agents (e.g., D-A, L-DOPA, apomorphine). Soon thereafter, it was found that these drugs also induced rotation, although at lower rates, in normal intact rats (Jerussi and Glick, 1976). Whereas rats with unilateral lesions of the nigrostriatal system would turn in circles at high rates (5–10 rotations/min) in response to D-A, a lower rate of turning (0.5–4 rotations/min) also took place in unlesioned rats (Glick and Jerussi, 1974; Jerussi and Glick, 1974). Subsequent studies demonstrated that, as with lesioned rats, the rotation in normal rats was consistent in direction and correlated in magnitude on repeated testing. Some rats rotate consistently to the left, whereas others rotate to the right. Furthermore, unilateral lesions of the striata of rats previously assessed for spontaneous rotation caused more rotation if the lesion was made ipsilateral rather than contralateral to the preoperative direction of rotation (Jerussi and Glick, 1975). Since the rotational behavior in lesioned rats was created by a large unilateral loss of nigrostriatal dopaminergic neurons, it was hypothesized that this behavior in normal rats was a manifestation of a smaller endogenous asymmetry of these same neurons (Glick et al., 1974; Jerussi and Glick, 1976).

The observations of rotation effects with dopaminergic drugs in nonlesioned animals led to the hypothesis suggesting that the afferent neurons of the nigrostriatal DA system would show morphological or biochemical asymmetry. These predictions have been generally confirmed. Endogenous biochemical asymmetries in the striatum have been identified for DA content (Zimmerberg et al., 1974), DA release and DA metabolism (Glick et al., 1988a), DA-stimulated adenylate cyclase activity (Jerussi et al., 1977), and DA receptors (Drew et al., 1986; Glick et al., 1988b). Furthermore, the strength and direction of these chemical asymmetries have been shown to be related to the magnitude and direction of behavioral asymmetries. Spontaneous side preferences as reflected by choice behavior in a T-maze (Zimmerberg et al., 1974) have been related to asymmetries in DA content. Rats have been also observed to circle at rates somewhat lower than their drugged rates during the active phase of their diurnal cycle. Variation in this "nocturnal" circling behavior has been shown to be predictive of asymmetries in striatal DA metabolism and DA-stimulated adenylate cyclase activity (Jerussi et al., 1977).

Side Preferences and Their Relationship to Biochemical and Morphological Asymmetries

Systemic injections of D-A also induce variations in circling behavior that have been related to hemispheric differences in striatal DA levels (Glick et al., 1974; Jerussi and Glick, 1976). Left and right striatal DA levels were determined in rats injected with various doses of amphetamine and tested for rotation. There were significant asymmetric differences in striatal DA that were related to the direction of rotation. Since D-A was found to be equally distributed to the two sides of the brain, the difference in striatal DA appeared to be the neurochemical substrate for rotation in normal rats. The results suggested that normal rats have asymmetrical levels of striatal DA. These asymmetries have since been confirmed (Robinson et al., 1980; Robinson and Becker, 1983) in other laboratories and under other conditions. Furthermore, since these early studies, other work has delineated other asymmetric characteristics of the striatal DA system. DA release (Robinson et al., 1982), DA uptake (Shapiro et al., 1986), and DA (D_2) receptor densities (Schneider et al., 1982; Drew et al., 1986) have been found to be asymmetrical. DA asymmetries also appear to exist in the mesocorticolimbic DA system. Differences in DA levels have been observed in the nucleus accumbens (Rosen et al., 1984) and frontal cortex (Slopsema et al., 1982).

Brain asymmetries are not limited to the nigrostriatal and mesoconticolimbic DA systems. Morphological asymmetries, for example, have been reported in the whole cortex (Diamond et al., 1981; Sherman et al., 1982; Fleming et al., 1986) and hippocampus (Diamond et al., 1982). Functional asymmetries, as evidenced by differential effects of lesions on circling behavior when performed on either side of the brain, have been seen in the cortex (Denenberg et al., 1978; Pearlson and Robinson, 1981; Crowne et al., 1987) and hippocampus (Therrien et al., 1982), as well as in the striatum (Glick and Cox, 1976, 1978; Rothman and Glick, 1976; Mittleman et al., 1985). Many neurochemical asymmetries have been reported for various regions of the rat brain. Some of these include side-to-side differences in norepinephrine content of the thalamus (Oke et al., 1980) and the striatum (Rosen et al., 1984), γ-aminobutyric acid (GABA) levels in substantia nigra, nucleus accumbens, thalamus, and striatum (Rosen et al., 1984); serotonin levels in the striatum (Knapp and Mandell, 1980; Rosen et al., 1984), nucleus accumbens (Rosen et al., 1984) and hippocampus (Knapp and Mandell, 1980); and luteinizing hormone-releasing hormone (LHRH) content in hypothalamus (Gerendai, 1984). From the foregoing, it would appear that multiple neurochemical asymmetries night contribute to the behavioral asymmetries elaborated above. Evidence that this is true came from studies that assessed glucose uptake in several brain structures (Glick et al., 1979). Since glucose is the brain's major energy source, variation in its uptake would relate directly to regional neuronal activity. Rats with known circling biases were injected with radiolabeled [1,2–3H]-deoxy-D-glucose, decapitated, and their brains dissected and assayed. Of the structures examined, most showed evidence of functional brain asymmetry. Greater glucose utilization on the side of the brain contralateral to the direction of rotation was seen in the midbrain and striatum. Differences in activity were also seen in the frontal cortex and hippocampus with the left side being greater than the right side regardless of circling bias. An absolute difference in activity between sides that was correlated to the rate of rotation was seen for the thalamus and hypothalamus, whereas an absolute asymmetry in the cerebellum was correlated with random movement. In other rats, D-A administered 15 min before the deoxyglucose injection altered asymmetries in the striatum, frontal cortex, and hippocampus but not in the midbrain, thalamus, hypothalamus, or cerebellum. Since functional brain activity has been held to be closely coupled to the rate of energy metabolism (Sokoloff, 1977), these findings suggest that multiple and interrelated functional asymmetries are present in the rat brain, and are related to the direction and intensity of turning behavior. The implication of this suggestion is that many of the neurotransmitter asymmetries noted above might contribute to this overall pattern of interrelated functional asymmetry. Thus, by measuring turning behavior, one may be assessing asymmetry not only in the nigrostriatal pathways but also in other regions that interact with the nigrostriatal system.

The Methodology of Circling

Circling behavior is normally assessed by placing subjects in a rotometer. The typical design of a rotometer is that of a cylindrical enclosure having a flat floor. The original design proposed by Ungerstedt and his colleagues (Ungerstedt, 1971b) used a concave or bowl-like floor that was so shaped as to "encourage turning." Originally, circling behavior in nonlesioned rats was assessed in the Ungerstedt-type bowl. However, it was subsequently demonstrated that circling behavior occurs independently of the shape of the rotometer (Glick et al., 1977a) and is simply the result of an intrinsic side preference.

The animal's movement must somehow be transmitted to a sensor. The most prevalent way of doing this is to use a semi-rigid link. In most applications, a thin steel wire is looped around the animal to form a flexible connector that harnesses the animal. The harness is connected to a shaft that activates a position-sensing device. The design of this device incorporates an array of four switches, either mechanical or photoelectric, that are arranged with one in each quadrant around the rotating shaft. A cam on this shaft sequentially operates one of the four switches in a manner that corresponds to the animal's angular movement. The four switches effectively divide the chamber into four areas or quadrants. Their outputs are read independently by either a discrete logic array or suitable computer software that will differentiate between incomplete and full turns in either direction.

The assessment of full turns is undertaken by a logic network that is either hard-wired or in the form of a computer program. Since the logic is similar whichever device is employed, only the hard-wired version will be described here. The hard-wired system employs standard discrete logic elements, such as gates, flip-flops, one-shots, and shift registers, and can be designed from TTL or CMOS components. Those interested in constructing such a device should consult

Horowitz and Hill (1989). A full schematic diagram is provided in Greenstein and Glick (1975). When the subject enters a quadrant (causing a switch to be tripped), a flip-flop representing that quadrant is set and, in turn, triggers a one shot. If a flip-flop representing a quadrant to the immediate left or right was also set (i.e., the animal was just in that quadrant), a gating network registers a "quarter turn" count for the respective direction, pulses a four-count shift register for the respective direction, and resets all quadrant flip-flops except the one just entered. This operation repeats every time the subject enters a new quadrant. If the subject enters four quadrants sequentially in the same direction (i.e., moves 360°), the four-count shift register enters a "full turn" count for the respective direction and reset to "0." If, however, a subject enters a quadrant from the opposite direction of the previous move, the shift register of the previous direction is reset to "0," and the shift register of the new direction counts "1." The resulting output consists of the following parameters: "quarter turns right," "full turns left," "quarter turns left, and "full turns right."

Although the apparatus described above has been shown to be accurate and provide reproducible turning data, it is not without its limitations. The major problem with this device is that it requires that the animal be constrained by the harness. The harness limits movement, exerts pressure on the subject's abdomen, and may be viewed as a mild stressor. Various attempts to circumvent this problem have appeared in the literature. The general approach is to monitor the direction of movement without in any way impeding that movement. However, it appears as though none of the approaches to this issue has been universally adopted, since each of these systems have its own problems ranging from expense to inaccuracy.

A rotometer has been described (Pons et al., 1990) that consists of a data-acquisition system made of several reed relays that are activated by a small moving magnet when the animal performs rotations. Angular discriminations are created from pulses that are produced when the reed relay contacts are closed. This output closes key contacts in the keyboard of a microcomputer that computes various turn fractions. Another apparatus (Etemadzadeh et al., 1989) uses digital pulses derived from an infrared photocell detector induced by animal rotations that are input directly to a microcomputer for on-line recording. A device described by Huston and his colleagues (Bonatz et al., 1987) employs video-image analysis by microcomputer. The system measures turning behavior of rats by microcomputer evaluation of a digitized video image. Its software uses algorithms that assess changes in angular movement by markings that are made on the animal's body. The system is able to discriminate 1/4, 1/2, 3/4, and complete turns that are classified in different diameter classes. Validity tests of the system indicate a correlation coefficient of $r = 0.93$ between its determination and those of a human rater. The major limitation of this approach would appear to be one of expense, since it requires the use of a video-acquisition system, video capture boards, and a computer capable of running fairly complex image analysis software.

In this laboratory, full turn data are summarized to describe two parameters—net rotations and percent preference. Each of these is a measure of lateral preference, but describes a different aspect of the response. Rotations in the dominant direction minus rotations in the opposite direction are referred to as "net rotations." They represent the magnitude of a rotational bias in absolute terms (i.e., the strength in a given direction). "Percent preference" is independent of the total number of rotations, and is calculated by dividing full rotations in the dominant direction by total full rotations and multiplying the quotient by 100. Thus, percent preference expreses the strength of a rotational bias as a proportion of total rotational movement. Both net rotations and percent preference together need to be derived in order to describe adequately a turning bias for a given subject.

As described earlier, normal intact rats circle both in response to various drugs and during the active phase of their diurnal cycle. This "nocturnal" circling behavior may be seen as providing a "truer" assessment of the animal's endogenous asymmetry than that obtained with drugs, since the former relies on undrugged brain activation. The "nocturnal" test session usually includes the 3 h before and the 3 h after the 12 h dark period (i.e., from 4:00 PM to 10:00 AM). In this laboratory, we have used a convention to identify the strength of lateralization on the basis of a nocturnal session. Rats referred to as "strong rotators" ("strongly lateralized") and "nonrotators" ("weakly rotating" or "nonlateralized") in nocturnal tests are classified according to the following criteria: > 60 net rotations and 85% preference vs < 5 net rotations and < 60% preference, respectively. Animals falling between these criteria are classified as moderately lateralized.

Individual Differences in Turning Behavior and in Cerebral Laterality

Through the use of the foregoing methods, it is possible to obtain a set of parameters that describe behavioral laterality for individual subjects. Turning differences may occur in populations or in individuals. Individual differences are not the same for all members of a population, but are consistently left for some individuals consistently right for others. A variety of studies (cf. Carlson and Glick, 1989, 1992) have indicated that differences in the direction and degree of laterality in animals are determined, in a complex way, by an interaction among genetic, hormonal, developmental, and experiential factors. One striking difference is between male and female rats of the same stock and strain. Early on, it was found that the female Sprague-Dawley rat exhibited a DA asymmetry in caudate-putamen (Zimmerberg et al., 1974). These findings have been extended (Robinson et al., 1980) to indicate that male rats do not show the same pattern of DA lateralization as do females. Other findings (Glick et al., 1977; Brass and Glick, 1981) have also reported evidence of sex differences in behavioral and neurochemical asymmetries. Morphological findings (Diamond et al., 1981, 1982; Diamond, 1985) have shown that the right cortex and right hippocampus are thicker than the corresponding left structures in male rats, whereas the left sides of these structures tend to be thicker than the right sides in female rats.

Genetic determinants have been shown to play a role in turning behavior and to be modulated by hormonal factors. The offspring of rats having opposite or same-sided turning biases were tested for turning biases as adults. Male off spring usually had the same turning bias as the male parent and the opposite bias as the female parent. There was also a significant tendency for female offspring to have biases that were different than those of the female parent in litters having more males than females (Glick, 1983). Testosterone has been hypothesized to delay preferentially the development of the left side of the brain (Geschwind and Galaburda, 1985) and to be responsible for the greater incidence of left-handedness in human males than females. Various reports have indicated that there is a relationship between the sex ratio of a litter and female levels of testosterone (e.g., vom Saal and Bronson, 1980; Meisel and Ward, 1981). It was suggested in the present study that exposure to testosterone reversed the coding of a heritable female influence and induced a tendency for offspring to have biases opposite to those of the female parent.

An attempt has also been made to breed strains of strongly (SL) and weakly rotating (WL) rats (Glick, 1985). A difference developed gradually over eight generations before reaching asymptote and was limited to females. Male and female rats bred for weak rotation developed a left-sided rotational bias. These findings were again consistent with the suggestion that testosterone modulates the expression of heritable influences by slowing the growth of the left hemisphere. In animals bred for weak rotation, this effect of testosterone on brain maturation may have been exerted independently of heritable influences, thus inducing a leftward rotational bias.

Further suggestion of the heritability of turning behavior has come from the development of the transgenic chakragati (**ckr**) mouse (Ratty et al., 1990; Fitzgerald et al., 1991, 1992, 1993). These mice display intense circling behavior as a result of insertional mutagenesis. The mutation is autosomal and recessive. Mice that are homozygous for the mutation express the phenotype, whereas heterozygous transgenic mice are normal. It has also been reported that DA D_2 receptor binding sites in the striata of the **ckr** mice are significantly elevated by about 31% compared to normal heterozygous transgenic mice. As a group, **ckr** mice had higher D_2 receptor levels on the side that was contralateral to the preferred direction of spontaneous nocturnal rotation. Striatal D_1 receptors and mesolimbic D_2 and D_1 receptors of **ckr** mice were neither elevated nor differentially asymmetric. Predictably, these mice increased their turning in response to the D_2–like agonist quinpirole. $GABA_A$ sites in the mediodorsal thalamus and superior colliculus were bilaterally and asymmetrically elevated in **ckr** mice. Further work with these mice offers the possibility of elucidating the molecular bases of the genetic control of turning behavior and brain DA receptor asymmetry.

Given the apparently complex determination of the magnitude of circling behavior, it might be concluded that, at the population level, there should be sizable variation among individuals. However, when genetic and environmental influences are more homogeneous, it might be expected that the magnitude of this variation should be diminished. A comprehensive study of rats from different strains and stocks has been undertaken (Glick et al., 1986). These animals were compared to the selectively bred SL and WL lines *(see above)*. The findings showed that among stocks obtained from various suppliers, there is a good deal of variation in the degree and direction of turning behavior. Not all commercially bred rats exhibit a sex difference in turning behavior. Although there was no sex difference in a WL strain of rats, an exaggerated sex difference occurred in SL rats. Two stocks of Sprague-Dawley rats—Taconic and Blue Spruce rats—clearly rotated less than the selectively bred WL rats.

Clearly, there is a genetic basis for turning behavior. The literature provides strong suggestions that the degree of cerebral late realization and thus the magnitude of turning are under strong genetic control. There also appears to be a strong influence from such variables as the endocrine environment during early development. It thus seems reasonable to conclude that strain and stock differences in turning behavior result, in part, from genetic factors, since they are influenced by endocrine factors. However, differences in turning behavior can also be affected more transiently. An important way that these changes can be brought about experimentally is through the manipulation of the organism's experience.

The Plasticity of Circling Behavior

Learned Rotational Behavior

A vast literature shows that most rats can learn to alter their spatial behavior in response to an appetitive or aversive stimulus. The demonstration of endogenous side preferences led to the questions of how endogenous and learned preferences might interact. Early studies using a T-maze showed that rats trained to reverse their initial intrinsic preferences gradually reverted back to their initial preferences when subsequently allowed to make spontaneous choices (Glick et al., 1977b; Zimmerberg et al., 1978). More recent work has employed an operant conditioning technique. Using these operant procedures, it has been possible to study conditioned changes in cerebral laterality. In conjunction with neurochemical techniques, the operant conditioning of turning can be used as a model to explore the plasticity of a well-defined neuronal system.

The Operant Control of Turning

Rats have been trained to turn in circles for water reinforcement using a continuous reinforcement schedule in which each 360° turn results in reward. Under such a contingency, rats typically make between 200 and 300 net rotations/1-h daily test session. In one experiment, different groups of rats were reinforced for turning either in the same or opposite direction as that elicited in a previous test of preferred direction with D-A. All 14 rats trained in the "same" direction readily acquired the task, whereas only 13 of 33 rats trained in the "opposite" direction showed evidence of learning. Rats were retested with D-A at 2 and 9 d after cessation of training. Initially, at 2 d, the effect of D-A was greater in rats trained in the "same" direction and decreased or reversed in rats successfully trained in the "opposite" direction (i.e., for the latter rats, D-A now elicited more turning in the trained than in the endogenous direction). A week later, however, these changes mostly disappeared—the reversed or "oppositely" trained rats inverted to their endogenous directions. This would suggest that a mechanism to reset the asymmetry is intrinsic to the organization of the nigrostriatal system. This might be predicated on an anatomical asymmetry, that is, for example, if the normally more active side of the system contained neurons than the other side. Training might reverse the functional asymmetry by selectively activating the side with fewer neurons; afterward, when normal inputs were restored, the system would gradually return to normal.

Since circling behavior has been associated with endogenous nigrostriatal DA asymmetry, a number of studies evaluated the hypothesis that the training-induced changes in rotational behavior are attributable to asymmetrical changes in the same system. The literature is somewhat conflicting on the tenability of this hypothesis. Studies have measured DA levels and metabolites in striatal homogenates of rats killed during the middle of an operant rotational test session. Yamamoto and Freed (Yamamoto et al., 1982; Yamamoto and Freed, 1982, 1984) reported that unidirectional training selectively increased DA and its metabolite 3,4-dihydroxyphenylacetic acid (DOPAC) in the striatum contralateral to the direction of circling. Szostak et al. (1986) found no changes in DA, DOPAC, or the metabolite homovanillic acid (HVA) in either striatum, although the ratio of HVA to DA and the ratio of serotonin to its metabolite 5-hydroxyindoleacetic acid (5-HIAA) were increased bilaterally in the striata as well as in the nuclei accumbens. Subsequent studies have found, however, that although conditioned circling was established and maintained by reinforcing the response with food, the food itself influenced DA metabolism and therefore precluded the detection of changes in DA metabolism specific to the circling response (Szostak et al., 1989). Thus, subtle asymmetric DA changes might have been overshadowed by those induced by the reinforcer. Schwarting and Huston (1987) also found no changes in DA, DOPAC, or HVA in either striatum; DOPAC/DA and HVA/DA ratios were higher bilaterally in the striata as well as in the amygdalae, but these effects also occurred, although to a lesser extent, in control rats that were not trained, but were simply water deprived. Nonetheless, it has been shown (Schwarting et al., 1987) that asymmetric changes in DA metabolism occur during conditioned lever pressing. Changes are asymmetrical with respect to the side of paw usage. The findings indicate that DA neurons in the two brain hemispheres are asymmetrically involved in such behavioral tasks.

Findings from this laboratory have shown that changes involving DA occur not in nigrostriatal areas, but in DA projection areas of the medial prefrontal cortex (PFC). In those animals where a conditioned turning bias was created, there was a significant increase in PFC DA concentration on the side ipsilateral to the direction of rotation. Thus, part of the chemical basis for the operant control of circling appears to reside in the DA projection areas of the medial prefrontal cortex (Glick and Carlson, 1989).

Pavlovian Conditioning
of D-A-Induced Circling

Amphetamine-induced rotation has also been classically conditioned in female rats using the test environment as the conditioned stimulus and 1.25 mg/kg D-A as the unconditioned stimulus (Drew and Glick, 1987). The goal of these studies has been to elucidate which environmental cues will, as a consequence of being associated with drug-induced rotation, subsequently elicit rotation in the absence of the drug. As a control measure, nonlateralized activity was also conditioned as extra quarter turns. Conditioned turning was found to extinguish more rapidly than nonlateralized activity, whereas a schedule of 50% partial reinforcement selectively decreased nonlateralized activity. Low doses of haloperidol blocked the conditioned turning response, suggesting that this response was mediated by DA. Thus, conditioned turning can be distinguished from conditioned changes in activity and, like the unconditioned response, appears to be mediated by DA. The turning response to D-A has also been found to sensitize or increase in magnitude with continued exposure to the drug. Conditioned turning behavior seems to play a role in this phenomenon. It was found that sensitization to D-A-induced circling behavior occurred only in rats that had received the drug previously in the test environment. This result occurred regardless of whether the interval between drug injections was 1 d or 1 wk (Drew and Glick, 1988a,c). Further studies (Drew and Glick, 1988b) showed that D-A-induced circling behavior could not totally account for the development of sensitization. Although both the conditioned response and the sensitized D-A response were environment-dependent and were both antagonized by DA antagonists (haloperidol, metoclopramide, and SCH 23390), only the conditioned response was found to be reduced by a schedule of partial reinforcement, by prior nonreinforced exposure to the test environment, and to be subject to extinction. Thus, conditioned turning behavior seems to be a unique form of a classically conditioned drug response. Conditioned DAergic activity (perhaps occurring in PFC) may underlie conditioned turning responses. The paradigm's future utility as a technique for understanding the dynamics of conditioned drug effects remains to be established.

The Effects of Stressors on Rotational Behavior

Perhaps related to learned alterations in cerebral lateralization and to turning behavior are changes brought about by a variety of stressors. Learning may be viewed as a brain process whereby the organism adapts to its environment. Similarly on exposure to most stressors, organisms make adaptive responses in order to maintain homeostasis. Generally, stressor effects on turning have been demonstrated for stressor administration either during the prenatal and neonatal period, or during adulthood. Changes brought about by stressors during the former tend to be relatively enduring even to the extent of persisting into adulthood, whereas stressor effects in adulthood may be more transitory.

Pre- and Postnatal Stressors

Asymmetrical brain function is strongly dependent on events that take place while the brain is developing. As noted above, testosterone has been suggested to play a major role in asymmetric brain development. There are also findings to suggest that stimuli during early development, such as handling during infancy (Zimmerberg et al., 1978), exposure to enriched environments (Glick and Cox, 1976) and exposure to prenatal stressors (Fleming et al., 1986), influence asymmetrical brain development. Sexually dimorphic cortical asymmetries are apparent in the neonatal rat (Van Eden et al., 1984). Random noise stress throughout gestation results in a shift of normal asymmetric differences in DA activity in the prefrontal cortex, nucleus accumbens, and striatum (Fride and Weinstock, 1988). It has been shown that the effects of early experience are asymmetrically distributed between the two hemispheres and that handling during infancy causes asymmetric changes in spatial behavior (Denenberg et al., 1978). Systematic studies using unilateral cortical at ablations (Sherman et al., 1980) established that handling eliminates a leftward turning bias normally seen in nonhandled rats. Intact handled rats were found to have a significant left bias, suggesting that, in nonhandled animals, behavioral symmetry in making spatial choices is owing to balanced brain asymmetry in which the right hemisphere biases the animal to move to the left, while the left hemisphere acts to inhibit this response.

Effects of Stressors in Adulthood

Various stressors experienced during early development alter cerebral laterality. This influence appears to interact with genetic determinants to determine the direction and magnitude of circling behavior. It is reasonable to assume that at least part of the sizable variation among rats in turning behavior might be the result of differences in early stressor exposure. In this sense, the direction and magnitude of turning behavior might be in part reflective of its past history of exposure to stressors. If the above assumption is correct, one might expect that variation in turning behavior might provide information about the animal's general responsiveness to stressors. One preliminary approach to testing this hypothesis was through the assessment of behavioral differences in response to stressors between left and right turning rats.

Cortical Asymmetry, Turning Biases, and Stress

Cortical Modulation of Turning

Several findings have suggested a left–right asymmetry in rat cortex. Differences in cortical thickness (e.g., Diamond et al., 1975), differences in behavioral effects of left and right cortical lesions (e.g., Denenberg et al., 1978; Robinson, 1979) and, based on measurements of labeled deoxyglucose uptake, a difference in frontal cortical energy metabolism (Glick et al., 1979) have been reported. Rats' turning preferences are stronger if deoxyglucose uptake is greater on the side contralateral to turning than on the ipsilateral side. These findings suggest that the left–right asymmetry in the frontal cortex modulates turning behavior. Since the left frontal cortex usually exhibits greater deoxyglucose uptake, it was postulated that, in a large population of rats, more rats would have right-turning preferences than left. Similarly, right preferences should be stronger than left preferences. Data of 602 rats tested for nocturnal or amphetamine-induced rotation showed a small (54.8%), but significant ($p < .025$) right population bias. The modulatory influence of an asymmetric frontal cortex on turning behavior was further shown by studies using bilateral cortical lesions (Ross and Glick, 1981). Such manipulations decreased turning preferences and activity in right-turning rats, and had opposite effects in left-turning rats.

Behavioral Responses in Left- and Right-Turning Rats

Behavioral differences have been observed between rats of differing rotational asymmetry in response to D-A. Right-turning rats were more active and had, as predicted, stronger side preferences than left-turning rats (Glick and Ross, 1981). It was also determined that right-sided rats, although displaying initially less activity to foot-shock stressors, learned to escape foot shock better than the left-turning rats. Furthermore, when Sprague-Dawley-derived rats of differing rotational direction were evaluated for stressor-induced behavioral disruptions using the learned helplessness model (Maier and Seligman, 1976), it was shown that nonbiased and right-biased subjects displayed a deficit, but left-biased rats did not (Carlson and Glick, 1989). These differences in stressor-related activity suggested that there might be differences in the way stressors affected asymmetric brain function. This hypothesis was supported when it was observed that stressors, such as foot shock (Carlson et al., 1987) and food deprivation (Carlson et al., 1988), systematically altered D-A-induced as well as nocturnal rotational behavior. It was suggested that stressors evoked these changes through their actions on asymmetric brain cortical systems.

The frontal cortex, and especially the mesocortical DA neurons, are known to be selectively activated by a variety of stressors (Thierry et al., 1976; Herman et al., 1982; Carlson et al., 1987b; Abercrombie et al., 1989). These cortical regions have been held to be important in governing the execution of anticipatory behaviors that are necessary for coping with stressors (Claustre et al., 1986; Le Moal and Simon, 1991). A large body of literature has also indicated that the activity of the frontal cortex modulates mesolimbic and nigrostriatal DA function (Glick and Greenstein, 1973; Carter and Pycock, 1980; Ross and Glick, 1981) through projections that use glutamate or aspartate as excitatory neurotransmitters (Fonnum, 1984; Christie et al., 1985). Thus, it was predicted that stressful stimuli that activate the cortex would also modulate the magnitude and direction of rotational behavior. Thus, for a given animal, aspects of the response to stressors might be predicted on the basis of turning behavior. Sprague-Dawley-derived rats were exposed to foot shock stress between two tests of D-A-induced rotational behavior. Foot shock induced a change in the direction and intensity of rotational behavior that was dependent on sex and pre-existing rotational bias. Right-rotating males and left-rotating females shifted their directional bias toward the opposite side, whereas left-rotating males and right-rotating females displayed increased rotation in their prestress direction (Carlson et al., 1987a). The data indicated that activation of the mesocortical DA system by stressors could induce a change in the direction of rotational behavior. The directional difference between sexes was in agreement with the reported sexually dimorphic nature of cortical asymmetry in Sprague-Dawley rats (Diamond et al., 1981).

Neurochemical Differences, Left and Right Turning, and the Response to Footshock

Studies have explored the underlying brain chemistry that is associated with the stressor-evoked behavioral differences described above. Evidence has indicated that activation of DA containing neuronal systems projecting to the medial prefrontal cortex (PFC) is asymmetrical. Food deprivation (Carlson et al., 1988), physical restraint (Carlson et al., 1991), and foot shock (Carlson et al., 1990) altered PFC DA utilization asymmetrically across the two hemispheres. These effects changed in magnitude and direction with increasing durations of stressor exposure, and were dependent on whether the stressor was controlled or not controlled by the animal. Experiments have been designed to assess specifically the effects of controlled and uncontrolled stressors on these DA systems (Carlson et al., 1993), in order to make comparisons with previously observed behavioral effects *(see above)*, rats of differing rotational direction were again evaluated using the procedures of the learned helplessness model (LH) (Maier and Seligman, 1976). Eighty trials of either a controllable (ESC) or identical uncontrollable foot-shock stressor (YOK) caused an activation, as indicated by increased DA metabolite DOPAC concentrations in the PFC, NAC, and striatum. In the PFC, YOK caused a bilateral DA depletion, relative to ESC and control animals, and a right > left increase in DOPAC/DA that was not seen in ESC animals. These findings suggested a preferential effect of YOK in the right PFC. A second experiment used rats that had been grouped according to their turning behavior. YOK right-biased rats showed an increase in DOPAC on the right side of the PFC, and YOK left-biased displayed a similar increase on the left side in response to a brief (5 min) controllable foot-shock stressor. Since right-turning rats had been shown to be more sensitive to the LH behavioral phenomenon, the data suggested that right PFC activation is responsible for the greater LH sensitivity. A final experiment evaluated the neurochemical and behavioral responses to a prolonged foot-shock stressor 24 h after uncontrolled foot shock. Right-turning YOK animals displayed depressed foot-shock escape behavior, and a right > left depletion in PFC DA and the metabolite HVA. Across groups, foot-shock escape performance was correlated with DA and HVA concentrations on the right, but not on the left side of the PFC. Thus, a disturbance of right PFC DA utilization was again associated with compromised coping behavior. The data suggest that the inability to control a stressor causes a lateralized alteration of PFC DA, which results in a disruption of the ability to respond to a new stressor. These findings indicate that the two sides of the PFC are differentially specialized for responding to a stressor. They also indicate that the increased tendency of rightward turning rats to show an LH effect may be dependent on differences in the way in which the right PFC is activated.

Stressors appear to exert a prominent role in the determination of turning behavior. The direction and intensity of circling behavior appear to reflect the organism's past history of stressor exposure as well as its susceptibility to future stressor-induced deficits. Turning behavior is in part reflective of lateralized functioning of the PPC and may provide important clues as to how this region functions in governing the behavioral response to stressors.

Circling as an Index of the Response to Drugs

Circling behavior is a useful response with which to assess the actions of various drugs. As is the case with the circling response and stressors, the drug not only plays a role in determining the parameters of circling, but circling behavior also provides information concerning the individual animal's response to the drug. For example, findings suggest that the magnitudes of several behavioral responses to stimulant drugs are dependent on the asymmetric nature of the neuronal systems on which they act. Lateralized rats have been shown to be more responsive to the locomotor stimulant effect of both D-A and low doses of morphine (Glick and Hinds, 1985). Conversely, nonlateralized rats have been shown to be more sensitive to the depressant effects of high doses of morphine. Various classes of drugs have been shown to induce turning behavior in rats. These include morphine (Morihisa and Glick, 1977), LSD (Fleisher and Glick, 1979), phencyclidine (Glick et al., 1980a), and cocaine (Glick et al., 1983). Whereas the effects of most drugs appear to be directly (D-A, morphine, cocaine) or indirectly (LSD) attributable to modulation of the striatal DA asymmetry, another mechanism involving the hippocampus appears to be largely responsible for the effect of phencyclidine (Glick et al., 1980a).

Circling and Drug Sensitization

An important aspect of the response to stimulant drugs is that of sensitization where, with repeated administration of stimulants, such as D-A or cocaine, rats display an enhanced behavioral response to the stimulant properties of the drug (Robinson and Becker, 1986). As was noted earlier, rats from different sources vary substantially in rotational behavior. Sprague-Dawley-derived rats from three sources were tested twice (a week between tests) for notation induced by D-A. Rats from two of these sources showed evidence of sensitization, there being significantly greater rotation in response to the second dose than in response to the first dose; the D-A-induced rotational behavior of rats from the third source did not change significantly from one week to the next. However, the latter rats had a greater initial response to the first dose of D-A than did rats from the other two breeders. Among rats from all three breeders, rats rotating weakly in response to D-A on the first test tended to rotate more on the second test. Rats rotating strongly in response to D-A on the first test tended to rotate less on the second test.

Long-Evans-derived rats exhibited no evidence of sensitization to D-A on repeated testing, and these rats differed in another important respect from the Sprague-Dawley-derived rats. Although population biases were sex-dependent for the latter, this was not the case for Long-Evans-derived rats. The majority (55%) of female Sprague-Dawley-derived rats had right-sided rotational (nocturnal and D-A-induced) biases (Ross and Glick, 1981), and males as a group tended (52%) to have left-sided biases (Robinson et al., 1985). In contrast, Long-Evans-derived rats of both sexes exhibited left-sided biases (62 and 56%, respectively for females and males). Long-Evans-derived rats also differed in cocaine-elicited rotation. Sprague-Dawley-derived rats had displayed sex-dependent differences in rotation between left- and right-biased rats (Glick et al., 1983). Right-sided females rotated more than left-sided females, whereas left-sided males rotated more than right-sided males. In Long-Evans-derived rats, cocaine (20 mg/kg, ip) elicited more intense rotation in left- than in right-sided rats of both sexes (Carlson and Glick, 1987). In both strains, females rotated more than males in response to D-A and cocaine as well as nocturnally. These differential interactions between sex and side variables in Long-Evans and Sprague-Dawley-derived rats illustrate how genetic variables can play an important role in modifying hormonal factors that determine turning behavior.

Circling and Drugs of Abuse

Circling behavior has provided some important insights into the relationship between brain asymmetry and the "euphoric" effects of various drugs. Studies of intracranial self-stimulation (ICS) have shown that the two sides of the rat brain are differentially sensitive to reinforcing brain stimulation (Glick et al., 1980b). Similarly, the effects of different classes of abused drugs have been distinguished on the basis of their predominant side of action (Glick et al., 1981; Fromm and Schopflocher, 1984; Glick and Badalamenti, 1986). Morphine and D-A were tested on self-stimulation rates in the two sides of the brain. Although both drugs caused an increase in self-stimulation responding, D-A preferentially affected the low-threshold (more rewarding) side of the brain, whereas morphine preferentially affected the high-threshold side. These deferential drug effects were also dependent on the individual animal's endogenous rotational asymmetry. For D-A, the shift in reward threshold occurred more for the side contralateral to the direction of rotation, whereas for morphine, the shift was more on the ipsilateral side. In addition, morphine had effects that depended on dose; low doses preferentially increased responding in the low-threshold side, and high doses preferentially decreased responding on the high-threshold side.

Differences in turning behavior are also predictive of differences in drug-seeking behavior among rats. In tests of drug self-administration, behaviorally lateralized and nonlateralized rats were found to be differentially sensitive to D-A and morphine. Lateralized rats self-administered more D-A than did non-lateralized rats, whereas nonlateralized rats self-administered more morphine than did lateralized rats (Glick and Hinds, 1985). Recent studies (Glick et al., 1992) using in vivo microdialysis have indicated possible lateralized neurochemical substrates for these differences. Extracellular levels of DA and its metabolites were measured bilaterally in the mesocorticolimbic and nigros-triatal systems of naive rats that were subsequently trained to self-administer morphine intravenously. There were several significant relationships between DA metabolite (DOPAC and HVA) levels and rates of morphine self-administration during both acquisition and asymptotic phases of testing. DA release in the nucleus accumbens appeared to be critically involved in supporting both the acquisition and maintenance of self-administration behavior, although no lateralization of release was seen. Hemispheric asymmetries in striatal metabolite levels were, however, inversely correlated with self-administration during the acquisition phase. DOPAC and HVA levels in the right, but not in the left side of the medial prefrontal cortex were positively correlated with self-administration rates during the acquisition phase. Similarly, right/left asymmetries in cortical metabolite levels were correlated with acquisition rates. There were no significant relationships between neurochemical indices and rates of bar pressing for water, suggesting that these effects were unique to morphine and not attributable to reinforcement *per se*. Lateralized release of striatal DA seems to be important in the initiation of bar pressing for morphine. DA release in the right, but not in the left medial prefrontal cortex also appears to be an important predictor of the initiation of morphine self-administration. More recent findings (Glick et al., 1994) have shown that DA metabolites in the left medial prefrontal cortex are positively correlated with asymptotic responding for cocaine. Thus, different stages of drug self-administration of morphine and cocaine appear to be related to activation of DA systems on opposite sides of the PFC. These results suggest that the normal variability in drug-seeking behavior is at least in part attributable to individual differences in the lateral organization and activity of brain DA systems. Furthermore, these findings are consistent with the differential effects of some of these drugs on turning behavior, especially in light of the specialized role that the cortex plays in modulating rotation.

Summary

Circling or turning may reflect hemispheric imbalances owing to unilateral lesions in one of several structures (e.g., substantia nigra, striatum, and so on), whereas circling in unlesioned animals indicates the presence of an endogenous functional lateralization of the brain. Normal circling behavior has been shown to covary with or be predictive of a number of biochemical and morphological measures of brain asymmetry. Circling may be measured by a number of different methods, most of which involve attaching the experimental subject to a device that discriminates angular from nonspecific movement. A number of electronic and computer-based instruments have been described that allow for the automated discrimination and summary of full and partial turns in either direction. Variation in circling behavior among subjects has been shown to be determined by an interaction among genetic, hormonal, and environmental factors, such as learning and stress. Turning behavior may be altered transiently by using either operant or classical conditioning procedures. Exposure to various stressors that occurs during early development or in adulthood can also alter turning behavior, and differences in turning behavior may be indicative of variation in the response to stressors and to various drugs. Variation among rats in the responses to foot-shock stress, drug sensitization, and drug reinforcement has been shown to be associated with differences in the magnitude and direction of measures of turning behavior. Circling behavior may thus be indicative of the differential influence of the two hemispheres in controlling a variety of behavioral functions.

Acknowledgments

This work was supported n part by NIMH Grant MH45539 (JNC) and NIDA Grant DA03817 (SDG).

References

Abercrombie, E. D., Keefe, K. A., DiFrischia, D. S., and Zigmond, M. J. (1989) Differential effect of stress on in vivo dopamine release in striatum, nucleus accumbens, and medial frontal cortex. *J Neurochem.* **52,** 1655–1658.

Annett, M. (1985) *Left, Right, Hand and Brain: The Right Shift Theory.* Erlbaum, Hillsdale, NJ.

Arnold, A. P. and Bottjer, S. W. (1985) Cerebral lateralization in birds, in *Cerebral Lateralization in Nonhuman Species* (Glick, S. D., ed.), Academic, Orlando, FL, pp. 11–39.

Bonatz, A E., Steiner, H., and Huston, J. P. (1987) Video image analysis of behavior by microcomputer: categorization of turning and locomotion after 6-ODA injection into the substantia nigra. *J. Neurosci. Methods* **22,** 13–26.

Brass, C. A. and Glick, S. D. (1981) Sex differences in drug-induced rotation in two strains of rats. *Brain Res.* **223,** 229–234.

Bryden, M. H. (1982) *Laterality, Functional Asymmetry in the Intact Brain.* Academic, New York.

Carlson, J. N. and Glick, S. D. (1987) Stress- and cocaine-like effects on amphetamine-induced rotational behavior caused by the anxiogenic ben-zodiazepine inverse agonist methyl B-carboline-3-carboxylate. *Soc. Neurosci. Abst.* **13,** 218 (abstract).

Carlson, J. N. and Glick, S. D. (1989) Cerebral lateralization as a source of interindividual differences in behavior. *Experientia* **45,** 788–798.

Carlson, J. N. and Glick, S. D. (1992) Behavioral laterality as a determinant of individual differences in behavioral function and dysfunction, in *Genetically Defined Animal Models of Neurobehavioral Dysfunctions* (Driscoll, P., ed.), Birkhauser, Boston, pp. 189–216.

Carlson, J. N., Glick, S. D., and Hinds, P. A. (1987a) Changes in d-amphetamine elicited rotational behavior in rats exposed to uncontrollable footshock stress. *Pharmacol. Biochem. Behav.* **26,** 17–21.

Carlson, J. N., Herrick, K. F., Baird, J. L., and Glick, S. D. (1987b) Selective enhancement of dopamine utilization in the rat prefrontal cortex by food deprivation. *Brain Res.* **400,** 200–203.

Carlson, J. N., Glick, S. D. Hinds, P. A., and Baird, J. L. (1988) Food deprivation alters dopamine utilization in the rat prefrontal cortex and asymmetrically alters amphetamine-induced rotational behavior. *Brain Res.* **454,** 373–377.

Carlson, J. N., Keller, R. M., and Glick, S. D. (1990) Individual differences in the behavioral effects of stressors attributable to lateralized differences in mesocortical dopamine systems. *Soc. Neurosci. Abstract* **16,** 438.

Carlson, J. N., Fitzgerald, L. W., Keller, R. W., and Glick, S. D. (1991) Side and region dependent changes in dopamine activation with various durations of restraint stress. *Brain Res.* **550,** 313–318.

Carlson, J. N., Fitzgerald, L. W., Keller, R. W., and Glick, S. D. (1993) Lateralized changes in prefrontal cortical dopamine activity induced by controllable and uncontrollable stress in the rat. *Brain Res.* **630,** 178–187.

Carter, C. J. and Pycock, C. J. (1980) Behavioral and neurochemical effects of dopamine and noradrenaline depletion within the medial prefrontal cortex of the rat. *Brain Res.* **192,** 163–176.

Christie, M. J., James, L. B., and Beart, P. M. (1985) An excitant amino acid projection from the medial prefrontal cortex to the anterior part of nucleus accumbens in the rat. *J Neurochem.* **45,** 477–482.

Claustre, Y., Rivy, J. P., Dennis, T., and Scatton, B. (1986) Pharmacological studies on stress-induced increase in frontal cortical dopamine metabolism in the rat. *J. Pharmacol. Exp. Ther.* **238,** 693–700.

Corballis, M. C. (1991) *The Lopsided Ape.* Oxford University Press, Oxford.

Cotzias, G. C., Van Woert, M. H., and Schiffer, L. M. (1967) Aromatic amino acids and modification of parkinsonism. *New Engl. J. Med.* **276,** 374–379.

Crowne, D. P., Richardson, C. M., and Dawson, K. A. (1987) Lateralization of emotionality in right parietal cortex of the rat. *Behav. Neurosci.* **101,** 134–138.

Davidson, R. J. (1992) Anterior cerebral asymmetry and the nature of emotion. *Brain Cogn.* **20,** 125–151.

Denenberg, V. H., Garbanati, J., Sherman, D. A., Yutzey, D. A., and Kaplan, R. (1978) Infantile stimulation induces brain lateralization in rats. *Science* **201,** 1150–1152.

Diamond, M. C. (1985) Rat forebrain morphology: right-left; male-female; young-old; enricheD-impoverished, in *Cerebral Lateralization in Nonhuman Species* (Glick, S. D., ed.), Academic, Orlando, FL, pp. 73–87.

Diamond, M. C., Dowling, G. A., and Johnson, R. E. (1981) Morphological cerebral cortical asymmetry in male and female rats. *Exp. Neurol.* **71,** 261–268.

Diamond, M. C., Johnson, R. E., and Ingham, C. A. (1975) Morphological changes in the young, adult and aging rat cerebral cortex, hippocampus, and diencephalon. *Behav. Biol.* **14,** 163–174.

Diamond, M. C., Murphy, G. M., Akiyama, K., and Johnson, R. E. (1982) Morphologic hippocampal asymmetry in male and female rats. *Exp. Neurol.* **76,** 553–565.

Drew, K. L and Glick, S. D. (1987) Classical conditioning of amphetamine-induced lateralized and nonlateralized activity in rats. *Psychopharmacology* **92,** 52–57.

Drew, K. L. and Glick, S. D. (1988a) Environment-dependent sensitization to amphetamine-induced circling behavior. *Pharmacol. Biochem. Behav.* **31,** 705–708.

Drew, K. L. and Glick, S. D. (1988b) Characterization of the associative nature of sensitization to amphetamine-induced circling behavior and of the environment dependent placebo-like response. *Psychopharmacology* **95,** 482–487.

Drew, K. L., Lyon, R. A., Titeler, M., and Glick, S. D. (1986) Asymmetry in D-2 binding in female rat striata. *Brain Res.* **363,** 192–195.

Etemadzadeh, E., Koskinen, L., and Kaakkola, S. (1989) Computerized rotometer apparatus for recording circling behavior. *Methods Find. Exp. Clin. Pharmacol.* **11,** 399–407.

Ferrier, D. (1876) *The Functions of the Brain.* Dawson, London.

Fitzgerald, L. W., Ratty, A. K., Miller, K. J., Ellsworth, M. K., Glick, S. D., and Gross, K. W. (1991) Ontogeny of hyperactivity and circling behavior in a transgenic insertional mutant mouse. *Behav. Neurosci.* **105**, 755–763.

Fitzgerald, L. W., Miller, K. J., Ratty, A. K., Glick, S. D., Teitler, M., and Gross, K. W. (1992) Asymmetric elevation of striatal dopamine D_2 receptors in the chakragati mouse: neurobehavioral dysfunction in a transgenic insertional mutant. *Brain Res.* **580**, 18–26.

Fitzgerald, L. W., Ratty, A. K., Teitler, M., Gross, K. W., and Glick, S. D. (1993) Specificity of behavioral and neurochemical dysfunction in the chakragati mouse: a novel genetic model of a movement disorder. *Brain Res.* **608**, 247–258.

Fleisher, L. N. and Glick, S. D. (1979) Hallucinogen-induced rotational behavior in rats. *Psychopharmacology* **62**, 193–200.

Fleming, D. E., Anderson, R. H., Rhees, R. W., Kinghorn, E., and Bakaitis, J. (1986) Effects of prenatal stress on sexually dimorphic asymmetries in the cerebral cortex of the male rat. *Brain Res. Bull.* **16**, 395–398.

Flor Henry, P. (1986) Observations, reflections and speculations on the cerebral determinants of mood and on the bilaterally asymmetrical distributions of the major neurotransmitter systems. *Acta Neurol. Scand. Suppl.* **109**, 75–89.

Fonnum, F. (1984) Glutamate: a neurotransmitter in mammalian brain. *J. Neurochem.* **42**, 1–11.

Fride, E. and Weinstock, M. (1988) Prenatal stress increases anxiety related behavior and alters cerebral lateralization of dopamine activity. *Life Sci.* **42**, 1059–1065.

Fromm, D. and Schopflocher, D. (1984) Neuropsychological test performance in depressed patients before and after drug therapy. *Biol. Psychiatry* **19**, 55–72.

Galaburda, A. M., Sherman, G. F., and Geschwind, N. (1985) Cerebral lateralization: historical note on animal studies, in *Cerebral Lateralization in Nonhuman Species* (Glick, S. D., ed.), Academic, Orlando, FL, pp. 1–10.

Gerendai, I. (1984) Lateralization of neuroendocrine control, in *Cerebral Dominance: The Biological Foundations* (Geschwind, N. and Galaburda, A. M., eds.), Harvard University Press, Cambridge, MA, pp. 167–178.

Geschwind, N. and Galaburda, A. M. (1985) Cerebral lateralization. Biological mechanisms, associations, and pathology: III. A hypothesis and a program for research. *Arch. Neurol.* **42**, 634–654.

Glick, S. D. (1973) Enhancement of spatial preferences by (+) -amphetamine. *Neuropharmacology* **12**, 43–47.

Glick, S. D. (1983) Heritable determinants of left-right bias in the rat. *Life Sci.* **32**, 2215–2221.

Glick, S. D. (1985) Heritable differences in turning behavior of rats. *Life Sci.* **36**, 499–503.

Glick, S. D. and Badalamenti, J. I. (1986) Sex difference in reward asymmetry and effects of cocaine. *Neuropharmacology* **25**, 633–637.

Glick, S. D. and Carlson, J. N. (1989) Regional changes in brain dopamine and serotonin metabolism induced by conditioned circling in rats: effects of water deprivation, learning and individual differences in asymmetry. *Brain Res.* **504**, 231–237.

Glick, S. D. and Cox, R. D. (1976) Differential effects of unilateral and bilateral caudate lesions on side preferences and timing behavior in rats. *J. Comp. Physiol. Psychol.* **90,** 528–535.

Glick, S. D. and Cox, R. D. (1978) Nocturnal rotation in normal rats: correlation with amphetamine-induced rotation and effects of nigrostriatal lesions. *Brain Res.* **150,** 149–161.

Glick, S. D. and Greenstein, S. (1973) Possible modulating influence of frontal cortex on function. *Br. J. Pharmacol.* **49,** 316–321.

Glick, S. D. and Hinds, P. A. (1985) Differences in amphetamine and morphine sensitivity in lateralized and nonlateralized rats: locomotor activity and drug self-administration. *Eur. J. Pharmacol.* **118,** 239–244.

Glick, S. D. and Jerussi, T. P. (1974) Spatial and paw preferences in rats: their relationship to rate-dependent effects of d-amphetamine. *J. Pharmacol. Exp. Theraput.* **188,** 714–725.

Glick, S. D. and Ross, D. A. (1981) Right-sided population bias and lateralization of activity in normal rats. *Brain Res.* **205,** 222–225.

Glick, S. D., Jerussi, T. P., Water, D. H., and Green, J. P. (1974) Amphetamine-induced changes in striatal dopamine and acetylcholine levels and relationship to rotation (circling behavior) in rats. *Biochem. Pharmacol.* **23,** 3223–3225.

Glick, S. D., Cox, F. D., Jerussi, T. P., and Greenstein, S. (1977a) Normal and amphetamine-induced rotation of rats on a flat surface. *J. Pharm. Pharmacol.* **29,** 51,52.

Glick, S. D., Zimmerberg, B., and Jerussi, T. P. (1977b) Adaptive significance of laterality in the rodent. *Ann. NY Acad. Sci.* **299,** 180–185.

Glick, S. D., Meibach, R. C., Cox, R. D., and Maayani, S. (1979) Multiple and interrelated functional asymmetries in the rat brain. *Life Sci.* **25,** 395–400.

Glick, S. D., Meibach, R. C., Cox, R. D., and Maayani, S. (1980a) Phencyclidine-induced rotation and hippocampal modulation of nigrostriatal asymmetry. *Brain Res.* **196,** 99–107.

Glick, S. D., Weaver, L. M., and Meibach, R. C. (1980b) Lateralization of reward in rats: differences in reinforcing thresholds. *Science* **207,** 1093–1095.

Glick, S. D., Weaver, L. M., and Meibach, R. C. (1981) Amphetamine enhancement of reward asymmetry. *Psychopharmacology* **73,** 323–327.

Glick, S. D., Hinds, P. A., and Shapiro, R. M. (1983) Cocaine-induced rotation: sex-dependent differences between left- and right-sided rats. *Science* **221,** 775–777.

Glick, S. D., Shapiro, R. M., Drew, K. L., Hinds, P. A., and Carlson, J. (1986) Differences in spontaneous amphetamine-induced rotational behavior, and in sensitization to amphetamine, among Sprague-Dawley derived rats from different sources. *Physiol. Behav.* **38,** 67–70.

Glick, S. D., Lyon, R. A., Hinds, P. A., Sowek, C., and Titeler, M. (1988b) Correlated asymmetries in striatal D1 and D2 binding: relationship to apomorphine-induced rotation. *Brain Res.* **455,** 43–48.

Glick, S. D., Merski, C., Steindorf, S., Wang, S., Keller, R. W., and Carlson, J. N. (1992) Neurochemical predisposition to self administer morphine in rats. *Brain Res.* **578,** 215–220.

Glick, S. D., Raucci, J., Wang, S., Keller, R. W., and Carlson, J. N. (1994) Neurochemical predisposition to self-administer cocaine in rats: individual differences in dopamine and its metabolites. *Brain Res.* 653, 148–154.

Greenstein, S. and Glick, S. D. (1975) Improved automated apparatus for recording rotation (circling behavior) in rats or mice. *Pharmacol. Biochem. Behav.* 3, 507–510.

Hellige, J. B. (1993) *Hemispheric Asymmetry: What's Right and What's Left.* Harvard University Press, Cambridge, MA.

Herman, J. P., Guillonneau, D., Dantzer, R., Scatton, B., Semerdjian Rouquier, L., and Le Moal, M. (1982) Differential effects of inescapable footshocks and of stimuli previously paired with inescapable footshocks on dopamine turnover in cortical and limbic areas of the rat. *Life Sci.* 30, 2207–2214.

Herrick, C. J. (1926) *Brains in Rats and Men.* Hafner, New York.

Horowitz, P. and Hill, W. (1989) *The Art of Electronics.* 2nd ed. Cambridge University Press, Cambridge, UK.

Jerussi, T. P. and Glick, S. D (1974) Amphetamine-induced rotation in rats without lesions. *Neuropharmacology* 13, 283–286.

Jerussi, T. P. and Glick, S. D. (1975) Apomorphine-induced rotation in normal rats and interaction with unilateral caudate lesions. *Psychopharmacologia* 40, 329–334.

Jerussi, T. P. and Glick, S. D. (1976) Drug-induced rotation in rats without lesions: behavioral and neurochemical indices of normal asymmetry in nigro-striatal function. *Psychopharmacology* 47, 249–260.

Jerussi, T. P., Glick, S. D., and Johnson, C. L. (1977) Reciprocity of pre- and postsynaptic mechanisms involved in rotation as revealed by dopamine metabolism and adenylate cyclase stimulation. *Brain Res.* 129, 385–388.

Kilshaw, D. and Annett, M. (1983) Right- and left-hand skill I: effects of age, sex and hand preference showing superior skill in left-handers. *Br. J. Psychol.* 74, 253–268.

Knapp, S. and Mandell, A. J. (1980) Lithium and chlorimipramine differentially alter bilateral asymmetry in mesostriatal serotonin metabolites and kinetic confirmations of midbrain tryptophan hydroxylase with respect to tetrahydrobiopterin cofactor. *Neuropharmacology* 19, 1–7.

Lashley, K. S. (1921) Studies on cerebral function in learning. No. III. Motor areas. *Brain* 44, 255–285.

LeMay, M. (1985) Asymmetries of the brains and skulls of nonhuman primates, in *Cerebral Lateralization in Nonhuman Species* (Glick, S. D., ed.), Academic, Orlando, FL, pp. 233–245.

LeMay, M. and Geschwind N. (1975) Hemispheric differences in the brains of great apes. *Brain Behav. Evolution* 11, 48–52.

LeMoal, M. and Simon, H. (1991) Mesocorticolimbic dopaminergic network: functional and regulatory roles. *Physiol. Rev.* 71, 155–234.

Levy, J. (1977) The mammalian brain and the adaptive advantage of cerebral asymmetry. *Ann. NY Acad. Sci.* 299, 264–272.

Maier, S. F. and Seligman, M. E. P. (1976) Learned helplessness: theory and evidence. *J. Exp. Psychol. Gen.* 105, 3–46.

Meisel, R. L. and Ward, I. L. (1981) Fetal female rats are masculinized by male litter mates located caudally in the uterus. *Science* 213, 239–242.

Mettler, F. A. and Mettler, C. C. (1942) The effects of striatal injury. *Brain* **65**, 242–255.

Mittleman, G., Fray, P. J., and Valenstein, E. S. (1985) Asymmetry in the effects of unilateral 6–OHDA lesions on eating and drinking evoked by hypothalamic stimulation. *Behav. Brain Res.* **15**, 263–267.

Morihisa, J. M. and Glick, S. D. (1977) Morphine-induced rotation (circling behavior) in rats and mice: species differences, persistence of withdrawal-induced rotation and antagonism by naloxone. *Brain Res.* **123**, 180–187.

Nottebohm, F. (1977) Asymmetries of neural control of vocalization in the canary, in *Lateralization in the Nervous System* (Harnad, S., Doty, R. W., Goldstein, L., Jaynes J., and Krauthamer, J., eds.), Academic, New York, pp. 23–44.

O'Boyle, M. W. and Hellige, J. B. (1989) Cerebral hemisphere asymmetry and individual differences in cognition. *Learning and Individual Differences* **1**, 7–35.

Oke, A., Lewis, R., and Adams, R. N. (1980) Hemispheric asymmetry of norepinephrine distribution in rat thalamus. *Brain Res.* **188**, 269–272.

Pearlson, G. D. and Robinson, R. G. (1981) Suction lesions of the frontal cerebral cortex in the rat induce asymmetrical behavioral and catecholaminergic responses. *Brain Res.* **218**, 233–242.

Pons, S., Lopez, J. A., Ramis, C., Planas, B., and Rial, R. (1990) A new precise microcomputer based rotometer. *J. Neurosci. Methods* **32**, 155–158.

Ratty, A. K., Fitzgerald, L. W., Titeler, M., Glick, S. D., Mullins, J. J., and Gross, K. W. (1990) Circling exhibited by a transgenic insertional mutant. *Brain Res. Mol. Brain Res.* **8**, 355–358.

Robinson, R. G. (1979) Differential behavioral and biochemical effects of right and left hemispheric cerebral infarction in the rat. *Science* **205**, 707–710.

Robinson, T. E. and Becker, J. B. (1983) The rotational behavior model: asymmetry in the effects of unilateral 6–OHDA lesions of the substantia nigra in rats. *Brain Res.* **264**, 127–131.

Robinson, T. E. and Becker, J. B. (1986) Enduring changes in brain and behavior produced by chronic amphetamine administration: a review and evaluation of animal models of amphetamine psychosis. *Brain Res. Rev.* **11**, 157–198.

Robinson, T. E., Becker, J. B., and Ramirez, V. D. (1980) Sex differences in amphetamine-elicited rotational behavior and the lateralization of striatal dopamine in rats. *Brain Res Bull.* **5**, 539–545.

Robinson, T. E., Becker, J. B., and Presty, S. K. (1982) Long-term facilitation of amphetamine-induced rotational behavior and striatal dopamine release produced by a single exposure to amphetamine: sex differences. *Brain Res.* **29**, 231–241.

Robinson, T. E., Becker, J. B., Camp, D. M., and Mansour, A. (1985) Variation in the pattern of behavioral and brain asymmetries due to sex differences, in *Cerebral Lateralization in Nonhuman Species* (Glick, S. D., ed.), Academic, Orlando, FL, pp. 185–231.

Rosen, G. D., Finklestein, B., Stoll, A. L., Yutzey, D. A., and Denenberg, V. H. (1984) Neurochemical asymmetries in the albino rat's cortex, striatum and nucleus accumbens. *Life Sci.* **34**, 1143–1148.

Ross, D. A. and Glick, S. D. (1981) Lateralized effects of bilateral frontal cortex lesions in rats. *Brain Res.* **210**, 379–382.

Rothman, A. H. and Glick, S. D. (1976) Differential effects of unilateral and bilateral caudate lesions on side preference and passive avoidance behavior in rats. *Brain Res.* **118**, 361–369.

Schneider, L. H., Murphy, R. B., and Coons, E. E. (1982) Lateralization of striatal dopamine (D2) receptors in normal rats. *Neurosci. Lett.* **33**, 281–284.

Schwarting, R. and Huston, J. P. (1987) Dopamine and serotonin metabolism in brain sites ipsi- and contralateral to directions of conditioned turning in rats. *J. Neurochem.* **48**, 1473–1479.

Schwarting, R., Nagel, J. A., and Huston, J. P. (1987) Asymmetries of brain dopamine metabolism related to conditioned paw usage in the rat. *Brain Res.* **417**, 75–84.

Shapiro, R. M., Glick, S. D., and Hough, L. B. (1986) Striatal dopamine uptake asymmetries and rotational behavior in unlesioned rats: revising the model. *Psychopharmacology* **89**, 25–30.

Sherman, G. F., Garbanati, J. A., Rosen, G. D., Yutzey, D. A., and Denenberg, V. H. (1980) Brain and behavioral asymmetries for spatial preference in rats. *Brain Res.* **192**, 61–67.

Sherman, G. F., Galaburda, A. M., and Geschwind, N. (1982) Neuroanatomical asymmetries in nonhuman species. *Trends Neurosci.* **5**, 429–431.

Slopsema, J. S., Van der Gugten, J., and De Bruin, J. P. C. (1982) Regional concentrations of noradrenaline and dopamine in the frontal cortex of the rat: dopaminergic innervation of the prefrontal subareas and lateralization of prefrontal dopamine. *Brain Res.* **250**, 197–200.

Sokoloff, L. (1977) Relation between physiological function and energy metabolism in the central nervous system. *J. Neurochem.* **29**, 13–26.

Springer, S. and Deutsch, G. (1981) *Left Brain, Right Brain.* Freeman, San Francisco, CA.

Szostak, C., Jakubovic, A., Phillips, A. G., and Fibiger, H. C. (1986) Bilateral augmentation of dopaminergic and serotonergic activity in the striatum and nucleus accumbens induced by conditioned circling. *J. Neurosci.* **6**, 2037–2044.

Szostak, C., Jakubovic, A., Phillips, A. G., and Fibiger, H. C. (1989) Influence of inherent directional biases on neurochemical consequences of conditioned circling. *Behav. Neurosci.* **103**, 678–687.

Therrien, B. A., Camp, D. M., and Robinson, T. E. (1982) Sex differences in the effects of unilateral hippocampal lesions on spatial learning. *Soc. Neurosci. Abst.* **8**, 312.

Thierry, A. M., Tassin, J. P., Blanc, G., and Glowinski, J. (1976) Selective activation of the mesocortical DA system by stress. *Nature* **253**, 242–244.

Ungerstedt, U. (1971a) Postsynaptic supersensitivity after 6-hydroxydopamine induced degeneration of the nigro-striatal dopamine system. *Acta Physiol. Scand. Suppl.* **367**, 69–93.

Ungerstedt, U. (1971b) Striatal dopamine release after amphetamine or nerve degeneration revealed by rotational behavior. *Acta Physiol. Scand.* **367**, 49–68.

Ungerstedt, U. and Arbuthnott, G. (1970) Quantitative recording of rotational behavior in rats after 6-hydroxydopamine lesions of the nigro-striatal dopamine system. *Brain Res.* **24**, 485–493.

Van Eden, C. G., Uylings, H. B. M., and Van Pelt, J. (1984) Sex-difference and left-right asymmetries in the prefrontal cortex during postnatal development in the rat. *Dev. Brain Res.* **12,** 146–153.

vom Saal, F. S. and Bronsol, F. H. (1980) Sexual characteristics of adult female mice are correlated with their blood testosterone levels during prenatal development. *Science* **208,** 597–599.

Wilson, S. A. K. (1914) An experimental approach to the anatomy and physiology of the corpus striatum. *Brain* **36,** 425–492.

Yamamoto, B. K. and Freed, C. R. (1982) The trained circling rat: a model for inducing unilateral caudate dopamine metabolism. *Nature* **298,** 467,468.

Yamamoto, B. K. and Freed, C. R. (1984) Asymmetric dopamine and serotonin metabolism in nigtostriatal and limbic structures of the trained circling rat. *Brain Res.* **297,** 115–119.

Yamamoto, B. K., Lane, R. F., and Freed, C. R. (1982) Normal rats trained to circle show asymmetric caudate dopamine release. *Life Sci.* **30,** 2155–2162.

Zimmerberg, B., Glick, S. D., and Jerussi, T. P. (1974) Neurochemical correlate of a spatial preference in rats. *Science* **185,** 623–625.

Zimmerberg, B., Strumpf A. J., and Glick, S. D. (1978) Cerebral asymmetry and left-right discrimination. *Brain Res.* **140,** 194–196.

CHAPTER 11

Asymmetrical Motor Behavior in Animal Models of Human Diseases

The Elevated Body Swing Test

Cesario V. Borlongan and Paul R. Sanberg

The essence of science lies in the methodology applied to answer the questions raised (Wallace, 1981).

Similarly, the validity of any scientific finding depends on the validity of the procedures employed in arriving at that finding (Chaplin and Krawic, 1976).

Introduction

This chapter will address primarily the behavioral characterization of lesion effects in the central nervous system (CNS). Animal models of Parkinson's (PD) and Huntington's disease (HD) will be discussed with emphasis on the novel "elevated body swing test" (EBST) in characterizing the behavioral deficits following unilateral CNS lesions. Advantages of the drug-free EBST paradigm over the conventional drug-induced rotational test in unilaterally lesioned rats will be presented. Because basal ganglia dysfunction underlies the pathophysiology of these two clincial disorders, it is appropriate to begin with an introduction of the basal ganglia.

From: *Motor Activity and Movement Disorders*
P. R. Sanberg, K. P. Ossenkopp, M. Kavaliers, Eds. Humana Press Inc., Totowa, NJ

The Basal Ganglia

Recent studies, ranging from molecular biological to human clinical studies, have implicated the basal ganglia as being involved in the fine tuning of behavioral selection (Hikosaka, 1991). It has been suggested that the basal ganglia contribute to the initiation of movement through the mechanism of disinhibition (Chevalier, 1990). The internal globus pallidus (GPi) and the substantia nigra pars reticulata (SNr), which are mediated by γ-aminobutyric acid (GABA), are the two major outputs of the basal ganglia. In turn, the major inputs to GPi and SNr originate from the striatum, and caudate nucleus (CD) and putamen (CPU). Both these output and input structures are GABA-mediated and inhibitory in function, suggesting that an excitatory input from the cerebral cortex or thalamus may lead to a disinhibition in the target structures of the basal ganglia (Hikosaka, 1990). Synergistic movements have been attributed to the CD-SNr-thalamic nuclei connection (Deniau and Chevalier, 1985). In addition, the basal ganglia has been demonstrated to have connection with the brainstem, such as the pedunculopontine nucleus (PPN), which in turn connects with the subthalamic nucleus (STN) (Kang and Kitai, 1990). Although the specific neural networks of the basal ganglia resulting in motor initiation and control remain speculative, the findings that the basal ganglia structures are characterized by inhibitory influences on target neurons, in contrast with most other brain structures, would imply a general role in selection of movement for the basal ganglia.

There is growing interest also in the influence of dopamine (DA) receptors that have distinct CNS distributions (in particular the basal ganglia) and pharmacological properties (Graybiel, 1991). DA, together with noradrenaline and adrenaline, are the three catecholamine neurotransmitters synthesized from the essential amino acid tyrosine. DA acts by binding to G protein-coupled receptors, the D_1 and D_2 receptors (Kebabian and Calne, 1979). In the case of the D_1 receptor stimulation, DA binding activates the adenylate cyclase, whereas in D_2 stimulation, production of cyclic-AMP (cAMP) is inhibited (Vallar and Meldolesi, 1989). Striatonigral target neurons contain D_1 receptors at their intrastriatal dendrites and presynaptically on their axon terminals in the SNr, whereas striatopallidal target neurons express mainly D_2 receptors at the presynaptic membrane of both dendrites and axon terminals (Gerfen, 1992). The concept of this CNS localization of DA receptors has been used to develop D_1 and D_2 receptor-specific pharmacological agonists and antagonists to investigate the functional role of the subdivisions of the nigrostriatal, as well as the mesocortiolimbic DA systems (Robertson, 1992).

The interaction between the DA system, in concert with the basal ganglia, and the initiation and control of motor behavior has received considerable attention. A model of how information is processed by the basal ganglia is presented by Mink and Thach (1993). As stated above, neocortical or cerebellar excitatory input initiates the motor program. The basal ganglia then selects a program and simultaneously suppresses other competing programs. From the GPi and SNr, acting as relay stations, the CPU and the STN receive the information. Through the GABA-ergic inhibitory influence, the excitatory activity of a selective center in the pallidum is suppressed, but the surrounding pallidal neurons continue to increase their inhibitory control over their respective target areas in the midbrain and thalamus. The coordination of the circuitry of the different nuclei involved in this disinhibition process of eliciting a motor behavior may be the role of the DA neurons.

The availability of experimental animal models for movement disorders has significantly improved our understanding of the relationship between the basal ganglia and movement disorders. Neuropharmacological studies using direct CNS administration of D_1 or D_2 receptor agonists and antagonists have revealed that both nigral and striatal DA systems actively mediate motor behavior (Pycock and Kilpatrick, 1989).

Movement Disorders

One of the many classifications of movement disorders involves site-specific vs. nonspecific aspects of basal ganglia lesions (Fahn, 1990). Examples of human disorders entailing selective involvement of the basal ganglia under this category include PD, HD, and paroxysmal dyskinesias, whereas examples of human disorders classified under nonspecific involvement of the basal ganglia include stroke, seizures, metastasis, and head injury. The authors decided to investigate PD and HD primarily because of the site-specific basal ganglia lesions associated with these diseases. The discussion of the animal models in this chapter has been limited to locomotor behavioral tests that directly relate to the nigrostriatal DA system. Also, a brief description of the accompanying neuropathological effects seen following the CNS insult has been included. For a complete report of behavioral tests and histopathological results in these animal models, the reader is referred to our original articles.

PD

PD is a progressive neurodegenerative disorder characterized by tremor, rigidity, hypokenesia, and postural instability. The pathologic feature is a loss of DA neurons in the substantia nigra pars compacta, which leads to degeneration of the mesostriatal pathway and depletion of DA in the striatum. Animal models of PD usually involve damaging the nigrostriatal dopaminergic pathway. The neuropathological effects following the nigrostriatal lesions have been demonstrated to resemble the neuropathology of the disease (Bernheimer et al., 1973; McGeer et al., 1977). 6-Hydroxydopamine (6-OHDA) is the widely used selective neurotoxin for inducing massive DA denervation of the nigrostriatal pathway in rodents (Ungerstedt and Arbuthnott, 1970). Bilateral 6-OHDA lesions most often result in death in animals, unless intragastric tube feeding is introduced to the animals (Dunnet et al., 1983). Because of this logistical problem, bilateral 6-OHDA lesions are rarely used for experimental purposes. Unilateral lesion models of PD (hemiparkinsonism) offer an alternative approach in investigations of PD. Unilateral stereotaxic injections of 6-OHDA into either the medial forebrain bundle or the substantia nigra destroy dopaminergic neurons on one side of the brain, thus creating a unilateral lesion of the nigrostriatal pathway (Ungerstedt, 1971a,b; Creese and Snyder, 1979).

Rotational Behavior in PD Animal Model

Unilaterally lesioned animals are known to develop drug-induced stereotypical rotational behavior (Ungerstedt, 1971a,b;

Silverman and Ho, 1981; Coward, 1983). Administration of apomorphine activates the supersensitive receptors on the lesioned side of the brain, causing the animal to rotate selectively in the direction contralateral to the lesioned side. In contrast, administration of amphetamine stimulates the release of DA from neurons on the intact side of the brain, inducing the animal to turn in the direction ipsilateral to the lesioned side. A pronounced turning rate owing to amphetamine or apomorphine injection requires a at least a 90% depletion of the nigral and striatal DA (Hudson et al., 1993). Rotation can thus be used as an index of the extent of the lesion-induced DA depletion. Accordingly, rotational behavior of unilateral 6-OHDA-lesioned animals in response to DA receptor agonists is the conventional method in assessing DA-mediated responses (Norman et al., 1990; Hudson et al., 1993b). A more detailed description of the rotational behavior has been discussed in Chapter 8.

HD

Huntington's disease (HD) is a progressive neurodegenerative disorder characterized by severe degeneration of basal ganglia neurons (Sanberg and Coyle, 1984). Behavioral symptoms of HD include abnormal, uncontrollable, and constant choreiform movements (Difliglia, 1990). Excitotoxic and the 3-nitropropionic acid (3-NP) animal models have been used to reproduce neurochemical, neuroanatomical, and behavioral sequelae of HD. Similar to the unilateral 6-OHDA-lesioned animal model for PD, animals with unilateral excitotoxic lesions exhibit asymmetrical rotational behavior in response to DA receptor agonists and antagonists.

Rotational Behavior in HD Animal Model

The rotational behavior has also been widely used as a behavioral test for the unilateral excitotoxin animal model of HD. The utility of the DA-induced rotational behavior in HD is founded on the observation that DA exerts a tonic influence on the striatal neurons, and drugs that alter dopaminergic transmission also modulate locomotor activity (Emerich et al., 1992). A more exhaustive discussion of the rotational behavior of HD animal models has been provided in Chapter 11 (Norman et al., 1991).

Limitation of the Drug-Induced Rotational Test: Sensitization

The neglect phenomenon has been documented in rats following either unilateral 6-OHDA destruction of the nigrostriatal dopaminergic pathway (Norman et al., 1988, 1990) or unilateral excitotoxic lesions of the striatum (Ungerstedt, 1973; Dunnett et al., 1988; Norman et al., 1988). Lesioned rats exhibit a stereotypical turning behavior in response to DA agonists and antagonists [32]. Heilman and Watson (1979), and Healton and colleagues (1982) suggested that the asymmetrical drug-induced rotational behavior in the lesioned animals may be the counterpart of the hemineglect motor functions in humans. The problem, however, in the drug-induced behavioral response is that the observed behavior may be a pharmacological reaction (Hattori et al., 1993; Klug and Norman, 1993). Furthermore, hemineglect in humans with unilateral CNS damage is characterized by an ipsilateral bias or increased movements toward the damaged side of the brain. In contrast, the bias in the drug-induced rotation test is dependent on the DA agonist/antagonist mode of action. For example, amphetamine stimulates DA release and blocks DA reuptake by the presynaptic terminals, whereas apomorphine triggers supersensitivity on the postsynaptic D_1 and D_2 receptors (Ungerstedt, 1971a,b). Furthermore, the locomotor topography of ipsilateral rotational behavior (following amphetamine and apomorphine challenge in nigral and striatal-lesioned animals, respectively) is not parallel with that of the human hemineglect phenomenon. In lesioned rats, the ipsilateral rotational behavior involves increased activity of the limbs contralateral to the lesioned side, whereas in humans with striatal damage, arm and leg movements are almost totally limited to the side of the body ipsilateral to the damaged brain. Although the resulting behavior in lesioned rats is an ipsilateral biased response, the limb movements accompanying this behavior actually demonstrate increased contralateral activity.

Sensitization owing to repeated drug administration may confound interpretation of drug-induced rotational behavior (Bevan, 1983; Coward, 1983; Kalivas and Weber, 1988), especially when assessing the efficacy of neural transplants (Norman et al., 1990). The problem of sensitization becomes more apparent when evaluating the efficacy of experimental therapies, such as neural transplantation, for human CNS disorders. The absence of functional recovery may be underestimated as a result of drug sensitization-induced increase in rotational behavior (Norman et al., 1991; Hattori et al., 1992).

Chronic use of the majority of stimulants (e.g., amphetamines) may result in the development of greater sensitivity to the drugs' effects (Kuczenski and Leith, 1981; Masur et al., 1986; Segal and Kuczenski, 1987a,b; Stewart and Vezina, 1988; Phillips et al., 1994). Of note, the neurochemical system implicated in drug-induced stimulation is the mesolimbic-striatal DA system (Dworkin and Smith, 1987; Wise, 1988; Phillips et al., 1994). Specifically, DA release and reuptake mediate the behavioral sensitization to amphetamine (Kalivas and Weber, 1988; Robinson et al., 1988; Phillips et al., 1994).

Similar behavioral sensitization effects have been observed in chronic administration of apomorphine. Repeated apomorphine injections lead to progressively greater increases in locomotor activity (Mattingly et al., 1991; Rowlett et al., 1991). Unilateral nigral 6-OHDA-lesioned rats were found to exhibit a significant increase in rotations during the second injection of apomorphine (Norman et al., 1990), which was quite robust in one study allowing a 5–6 wk interval between injections (Klug and Norman, 1993).

These studies would preclude occurrence of sensitization to drugs that induce rotational behavior in unilaterally lesioned animals. Thus, a behavioral parameter that can evaluate asymmetrical motor behavior of lesioned animals in a drug-free state may reflect a more natural response of the animals to the lesion effects.

EBST: A Drug-Free Behavioral Test for Unilateral Lesioned Animal Models

In a series of experiments using PD and HD animal models, the feasibility of EBST as a measure of asymmetrical motor behavior was evaluated (Fig. 1) (Borlongan and Sanberg, 1995; Borlongan et al., 1995a,b). The EBST involves measurement of frequency and direction of the swing behavior of the rodent when it is held by the tail. It is hypothesized that an animal with unilat-

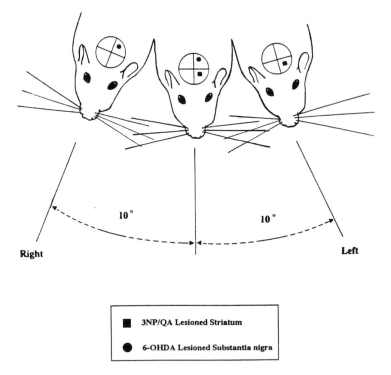

Fig. 1. Vertical axis **(middle)**: The animal is shown positioned at the vertical axis, which was defined as no more than 10° to either left or right side. The animal is held approx 1 in. from the base of its tail. The height between the animal and the surface is about 1 in. Right-biased swing **(left)**: The animal is shown at the right-biased position. A right swing is counted when the head of the animal moved more than 10° from the vertical axis to the right side. This right-biased swing is exhibited by the 6-OHDA substantia nigra unilaterally lesioned animal. Left-biased swing **(right)**: The animal is shown at the left-biased position. A left swing is counted when the head of the animal moved more than 10° from the vertical axis to the left side. This left-biased swing is exhibited by the excitotoxin striatal unilaterally lesioned animal (from Borlongan et al. (1995) *J. Neurosci. Protocols*, submitted).

eral nigral 6-OHDA lesions will exhibit biased swings contralateral to the lesioned side, whereas an animal with excitotoxin lesions will display biased swings ipsilateral to the lesioned side.

The swing test is a simple and easy behavioral test that only requires handling the animal by its tail and recording the direction of swings made by the animal for a certain period of time. Initially, the animal is placed into a Plexiglas™ box (40 × 40 × 35.5 cm), and allowed to habituate for 2 min and attain a neutral position, defined as having all four paws on the ground. The animal is held approx 1 in. from the base of its tail. It is then elevated to an inch above the surface on which it has been resting (Fig. 1). The animal is held in the vertical axis, defined as no more than 10° to either the left or the right side. A swing is recorded whenever the animal moves its head out of the vertical axis to either side. Before attempting another swing, the animal must return to the vertical position for the next swing to be counted. In cases when the animal swings and redoubles its efforts to move toward one side without returning to the vertical position, only one swing is counted. In few cases when the animal refuses to return to the vertical position for more than 5 s or when it grabs its tail, time is stopped and the animal is momentarily placed back on the ground. Once in a neutral position, the animal is then resuspended and time is restarted. Swings are usually exhibited in less than a second. Thus, frequency, and not amount of time, of swings is counted. When the animal does not commence swing behavior when it is elevated for more than 5 s, a gentle pinch to the tail induces the behavior. Swings are recorded using a hand counter. The total number of swings made to each side is divided by the overall total number of swings made to both sides to obtain percentages of left and right swings. The criterion for biased swing behavior is set at 70% or higher.

EBST for PD Animal Model

We observed that 6-OHDA-lesioned animals exhibited 70% or higher contralateral (to the lesion) biased swings (Fig. 2). This significant right-biased swing behavior persisted from 7 d postlesion, the earliest time the EBST was conducted, and each week thereafter up to 2 mo postlesion. Sham-lesioned animals did not display any biased swing behavior throughout the experiment.

Similar behavioral tests with 6-OHDA-lesioned animals in a drug-free state have been reported. Using a treadmill apparatus to measure motor function of 6-OHDA-lesioned rats, animals were induced to run uphill on the treadmill by

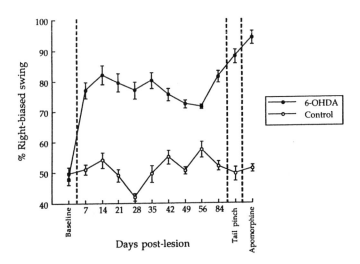

Fig. 2. EBST for animals with unilateral substantia nigra 6-OHDA lesions. Only normal, naive animals showing unbiased swings (prelesion) were used in the study. 6-OHDA-lesioned animals exhibited significant right-biased swings throughout the entire postlesion test period, but note a slight decline at 42, 49, and 56 d postlesion. A restored high level of biased activity, however, was observed after the 1-mo rest period (84 d postlesion). Similar high levels of biased swing activity were noted when animals were subjected to tail pinch or apomorphine challenge. A new set of animals was introduced to these tail pinch- and apomorphine-induced EBST at 56 d postlesion. Sham-lesioned animals did not display any biased swing on all test sessions. Data points were computed as means ± SEM. Numbers on top of each data point correspond to mean total frequency of swings recorded for striatal-lesioned animals (modified from Borlongan and Sanberg, *J. Neurosurg.,* 1995, submitted).

electrostimulation (Hattori et al., 1992). Although this method is free from confounding effects of sensitization to drugs, sensitivity to electrical stimulation becomes the apparent problem. Another suggested technique is an automated recording of open-field locomotor activities (Sanberg et al., 1987; Hudson et al., 1993a). Although this technique ensures quantification of a variety of locomotor variables, the cost of the activity monitors seems expensive. A more common drug-free test is the paw-reaching test, which involves measurement of the skilled reaching of the paw contralateral to the lesioned side (Dunnett et al., 1988; Abrous et al., 1992). The concern here is the inherent subjective scaling (0 = weak, 1 = moderate, 2 = exaggerated) and the complexity of the paw-reaching task. In addition, the paw-reaching task and apomorphine-induced rotations are negatively correlated, which might imply that different types of behaviors are mediated by the dopaminergic pathway. Another drug-free behavioral test that is also negatively correlated to the rotational behavior employs recording of asymmetrical orientation to edges of an open field (Sullivan et al., 1994). In this test, undrugged 6-OHDA-lesioned animals were found to align preferentially with the edge of a platform with the intact striatum contralateral to the edge. The negative correlation between edge and rotational behavior may be the result of the former behavior being a sensorimotor response and the latter behavior being a motor response (Sullivan et al., 1994). In contrast, the swing behavior was noted to be positively correlated with the rotational behavior and may be considered a motor response. Recently, Schallert and colleagues (1994) also described a preferential use of the ipsilateral (to the lesion) forelimb for postural and exploratory movements in 6-OHDA-lesioned animals, which supports our observation that the ipsilateral forelimb was consistently used in making the contralateral swing. Further characterization of the swing behavior may reveal interesting correlations with those behaviors reported by Schallert and colleagues (1994).

EBST for HD Animal Model

Significant biased swing rotations were observed in 3-NP and QA lesioned animals using the EBST (Fig. 3). Animals with striatal lesions exhibited a significant right-biased swing behavior on all test sessions compared with control animals. These biased swing data on unilateral excitotoxin animal models of HD and those of 6-OHDA-induced hemiparkinsonian rats, taken together, indicate that the striatal-lesioned animals displayed a biased swing ipsilateral to the lesioned side of the brain, whereas the nigral-lesioned animals exhibited a biased swing contralateral to the lesioned side. These findings may indicate that the underlying brain mechanism of the swing behavior is related to the location of lesions in the striatonigral dopaminergic pathway. Of note, the direction of biased swing

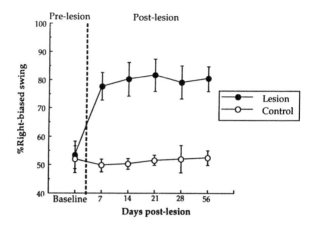

Fig. 3. EBST for animals with unilateral excitotoxin striatal-lesions.
Only normal, naive animals showing unbiased swings (prelesion)
were used in the study. Striatal excitotoxin-lesioned animals exhibited significant right-biased swings (70% or higher) throughout the
entire postlesion test period. Sham-lesioned animals did not display
any biased swing on all test sessions. Data points were computed as
means ± SEM. Numbers on top of each data point correspond to mean
total frequency of swings recorded for striatal-lesioned animals (from
Borlongan et al., *Brain Res.*, 1995, in press).

taken by either the striatal- or the nigral-lesioned animal corresponded with the direction of the asymmetrical apomorphine-
induced rotation.

The biased swing behavior in unilateral excitotoxin-
lesioned animal models of HD may be a correlate of a similar
bias or "neglect" phenomenon observed in humans with CNS
damage limited to the striatum and surrounding white matter
tracts (Hamilton and Gould, 1987; Dunnett et al., 1988; Keefe et
al., 1989). Humans with CNS damage to the right cerebral cortex, including the parietal lobe (Denny-Brown et al., 1952;
Heilman and Watson, 1979), or to the basal ganglia (Damasio et
al., 1980), especially the striatum, and the surrounding white
matter tracts (Healton et al., 1982; Mesulam, 1985; Ferro et al.,
1987) may display a unilateral neglect or hemineglect phenomenon. Failure to recognize, respond to, or explore stimuli to one-
half of extrapersonal space is manifested when hemineglect
occurs (Healton et al., 1982; Mesulam, 1985; Daffner et al., 1990).
Three mechanisms may underlie the hemineglect phenomenon

(Healton et al., 1982; Daffner et al., 1990). First, deafferentation of the right primary sensory pathway may result in a reduced flow of sensory information to the ipsilateral hemisphere. Second, although there are sufficient sensory stimuli, destruction of sensory association and limbic cortical areas may interfere with selective attention, which is needed to activate motor responses. Third, the striatum, which receives several projections from thalamic intralaminar nuclei and parietal, anterior frontal and cingulate cortex, may direct the orienting response. Indeed, hemineglect has been observed in human patients with unilateral CNS damage restricted to the striatum and deep white matter, but with intact cortex or thalamus (Damasio et al., 1980; Healton et al., 1982; Daffner et al., 1990).

EBST Correlation with Drug-Induced Rotational Test

Apomorphine-induced rotations appear to demonstrate the same side of biased swing activities observed in lesioned animals. Indeed, analyses of data from swing behavior and apomorphine-induced rotations revealed high positive correlations. Rats with unilateral 6-OHDA substantia nigra lesions exhibited a biased swing contralateral to the lesioned side, whereas rats with excitotoxin intrastriatal-lesions displayed a biased swing ipsilateral to the lesioned side. These findings may indicate that the underlying brain mechanism of the swing behavior is related to the location of lesions in the striatonigral dopaminergic pathway. Of note, the direction of biased swing taken by either the striatal- or the nigral-lesioned animal corresponded with the direction of the asymmetrical apomorphine-induced rotation.

Possible Mechanisms Underlying EBST

The possible mechanism involved in the display of biased swing behavior may be the stress caused by handling the animal by its tail. Past studies have reported that tail pinch and stressor effects may result in changes in locomotor activity (Boutelle et al., 1990; Tanaka et al., 1991; Rouge-Pont et al., 1993). Furthermore, stressors may interact with the dopaminergic pathway (Pei et al., 1990; Cenci et al., 1992). For the biased swing activity in hemiparkinsonian rats, an imbalance in DA levels between the two sides of the brain, coupled with stressor effects, may explain the observed biased swing activity. Neurochemical studies on tail pinch and stressors have revealed increases in DA metabolite, DOPAC, and [³H]SCH 23390 binding (Morelli et al., 1987; Rodriguez and Castro, 1991; Rowlett et al., 1991, 1993). Of note, a higher level of biased swing actvity was observed when the animal's tail was pinched or when the animal was injected with apomorphine. Taken together, these results would suggest that the swing activity is a DA-mediated motor response. A similar mechanism is also implicated in apomorphine-induced rotational behavior (Ungerstedt, 1971a,b; Silverman and Ho, 1981; Coward, 1983). Future investigations into the neurochemical alterations during swing activity of animals with lesions in the dopaminergic pathway may reveal further evidence of interaction between DA and stressor effects.

For the unilateral excitotoxin-lesion model of HD, the nigrostriatal dopaminergic system coupled with stress factors/ stressors may also explain the occurrence of asymmetrical motor behaviors. In animals with unilateral nigral lesions, the denervated striatum has more supersensitive postsynaptic DA receptors than the intact striatum. Injection of apomorphine, being a DA receptor agonist, leads to rotational behavior contralateral to the lesioned striatum (Hudson et al., 1993). In animals with striatal lesions, DA axons of passage and terminals are spared (Isacson et al., 1984), which would indicate that DA receptor levels in the bilateral striata remain balanced. This sparing of DA fibers-of-passage would seem, at first, not to support an imbalance in bilateral DA receptor levels as the underlying mechanism of turning behavior in striatal-lesioned animals. However, DA exerts some tonic influence on striatal neurons (Sanberg and Coyle, 1984; Emerich et al., 1992). Thus, an imbalance in striatal DA levels may still affect stress-evoked DA release and reuptake in the striatum (Abercrombie et al., 1989; Keefe et al., 1989) which, in turn, may alter locomotor activity. Indeed, direct injections of DA agonists into the intact striatum stimulate DA function and result in increased stereotypic behaviors (Sharp et al., 1987). Alternatively, temporal patterns between DA release and stereotypy are not parallel, suggesting that the stress-DA-evoked behavioral effect may be mediated by other neurotransmitters (Ungerstedt, 1973). Since neurotransmitter synthesizing enzymes, primarily choline acetyltransferase (ChAT) and glutamic acid decarboxylase (GAD), are markedly reduced in animals with striatal lesions (Schmidt et al., 1981; Isacson et al., 1984), acetylcholine and GABA may be factors that interact with the motor behavior asymmetry. Measurements of striatal DA receptor binding and other neurotransmitters during EBST may provide direct evidence of a DA-mediated biased swing behavior.

In EBST sessions toward the end of the 2-mo period, the lesioned animals displayed a decreasing trend of biased swing activity. This could be the result of lower stress levels owing to repeated handling. However, the 2-mo weekly test period seems to be a sufficient time to evaluate any behavioral alterations following brain insults. Furthermore, animals showed a restored high level of biased swing activity when tested 1 mo following the last swing test (84 d postlesion) or when their tails were pinched.

The contention that stress factors may influence the direction of the biased swing in the EBST is not supported by previous observations on stress-induced rotations reported by Ungerstedt and Arbuthnott (1970) and Nikkhah and colleagues (1994). These investigators found that stress, which was accomplished by attaching a paper clip at the tail, induces ipsilateral rotations in unilaterally 6-OHDA lesioned rats, whereas in the EBST, a contralateral biased swing activity, paralleling the direction of the apomorphine-induced rotational behavior, was observed. Thus, although stress factors may contribute to the display of asymmetrical motor behavior, other factors may influence the direction of this biased behavior. Further investigations are warranted to elucidate the mechanisms involved in the motor asymmetry displayed in the EBST.

In summary, the EBST has been shown to be a rapid, easy, inexpensive, and accurate measure of a DA-mediated motor function. The biased swing behavior is good estimate of true effects of unilateral lesions in the nigrostriatal pathway. Repeated behavioral assessment is very important when evaluating effects of neural transplants on rat models with hemiparkinsonism or unilateral HD motoric symptoms. The EBST circumvents the sensitization problem and provides an alternative tool for studies of animal models of neurodegenerative disorders characterized by asymmetrical brain lesions. In addition, EBST can be used in assessment of behavioral recovery following experimental therapies, such as neural transplantation.

Conclusion

Asymmetrical motoric deficits following CNS insults can be characterized by a number of behavioral tests. However, caution should be exercised in the procedural controls of the test and interpretation of the data. The EBST offers an alternative approach to investigating the motor deficits in animal models with asymmetrical CNS lesions. Although it has advantages over the drug-induced tests, the EBST should in no way be used as a sole index of motor deficits. We believe that intermittent use of EBST together with other behavioral tests would provide a more exhaustive characterization of the neuropathological damage in the CNS.

> The methods/behavioral tests used by the neuroscientists must be adapted to the task, and when the human brain is concerned, they cannot be too refined (adapted from Lars Leksell, 1971).

References

Abercrombie, E. D., Keefe, K. A., Di Frischa, D. S., and Zigmond, M. J. (1989) Differential effect of stress on in vivo dopamine release in striatum, nucleus accumbens and medial frontal cortex. *J. Neurochem.* **52**, 1655–1658.

Abrous, D. N., Wareham, A. T., Torres, E. M., and Dunnett, S. B. (1992) Unilateral dopamine lesions in neonatal, weanling and adult rats: comparison of rotation and reaching deficits. *Behav. Brain Res.* **51**, 67–75.

Bernheimer, H. W., Birkmayer, W., Hornykiewicz, O., Jellinger, K., and Seitelberger, F. (1973) Brain dopamine and the syndromes of Parkinson and Huntington. Clinical, morphological and neurochemical correlations. *J. Neurol. Sci.* **20**, 415–455.

Bevan, P. (1983) Repeated apomorphine treatment causes behavioral supersensitivity and dopamine D_2 receptor hyposensitivity. *Neurosci. Lett.* **35**, 185–189.

Borlongan, C. V. and Sanberg, P. R. (1995) Elevated body swing test: a new behavioral parameter for rats with 6-hydroxydopamine-induced Hemiparkinsonism. *J. Neurosci.* (submitted).

Borlongan, C. V., Cahill, D. W., and Sanberg, P. R. (1995a) Motor behavior assymetry in rats with striatal-lesion: a correlate of human neglect phenomenon. *Restor. Neurol. Neurosci.* (submitted).

Borlongan, C. V., Randall, T. S., Cahill, D. W., and Sanberg, P. R. (1995b) Asymmetrical motor behavior in rats with unilateral striatal excitotoxic lesion as revealed by the elevated body swing test. *Brain Res.* (in press).

Boutelle, M. G., Zetterstrom, T., Pei, Q., Svensson, L., and Fillenz, M. (1990) In vivo neurochemical effects of tail pinch. *J. Neurosci. Meth.* **34**, 151–157.

Cenci, M. A., Kalen, P., Mandel, R. J., and Björklund, A. (1992) Regional differences in the regulation of dopamine and noradrenal release in the medial frontal cortex, nucleus accumbens and caudate-putamen: a microdialysis study in the rat. *Brain Res.* **581**, 217–228.

Chevalier, G., Vacher, S., Deniau, J., and Desban, M. (1985) Disinhibition as a basic process of the expression of striatal functions. I. The striato-nigral influence of tecto-spino/tecto-diencephalic neurons. *Brain Res.* **334**, 215–226.

Coward, D. M. (1983) Apomorphine-induced circling behavior in 6-hydroxy-dopamine-lesioned rats. *Arch. Pharmacol.* **323**, 49–53.

Creese, I. and Snyder, S. H. (1979) Nigrostriatal lesions enhance striatal [^{3}H]apomorphine and [^{3}H]spiroperidol binding. *Eur. J. Pharmacol.* **56**, 277–281.

Damasio, A. R., Damasio, H., and Cjui, H. C. (1980) Neglect following damage to frontal lobe or basal ganglia. *Neuropsychologia* **18**, 123–132.

Deniau, J. M. and Chevalier, G. (1985) Disinhibition as a basic process of striatal functions. II. The striato-nigral influence on the thalamocortical cells of the ventromedial thalamic nucleus. *Brain Res.* **334**, 227–233.

Denny-Brown, D., Meyer, J. S., and Horenstein, S. (1952) The significance of perceptual rivalry resulting from parietal lesion. *Brain* **75**, 433–471.

Dunnet, S. B., Björklund, A., Schmidt, R. H., Stenevi, U., and Iversen, S. D. (1983) Intracerebral grafting of neruonal cell suspensions. IV. Behavioural recovery in rats with unilateral 6-OHDA lesions following implantation of nigral cell suspensions in different forebrain sites. *Acta Physiol. Scand.* **522**, 29–37.

Dunnett, S. B., Isacson, O., Sirinathsinghji, D. J., Clarke, D. J., and Björklund, A. (1988) Striatal grafts in rats with unilateral neostriatal-lesions-III. Recovery from dopamine-dependent motor asymmetry and deficits in skilled paw-reaching. *Neuroscience* **24**, 813–820.

Dworkin, S. I. and Smith, J. E. (1987) Neurobiological aspects of drug-seeking behaviors, in *Neurobehavioralpharmacology: vol. 6. Advances in Behavioral Pharmacology* (Thompson, T., Dews, P. B., and Barrett, J. E., eds.), Erlbaum, Hillsdale, NJ, pp. 1–43.

Emerich, D. F., Ragozzino, M., Lehman, M., and Sanberg, P. R. (1992) Behavioral effects of neural transplantation. *Cell Transplant* **1**, 401–427.

Fahn, S. (1990) What are the basal ganglia diseases. *Rev. Neurosci.* **2**, 165–172.

Ferro, J. M., Kertesz, A., and Black, S. E. (1987) Subcortical neglect: quantification, anatomy and recovery. *Neurology* **37**, 1487–1492.

Gerfen, C. R. (1992) The neostriatal mosaic: multiple levels of compartmental organization. *TINS* **15**, 133–139.

Graybiel, A. M. (1991) Basal ganglia-input, neural activity, and relation to the cortex. *Curr. Opinion Neurobiol.* **1**, 644–651.

Hamilton, B. F. and Gould, D. H. (1987) Nature and distribution of brain lesions in rats intoxicated with 3-nitropropionic acid: a type of hypoxic (energy deficient) brain damage. *Acta Neuropathol.* **72**, 286–297.

Hattori, S., Li, Q., Matsui, N., and Nishino, H. (1993a) Treadmill running test for evaluating locomotor activity after 6-OHDA lesions and dopaminergic cell grafts in the rat. *Brain Res. Bull.* **31**, 433–435.

Hattori, S., Li, Q., Matsui, N., and Nishino, H. (1993b) Treadmill running test for evaluating locomotor activity after 6-OHDA lesions and dopaminergic cell grafts in the rat. *Brain Res. Bull.* **31**, 433–435.

Healton, E. B., Navarro, C., Bressman, S., and Brust, J. C. M. (1982) Subcortical neglect. *Neurology* **32**, 776–778.

Heilman, K. M. and Watson, R. T. (1979) Thalamic neglect. *Neurology* **29**, 690–694.

Hikosaka, O. (1991) Basal ganglia-possible role in motor coordination and learning. *Curr. Opinion Neurobiol.* **1**, 638–643.

Hudson, J. L., Levin, D. R., and Hoffer, B. J. (1993a) A 16-channel automated rotometer system for reliable measurement of turning behavior in 6-hydroxydopamine lesioned and transplanted rats. *Cell Transplant* **2**, 507–514.

Hudson, J. L., Van Horne, C. G., Stromberg, I., Brock, S., Clayton, J., Masserano, J., Hoffer, B. J., and Gerhardt, G. A. (1993b) Correlation of apomorphine- and amphetamine-induced turning with nigrostriatal dopamine content in unilateral 6-hydroxydopamine lesioned rats. *Brain Res.* **626**, 167–174.

Isacson, O., Brundin, P., Kelly, P. A. T., Gage, F. H., and Björklund, A. (1984) Functional neuronal replacement by grafted striatal neurons in the ibotenic-acid-lesioned rat striatum. *Nature* **311**, 458–460.

Kalivas, P. W. and Weber, B. (1988) Amphetamine injection into the ventral mesencephalon sensitizes rats to peripheral amphetamine and cocaine. *J. Pharmacol. Exp. Ther.* **245**, 1095–1102.

Kang, Y. and Kitai, S. T. (1990) Electrophysiological properties of pendunculopontine neurons and their postsynaptic responses following stimulation of substania nigra reticulata. *Brain Res.* **553**, 79–95.

Kebabian, J. W. and Calne, D. B. (1979) Multiple receptors for dopamine. *Nature* **277**, 93–96.

Keefe, K. A., Stricker, E. M., Zigmond, M. J., and Abercrombie, E. D. (1989) Environmental stress increases extracellular dopamine in striatum of 6-hydroxydopamine-treated rats: in vivo microdialysis studies. *Brain Res.* **527**, 350–353.

Klug, J. M. and Norman, A. B. (1993) Long-term sensitization of apomorphine-induced rotation behavior in rats with dopamine deafferentation or excitotoxin-lesions of the striatum. *Pharmacol. Biochem. Behav.* **46**, 397–403.

Klug, J. M. and Norman, A. B. (1993) Long-term sensitization of apomorphine-induced rotation behavior in rats with dopamine deafferentation or excitotoxin lesions of the striatum. *Pharmacol. Biochem. Behav.* **46**, 397–403.

Kuczenski, R. and Leith, N. J. (1981) Chronic amphetamine: is dopamine a link in or a mediator of the development of tolerance and reverse tolerance? *Pharmacol. Biochem. Behav.* **15**, 405–413.

Masur, J., Oliveira de Souza, M. L., and Zwicker, A. P. (1986) The excitatory effect of ethanol: absence in rats, no tolerance and increased sensitivity in mice. *Pharmacol. Biochem. Behav.* **24**, 1225–1228.

Mattingly, B. A., Rowlett, J. K., Graff, J. T., and Hatton, B. J. (1991) Effects of selective D_1 and D_2 dopamine antagonists on the development of behavioral sensitization to apomorphine. *Psychopharmacology* **105**, 501–507.

McGeer, P. L., McGeer, E. G., and Suzuki, J. S. (1977) Aging and extrapyramidal function. *Arch. Neurol.* **34**, 33–35.

Mesulam, M. M. (1985) Attention, confusional states, and neglect, in *Principles of Behavioral Neurology* (Mesulamm, M. M., ed.), FA Davis, Philadelphia, PA, pp. 125–168.

Mink, J. W. and Thach, W. T. (1993) Basal ganglia intrinsic circuits and their role in behavior. *Curr. Opinion Neurobiol.* **3**, 950–957.

Morelli, M., Fenu, S., and Di Chiara, G. (1987) Behavioral expression of D-1 receptor supersensitivity depends on previous stimulation of D-2 receptors. *Life Sci.* **40**, 245–251.

Nikkah, G., Cunningham, M. G., McKay, R., and Björklund, A. (1995) Dopaminergic microtransplants into the substantia nigra with bilateral 6-OHDA lesions. II. Transplant-induced behavioral recovery. *J. Neurosci.* (in press).

Norman, A. B., Calderon, S. F., Giordano, M., and Sanberg, P. R. (1988) A novel rotational behavior model for assessing the restructuring of striatal dopamine effector systems: are transplants sensitive to peripherally acting drugs? in *Transplantation into the Mammalian CNS, Progress in Brain Research*, vol. 78 (Gash, D. M. and Sladek, J. R., Jr., eds.), Elsevier, Amsterdam, pp. 61–67.

Norman, A. B., Wyatt, L. M., Hildebrand, J. P., Kolmonpunporn, M., Moody, C. A., Lehman, M. N., and Sanberg, P. R. (1990) Sensitization of rotation behavior in rats with unilateral 6-hydroxydopamine or kainic acid-induced striatal lesions. *Pharmacol. Biochem. Behav.* **37**, 755–759.

Pei, Q., Zetterstrom, T., and Fillenz, M. (1990) Tail pinch-induced changes in the turnover and release of dopamine and 5-hydroxytryptamine in different brain regions of the rat. *Neuroscience* **35**, 133–138.

Phillips, T. J., Dickenson, S., and Burkhart-Kasch, S. (1994) Behavioral sensitization to drug stimulant effects in C57BL/6J and DBA/2J inbred mice. *Behav. Neurosci.* **108**, 789–803.

Pycock, C. J. and Kilpatrick, I. C. (1989) Motor asymmetries and drug effects, in *Neuromethods: Psychopharmocology*, vol. 13 (Boulton, A. A., Baker, G. B., and Greenshaw, A. J., eds.), Humana, Clifton, NJ, pp. 1–93.

Robertson, H. A. (1992) Dopamine receptor interactions: some implication for the treatment of Parkinson's disease. *Trends Neurosci.* **15**, 201–206.

Robinson, T. E., Jurson, P. A., Bennett, J. A., and Bentgen, K. M. (1988) Persistent sensitization of dopamine neurotransmission in ventral striatum (nucleus accumbens) produced by prior experience with (+) amphetamine: a microdialysis study in freely moving rats. *Brain Res.* **462**, 211–222.

Rodriguez, M. and Castro, R. (1991) Apomorphine lowers dopamine synthesis for up to 48 h: implications for drug sensitization. *NeuroReport* **2**, 365–368.

Rouge-Pont, F., Piazza, P. V., Kharouby, M., Le Moul, M., and Simon, H. (1993) Higher and longer stress-induced increase in dopamine concentrations in the nucleus accumbens of animals predisposed to amphetamine self-administration. A microdialysis study. *Brain Res.* **602**, 169–174.

Rowlett, J. K., Mattingly, B. A., and Bardo, M. T. (1991) Neurochemical and behavioral effects of acute and chronic treatment with apomorphine in rats. *Neuropharmacology* **30**, 191–197.

Rowlett, J. K., Mattingly, B. A., and Bardo, M. T. (1993) Neurochemical correlates of behavioral sensitization following repeated apomorphine treatment: assessment of the role of D_1 dopamine receptor stimulation. *Synapse* **14**, 160–168.

Sanberg, P. R. and Coyle, J. T. (1984) Scientific approaches to Huntington's disease. *CRC Crit. Rev. Clin. Neurobiol.* **1**, 1–44.

Sanberg, P. R., Zoloty, S. A., Willis, R., Ticarich, C. D., Rhoads, K., Nagy, R. P., Mitchell, S. G., Laforest, A. R., Jenks, J. A., Harkabus, L. J., Gurson, D. B., Finnefrock, J. A., and Bednarik, E. J. (1987) Digiscan activity: automated measurement of thigmotactic and stereotypic behavior in rats. *Pharmacol. Biochem. Behav.* **27**, 569–572.

Schallert, T., Norton, D., Razcok, S. E., Johnston, R. E., and Becker, J. B. (1994) Recovery of spontaneous behaviors after grafts of fetal ventral mesencephalic tissue into the dopamine-denervated rat striatum. *Int. Behav. Neurosci. Soc. Abstract* **3**, 78.

Schmidt, R. H., Björklund, A., and Stenevi, U. (1981) Intracerebral grafting of dissociated CNS tissue suspensions: a new approach for neuronal transplantation to deep brain sites. *Brain Res.* **218**, 347–356.

Segal, D. S. and Kuczenski, R. (1987a) Behavioral and neurochemical characteristics of stimulant-induced augmentation. *Psychopharmacol. Bull.* **23**, 417–424.

Segal, D. S. and Kuczenski, R. (1987b) Individual differences in responsiveness to single and repeated amphetamine administration: behavioral characteristics and neurochemical correlates. *J. Pharmacol. Exp. Ther.* **242**, 917–926.

Sharp, T., Zetterstrom, T., Ljungerg, T., and Ungerstedt, U. (1987) A direct comparison of amphetamine induced behaviours and regional brain dopamine release and metabolism in rat brain regions using intracerebral dialysis. *Brain Res.* **401**, 322–330.

Silverman, P. B. and Ho, B. T. (1981) Persistent behavioral effect of apomorphine in 6-hydroxydopamine-lesioned rats. *Nature* **294**, 475–477.

Stewart, J. and Vezina, P. (1988) Conditioning and behavioral sensitization, in *Sensitization in the Nervous System* (Kalivas, P. W. and Barnes, C. D., eds.), Telford, Caldwell, NJ, pp. 207–224.

Sullivan, R. M., Fraser, A., and Szechtman, H. (1994) Asymmetrical orientation to edges of an openfield: modulation by striatal dopamine and relationship to motor asymmetries in the rat. *Brain Res.* **637**, 114–118.

Tanaka, T., Yokoo, H., Mizoguchi, K., Yosiha, M., Tsuda, A., and Tanaka, M. (1991) Noradrenaline release in the rat amygdala is increased by stress: studies with intracerebral microdialysis. *Brain Res.* **544**, 174–176.

Ungerstedt, U. (1971a) Striatal dopamine release after amphetamine or nerve degeneration revealed by rotational behavior. *Acta Physiol. Scand.* **367**, 49–66.

Ungerstedt, U. (1971b) Post synaptic supersensitivity after 6 hydroxydopamine induced degeneration of the nigro-striatal dopamine system. *Acta Physiol. Scand.* **367**, 69–93.

Ungerstedt, U. (1973) Selective lesions of central catecholamine pathways: application in functional studies, in *Neuroscience Research*, vol. 5 (Ehrenpreis, S. and Kopin, I., eds.), Academic, New York, pp. 73–96.

Ungerstedt, U. and Arbuthnott, G. W. (1970) Quantitative recording of rotational behavior in rats after 6-hydroxydopamine lesions of the nigrostriatal dopamine system. *Brain Res.* **24**, 485–493.

Vallar, L. and Meldolesi, J. (1989) Mechanisms of signal transduction at the dopmine D_2 receptor. *Trends Pharmacol. Sci.* **10**, 74–77.

Wallace, R. B. (1983) Behavioral analysis of the transplantation phenomenon withiing a motor and sensory system, in *Neural Tissue Transplantation Research* (Wallace, R. B. and Das, G. D., eds.), Springer-Verlag, New York, pp. 217–237.

Wise, R. A. (1988) Psychomotor stimulant properties of addictive drugs. *Ann. NY Acad. Sci.* **537**, 228–234.

CHAPTER 12

Measuring Spontaneous Turning Behaviors in Children and Adults

H. Stefan Bracha and Jeffrey W. Gilger

Introduction

Asymmetrical spontaneous turning behavior (rotational preference or circling bias) has been observed in humans and animals, and it is related to clinical hemineglect (Glick and Ross, 1981; Bradshaw and Nettleton, 1987). When a rotational preference is observed, it is typically in a direction away from the brain hemisphere with higher striatal dopaminergic transmission (Glick and Ross, 1981; Bradshaw and Nettleton, 1987).

Over the past decade, Bracha and coworkers have studied rotational asymmetries in a variety of psychiatric and normal populations (Bracha, 1987; Bracha et al., 1987a,b, 1989, 1993; Gordon et al., 1992; Lyon et al., 1992). The focus has been on populations where the brain hemispheres are thought to be asymmetrically involved, and/or the dopaminergic system is considered part of the etiology (e.g., schizophrenia). The paradigm utilizes a technology for humans that has been adapted from circling studies of rodents (Bracha et al., 1987a,b). In this chapter, we will describe the logic and methodology as they have been applied to normal and psychiatric populations. The ultimate objective of these studies has been to assess the usefulness of measuring spontaneous human circling behavior as

From: *Motor Activity and Movement Disorders*
P. R. Sanberg, K. P. Ossenkopp, M. Kavaliers, Eds. Humana Press Inc., Totowa, NJ

a diagnostic and cost-effective research tool that is less intrusive, less demanding, and less labor-intensive than many methods commonly used in psychiatry.

Theoretical Underpinnings of Asymmetrical Turning Behavior in Humans

Animal Research

A theoretical understanding of human turning behavior begins with animal studies that are reviewed comprehensively elsewhere in this book (e.g., *see* Chapter 8). Animal studies suggest that rotational preference is determined by a dopamanergic predominance in one hemisphere compared to the other (Ungerstedt and Arbuthnott, 1970; Glick and Cox, 1978; Glick and Ross, 1981; Crowne and Pathria, 1982; Collins, 1985; Stewart et al., 1985). As a rule, animals rotate toward the hemisphere with lower striatal dopaminergic activity (away from the hemisphere with higher striatal dopaminergic activity). Although the dopaminergic imbalance has most commonly been produced by electrical stimulation or lesions, an endogenous asymmetry in dopaminergic function has also been identified (e.g., Zimmerberg et al., 1974) and related to spontaneous and drug-induced circling (Glick and Shapiro, 1985).

In normal rodents, the nigrostriatal asymmetry has been shown to be reflected in both pre- and postsynaptic properties. Differences in the levels of dopamine metabolites, indicative of differences in presynaptic activity, and an opposite imbalance in striatal dopamine-stimulated adenylate cyclase activity, indicative of a difference in postsynaptic receptor sensitivity, have been observed. The implication is that the brain hemisphere with more active terminals had fewer or less sensitive receptors (Donaldson et al., 1976). Typically, the presynaptic component of this asymmetry appears to be dominant. Normal unmedicated rodents spontaneously rotate away from the hemisphere that has higher striatal dopamine levels. Following a direct agonist, such as apomorphine, however, normal rats reverse their normal rotation and rotate away from the hemisphere that has more striatal dopamine receptors.

Nonlesioned, unmedicated rodents rotate spontaneously in a consistent direction when placed in open spaces (Ungerstedt and Arbuthnott, 1970; Glick and Cox, 1978; Glick and Ross, 1981; Crowne and Pathria, 1982; Collins, 1985; Glick and Shapiro, 1985; Stewart et al., 1985). This rotation is thought to be related to the normal small asymmetry in the dopamine contents of the right and left basal ganglia: Dopamine content is signifi-cantly higher in the striatum contralateral to a naive rat's domi-nant paw. High doses of amphetamine (20 mg/kg) can increase this difference and cause the animal to rotate in tight circles away from the side with the higher dopamine levels, regard-less of the environmental architecture.

Human Research

Several studies have noted or hypothesized a role for the dopaminergic system in a variety of clinical disorders, includ-ing schizophrenia (e.g., Bracha, 1987, 1991; Bracha et al., 1993; Seeman and Guan, in press), Parkinson's disease (Bracha et al., 1987b), and depression (Poschel and Ninteman, 1964; Lyon et al., 1992). Some percentage of dopaminergic asymmetry has also been noted in the brains of normals, and hemispheric laterality for a variety of cognitive and motor tasks in normals is a well established fact (McGlone, 1980; Springer and Deutsch, 1981; Bracha et al., 1987a). Subtle asymmetry in dopamine levels (L > R) have also been demonstrated in postmortem brains of normal humans (Glick et al., 1982).

Perhaps one of the better known psychiatric disorders where hemispheric asymmetry and dopamine-related processes have been described is schizophrenia. The disorder therefore serves as a good example of a testable model for preferential turning in humans (Bracha, 1987, 1991; Bracha et al., 1993), and it deserves special attention.

There is a vast, although conflicting, clinical literature sug-gesting that in some patients with schizophrenia, the cerebral hemispheres may be asymmetrically involved. Support for this hypothesis comes from a variety of electrophysiological stud-ies (Flor-Henry, 1969; Serafentinides, 1972), hemiretinal and tachistoscopic studies (Connolly et al., 1979; Posner et al., 1988), regional cerebral blood flow studies (Gur et al., 1985), eye move-ment studies (Gur, 1977), EEG and auditory and visual evoked potential studies (Serafinides, 1972; Gruzelier, 1973; Gruzelier et al., 1974; Connolly et al., 1979), handedness studies (Gur,

1977), skin conductance studies (Gruzelier, 1973), BEAM studies (Morihisa et al., 1983), CT-scan studies (Losonczy et al., 1986),
and Positron Emission Tomography studies (Buschbaum et al.,
1984). Either a dominant hemisphere dysfunction or a nondo-
minant hemisphere dysfunction in schizophrenia has been suggested as the underlying pathology, and genetic as well as
nongenetic explanations for the dysfunction have been given
(e.g., Crow, 1990; Bracha, 1991).

Increasingly some common ground has been established
between the asymmetry hypothesis and the dopamine hypotheses in schizophrenia research. Several studies using varying
methodologies have reported significantly higher levels of
dopamine or dopaminergic striatal receptors in the right hemispheres of persons with schizophrenia (Reynolds, 1983; Reynolds
et al., 1987; Seeman and Guan, in press). The laterality effects noted
for schizophrenia may be a manifestation of this contralateral
hemispatial inattention owing to differential dopaminergic system functioning. It is noteworthy that neuroleptics block
amphetamine-induced rotation in rats (Pijnenburg et al., 1975;
Jerussi and Glick, 1976), and this rodent circling paradigm has
become a standard tool for screening new neuroleptic and anti-
Parkinsonian drugs (Glick and Shapiro, 1985).

Using the rotametric technique described below, Bracha
(1987) in fact demonstrated subclinical right-sided hemineglect
in unmedicated schizophrenics as assessed by a preference for
spontaneous turning behaviors toward the left side (i.e., counterclockwise). This is consistent with right anterior subcortical
structures manifesting an overactivity of dopaminergic systems
relative to those of the left side, as would be predicted by both
animal and human models of neurobiology and psychosis. Interestingly, later research on autopsied brains of patients with schizophrenia supported the differential hemispheric functioning of
dopaminergic systems (e.g., Seeman and Guan, in press) suggested by the unintrusive and easily applied rotametric technique.

The Need for a Tool
to Study Human Circling Behavior

Starting in 1985 the search for human analogs of the well-
understood animal model of circling behavior began in the labs
of Bracha and coworkers. Although circling had been for more

than two decades one of the best understood behaviors in psychopharmacology, the technology to measure the human analogs of circling was not available.

Bracha and colleagues (1987) developed a device to quantify in humans the kinds of subtle rotational movements observed in unmedicated, unlesioned rats. Subjects wear the device (the "human rotameter" or "turn counter") for a specified period of time, and are completely unaware of the type of counts it records. The device has provided a means to investigate inexpensively and unintrusively a human model of a dopamine-related hemispheric inattention, similar to rodent behavior.

The Human Rotameter

The rotameter is an automated hip-mounted device developed by Bracha and colleagues while at the Neuropsychiatry branch of the National Institute of Mental Health.* The rotameters are worn in a belt-mounted calculator case, are lightweight, and use rechargeable batteries (*see* Fig. 1). The model currently in use is a modified version of a prototype originally designed by S. D. Glick and S. Green.

The devices can operate up to 100 h at a time and register up to 9999 90, 180, and 360° turns in each direction. Two major functional components of the rotameters are the position sensor and the electronic processing circuit. The position sensor monitors changes in the orientation of the dorsal–ventral axis of the subject. Magnetic north is used as a necessary external reference, and a compass is used to track this reference. A compass transducer system moves with the subject relative to the needle. When the subject turns left or right, the needle maintains its fix on magnetic north, while the compass casing moves with the subject relative to the needle. The advantages of the computerized rotameter over the clinical tests of rotational or hemispheric preference used in other work (e.g., Heilman and Valenstein, 1972; Posner et al., 1988; Lyon et al., 1992; Harvey et al., 1993) are that it is much less laborious, allows continuous monitoring rather than spot checks of rotational preference, and is less prone to breech of rater blind.

*Contact the authors for more detail about the device and its availability.

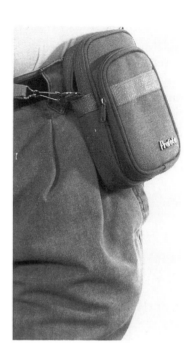

Fig. 1. The electronic rotameter and body attachment apparatus.

The apparatus utilizes an Intel 8051 microchip that is programmed with algorithms for generating "full turn" counts and accumulates and stores the six data types: right and left 90°, right and left 180°, and right and left 360° turns. These data can be displayed on demand and then recorded for future study.

The rotameter was designed such that its output would be directly analogous to that used in rodent studies. Although a variety of indices have been examined (*see* Table 1), one typically serves as the dependent measure of interest: the 360° percent right turns (PRT). This is the principal index of rotational asymmetry used in our lab and the labs of others, and it is defined as the number of full (360°) turns to the right (clockwise) divided by the number of total (right plus left) 360° turns. It is important to recognize that this measure reflects a behavioral asymmetry that is independent of the absolute amount of spontaneous circling in either direction.

Table 1
Four Common Rotameter Variables and Their Definitions

Number of 90, 180, and 360° turns to the left and right. This index can be used as an indicator of the overall circling activity level of the subject.
Percent of 90, 180, and 360° turns in either direction. Obtained by dividing the number of interest by the total number of left and right turns. These are the primary variables of interest for asymmetry and turning preference studies *(see text)* and are independent of the total number of turns made by the subject.
The absolute difference between full 360° turns in the preferred and nonpreferred directions. This is another measure of asymmetry of preferential circling behavior, although it is not independent of the total amount of turning activity engaged in.
Extra quarter turns: This is obtained by subtracting four times the number of full 360° turns from the number of 90° turns. It is a human analog to the index of "nonlateralized activity" used in rodent studies.[a]

[a]*See* Greenstein and Glick (1975).

Application of the Rotameter to the Study of Brain Disorders

The rotameter has proven quite useful since its development, and more than a dozen independent labs in the United States, Australia, Asia, and Europe have begun to study hemispheric asymmetries using the device. In terms of our own work, the rotameter has been applied to study individual, gender, and cognitive ability differences in normal adults (Bracha et al., 1987a; Gordon et al., 1992), asymmetrical movements in patients with anterolateral ischemic cortical lesions (stroke; Bracha et al., 1989), adults with hemi-Parkinsonism (Bracha et al., 1987b), children with autism and other psychiatric disturbances (Bracha et al., in review), depressed patients (Bracha et al., unpublished manuscript), and medicated and unmedicated patients with schizophrenia (Bracha, 1987; Richmond et al., 1992; Bracha et al., 1993). Unpublished work in progress includes the study of children with learning and attention deficits.

The research completed thus far has facilitated an understanding of the possible neurobiological underpinnings of group and individual differences in normal and disordered populations, as well as help clarify how gross motor functional and active laterality is related to more common and static indices of laterality, such as handedness, eye preference, and performance on focused attention and decision tasks (e.g., Bracha et al., 1987b). As an example of the potential of the rotameter technique, we summarize an ongoing study of the rotational behaviors of autistic children below (for more detailed reports on preliminary data *see* Bracha et al., in review).

An Example of the Rotametric Method: Do Autistic Children Preferentially Rotate to the Left or Right Side?

Autism is a neurodevelopmental disorder that is first manifested in the early childhood years. Diagnostic criteria are predominantly behavioral, with one of the typical symptoms being stereotypical and repetitive body movements. Hand stereotypes and mannerisms, for example, are widely recognized. Finger flicking, head banging, rocking, hand twisting, whirling, twirling, and body spinning are other clinical motor signs of severe autism (Sorosky et al., 1968; APA, 1987; Edwards and Bristol, 1991).

Here we report some preliminary data where we employed the rotameter to measure spontaneous circling behavior in children with autism of unknown etiology. It has already been shown that normal and psychiatrically disordered subjects exhibit rotational biases, possibly in response to dopaminergic activational asymmetries (e.g., Bracha, 1987). For example, as was mentioned earlier, a leftward circling preference was demonstrated in adults with schizophrenia, suggesting that the dopamine system may be involved in the disorder and that the right brain hemisphere may be more intact than the left (Bracha, 1987, 1991). Similar to previous research on persons with schizophrenia, the current study, although preliminary in nature, may provide clues to some of the neurologic abnormalities affecting autistic individuals.

Method

The group of nine autistic subjects had been referred to the Arkansas Children's Hospital for diagnosis and assessment. All autistic subjects entering the study had a confirmed diagnosis of autism of unknown etiology using the DSM-III-R crite-

Table 2
Descriptive Statistics and Mean Comparisons
for Autistic and Control Subjects

| Variable | Group | | *t* Value | *p*[a] |
	Autistic	Controls		
N	9	27		
Males[b]	5	13		
Females	4	14		
Mean age (yr)	8.33	5.96	2.75	<.05
SD[c]	3.64	1.58		
Mean total[d]	10,148.22	24,908.85	−2.83	<.01
SD	5150.39	15,233.16		
Mean 360° PRT	33.0	50.0	−5.36	<.00
SD	1.3	0.7		
Mean 180° PRT	40.0	49.0	−4.99	<.00
SD	0.9	0.3		
Mean 90° PRT	44.0	49.0	−4.74	<.00
SD	0.6	0.2		

[a]Two-tailed probability.
[b]Chi-square test of male–female proportions across groups =
0.15, 1 df, *p* > .05.
[c]SD = Standard deviation.
[d]Total = Sum of all left and right turns made for 360, 180,
and 90° rotations.

ria (APA, 1987), and had scores on the Childhood Autism Rating Scale (Schopler et al., 1985) in excess of 30 points.

Autistic participants were free of any physical (e.g., orthopedic) problems that would interfere with proper acquisition of turning data and were not on psychotropic medication at the time of the study. Table 2 displays some descriptive statistics for the 9 autistic subjects.

Twenty-seven control children were recruited from an elementary school in Little Rock, AR. All children were required to be functioning normally in school and be free of emotional or behavioral disturbance. Descriptive data on the controls are also shown in Table 2.

Procedure

We employed the standard data collection methods utilized in previous work and described in detail above and elsewhere (e.g., Bracha et al., 1993). For the autistic subjects, a parent or guardian was contacted and asked permission for his/her child to be involved in a 1-d study during which data on activity level would be collected. Parents were instructed to bring their children to a designated daycare facility, and on arrival, they were given a description of the project.

After parental consent had been obtained, a rotameter was placed on the child. Data on all autistic subjects were acquired during this 1-d, 8-h period, during which all subjects were allowed to play independently at the daycare facility. Parents/guardians were present throughout the data collection period.

For controls, consent was first obtained from parents or guardians, and then the use and operation of the monitoring device were described. Arrangements were then made for a research assistant to go to a control subject's school and attach the device to the child's belt. Children were told to wear the device on their belts during waking hours, on two consecutive days, and to go about their usual activities. An experimenter followed up with each control child and parent or guardian to ensure proper use of the monitor occurred. As it turned out, most of the activity recordings for controls occurred during school recess and after school playtime. This was appropriate in that the autistics were also recorded during similar activities and situations. Like the autistic group, the control children were unaware of the nature of the study and the monitor.

Analysis

As described above and in Table 1, there are several potential variables of interest that are yielded by the rotameter. However, the principal index of asymmetric turning behavior we shall discuss here is the percent of 360° PRT as defined previously. Again, PRT reflects hemispatial preference asymmetry independent of total activity. The further PRT is from 50%, the more asymmetric the behavior: Values between 0 and 40% reflect varying degrees of left-sided turning preference (i.e., right hemineglect), whereas values between 60 and 100% reflect varying degrees of right-sided turning preference (i.e., left hemineglect). The primary question of this study is whether autistics exhibit more or less frequent turns to the right relative to controls. Because of the small and disproportionate sizes of our two samples, both parametric (i.e., *t*) and nonparametric (i.e., Mann-Whitney U) tests were used (Hays, 1981).

Results

Before discussing the group differences on the PRT, some preliminary analyses needed to be dispensed with. First, in all cases, parametric and nonparametric test results agreed. Thus, we are confident that the statistical conclusions we draw from our data are robust and not simply owing to distributional peculiarities that may exist in our data. For ease of interpretation, however, we will present only the *t*-test results. Second, we noted (*see* Table 2) that autistics were significantly older than controls and displayed a lower mean number of total moves (irrespective of direction). The lower mean number of moves is in large part owing to group differences in the method by which data were recorded: For autistics, measurement occurred during a single 6–8 h time period, whereas for controls, the measurement period lasted however many waking hours there were over a 2-d period. As expected, however, there was not a significant relationship between the total number of moves and the percent of right handed turns ($r = .23, p > .05$). Consequently, we did not consider the total number of moves an important covariate in our main analyses. On the other hand, age was nearly significantly correlated with the PRT ($r = - .28, p < .09$, two-tailed). Although the specifics will not be presented here, using age as a covariate in our group analyses did not diminish the statistical differences between autistic and control children on the PRT.

As can be seen in Table 2, autistic children exhibited significantly lower percentages of 360° turns (i.e., significantly higher percentages of left turns) than controls. For completeness, we show data for the percent of 90 and 180° right-hand turns in Table 2 as well.

As expected for a normal population, control children's propensity to turn right or left was around 50/50. In fact, group membership was a moderately powerful predictor of turning bias and accounted for approx 40% of the variance observed in right-turning preferences. Specifically, η-squared values (Hays, 1981), a measure of group effect size, were .46, .42, and .39 for 360, 180, and 90° PRT indices, respectively. That these group

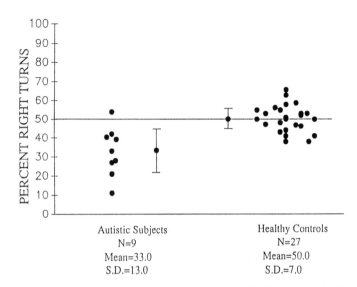

Fig. 2. Percent of spontaneous 360° right-sided turning in 9 autistic and 27 control children.

differences were so large and significant in spite of the small autistic sample is noteworthy. As a graphic illustration, the data for 360° PRT are displayed in Fig. 2.

It is also of some interest that for all three degree categories, the variance for the autistic group was significantly smaller than that of the control group (Hays, 1981). Nonhomogeneous variances are often of more than a simple statistical interest, and may suggest something about the comparative range of individual differences across groups of neurodevelopmentally impaired children and normals. Significantly smaller variances in the autistics, for instance, are consistent with a neurological abnormality resulting in restricted turning behavior and development, and are another index affirming the autistic stereotypies and rigidity reported in the clinical literature.

Why Might Autistics Show a Left-Sided Turning Preference?

Our preliminary study of autistics confirms the numerous observations made by clinicians and noted in DSM-III-R that the child with autism often "spins" or "twirls" (Sorosky et al.,

1968; APA, 1987). Yet this study goes beyond the previous clinical reports by:

1. Demonstrating measurable turning behavior in children with autism, in whom clinical turning may not have been identified;
2. Providing a means to quantify this behavior; and
3. Demonstrating that this behavior is asymmetric. Specifically, autistic children relative to controls demonstrated spontaneous rotational behaviors preferentially toward the left hemispace.

Additional work is in progress to replicate and extend our findings. There are also other, intuitively plausible concerns about the interpretation of the group differences in these data that may have come to the reader's mind. Namely, were the controls appropriately matched to the autistics for the seemingly relevant variables of handedness and IQ, and would the use of a psychiatric control group of children make a difference in our conclusions about rotational preferences in autistics?

Unfortunately, subject availability precluded a thorough matching procedure, and adequate laterality and IQ information was not available for the child participants. We are currently expanding the study in response to these weaknesses. However, we can address the importance of handedness, IQ, and psychiatric controls indirectly.

First, contrary to popular expectation, several recent studies have shown that a subject's handedness has surprisingly little effect on turning behavior (Bracha, 1987; Bracha et al., 1989; Lyon and Satz, 1991; Gordon et al., 1992). Handedness differences, therefore, probably do not explain our rotational asymmetry finding in autism (or schizophrenia), even if there were handedness differences across study groups. Second, in regard to the need for a psychiatric control group and IQ effects, data from a study in progress suggest that the autistic turning preference is not a reflection of IQ or simply having some/any psychiatric disorder. Specifically, the correlation between IQ and the percent of right-sided turns was nonsignificant ($r = .20$) in a sample of 18 psychiatrically ill adolescents, and a left-sided turning preference was not shown in these adolescents or in a sample of five children with Attention Deficit Disorder with Hyperactivity (Bracha, unpublished manuscript). It is noteworthy, however, that what information we have on our autistics indicates that they were severely affected and had IQs (at least as very young children at first assessment) in the mild to moderately retarded range.

We also have some data relevant to the need for different psychiatric controls: Not only was a left-turning preference not demonstrated in the group of nonautistic but psychiatrically ill adolescents mentioned above, but it was also not evident in adult inpatients with depression (Bracha et al., unpublished manuscript), and a heterogeneous group of child psychiatric inpatients (Bracha, unpublished manuscript). Moreover, the observed mean 360° PRP in autistics (33.0%) is significantly lower than the 50% we would expect in nonpreferential turners. That is, the two-tailed .05 confidence limits for the mean autistic 360° PRT ranges from 32.15–33.85%, and does not include 50% as a value. This indicates a within-subject bias toward left-sided turns independent of a comparison group.

However, similar to the autistics, severely ill and unmedicated patients with schizophrenia do manifest a pattern of subtle left-circling preference. Although the etiology of autism is unknown in the majority of cases (APA, 1987; Edwards and Bristol, 1991), it is quite clear that autism is genetically unrelated to schizophrenia (e.g., Volkmar and Cohen, 1991). These pilot data merely suggest that schizophrenia shares the phenomenon of left-turning behavior with a neurodevelopmental disorder manifested in children. Both striatal and cerebellar lesions may produce left-turning behavior, and such cerebellar abnormalities may exist in some (Courchesne et al., 1988), but not all (Garber and Ritvo, 1992), autistic children. Recent neurological literature, however, associates persistent motor preference for the left hemispace with the immature brain (Gospe et al., 1990) or with severe bilateral cortical pathology both in rats (Crescimanno et al., 1982) and in humans (Schneider et al., 1968; Quesney, 1986; Weintraub and Mesulam, 1989).

We and others (e.g., Campbell and Palij, 1985) advocate the potential use of automated devices as a source of objective measures of behavior in child psychiatric research. For example, this rotameter may allow measurement of drug effects on one objective sign of autism, circling. Also, animal studies suggest (e.g., Christie and Crow, 1971; Pycock, 1980) that very marked turning behavior may identify the few patients with autism who would gain some benefit from dopamine antagonists. Further automated studies of circling behavior of autistic children and carefully matched psychiatric controls are indicated.

Conclusions and Future Directions Regarding the Study of Human Motoric Asymmetry

Conclusions drawn from the data presented here as an example of a study using the rotameter must be viewed as tentative, and in need of replication and expansion. The vast animal literature on rotational behaviors, their connection to dopamine, and our ability to quantify it in humans may make spontaneous rotational behavior one of the more useful asymmetries and signs assessed and studied in psychiatry, psychology, and neurology, especially with reference to child and adolescent populations.

Biological approaches to disorders, such as schizophrenia, are likely to be cost-effective in the long run. Elucidating the pathophysiology of such devastating brain dysfunctions is imperative. Novel approaches to the problem may help us to identify in advance subgroups of patients, say, with schizophrenia, that may respond favorably to treatment with dopamine blockers vs those who will not. In addition to turning behavior, other animal models of drug effects exist, and it may be a worthwhile endeavor to apply these models directly to humans, as we have started to do with the human rotameter.

Acknowledgments

Portions of the data reported in this chapter were obtained through work funded in part by grants from the NIMH (MH43537), and the United States Veteran's Administration.

References

APA. (1987) *Diagnostic and Statistical Manual of Mental Disorders*, 3rd ed. rev. American Psychiatric, Washington, DC.
Bracha, H. S. Asymmetric turning behaviors in adolescent psychiatric population, unpublished.

Bracha, H. S., Gillin, N. R., Dupont, J. A., and Rapaport, J. L. Asymmetric turning behaviors in a sample of patients with idiopathic depression, unpublished.

Bracha, H. S. (1987) Asymmetric rotational (circling) behavior, a dopamine-related asymmetry: preliminary findings in unmedicated and never-medicated schizophrenic patients. *Biol. Psychiatry* **22**, 995–1003.

Bracha, H. S. (1991) Etiology of structural asymmetry in schizophrenia: an alternative hypothesis. *Schizophr. Bull.* **17**, 551–553.

Bracha, H. S., Edwards, D. R., Balkozar, A., Dykman, K., Livingston, R., and Gilger, J. W. Do autistic children circle left or right? in review.

Bracha, H. S., Seitz, D., Otemaa, J., and Glick, S. D. (1987a) Rotational movement (circling in normal humans): sex difference and relationship to hand, foot and eye preference. *Brain Res.* 231–235.

Bracha, H. S., Shults, C., Glick, S. D., and Kleinman, J. E. (1987b) Spontaneous asymmetric circling behavior in hemiparkinsonism: a human equivalent of the lesioned-circling rodent behavior. *Life Sci.* **40**, 1127–1130.

Bracha, H. S., Livingston, R., Clothier, J., Linington, B. B., and Karson, C. (1993) Correlation of severity of psychiatric patient's delusions with right hemispatial inattention (left-turning behavior) *Am. J. Psychol.* **150(2)**, 330–332.

Bracha, H. S., Lyden, P. D., and Khansarinia, S. (1989) Delayed emergence of striatal dopaminergic hyperactivity after anterolateral ischemic cortical lesions in humans: evidence from turning behavior. *Biol. Psychiatry* **25**, 265–274.

Bracha, H. S., Livingston, R., Dykman, K., Edwards, D. R., and Adam, B. (1995) An automated electronic method for quantifying spinning (circling) in children with autistic disorder. *J. Neuropsychiat. Clin. Neurosci.* **7**, 1–5.

Bradshaw, J. L. and Nettleton, N. C. (1987) Coordinates of extracorporeal space, in *Neurophysical and Neuropsychological Aspects of Spatial Neglect* (Jeannerod, M., ed.), Elsevier, North Holland, pp. 41–67.

Buschsbaum, M. S., Delisi, L. E., Holcomb, H. H., Cappelletti, J., King, A. C., Johnson, J., Hazlett, E., Dowling-Zimmerman, S., Post, R., and Morihisa, J. (1984) Anteroposterior gradients in cerebral glucose use in schizophrenia and affective disorders. *Arch. Gen. Psychiatry* **41**, 1159–1166.

Buschsbaum, M. S., Ingvar, D. H., Kessler, R., Waters, R. N., Cappelletti, J., van Kammen, P., King, A. C., Johnson, J. L., Manning, R. G., Flynn, R. W., Mann, L. S., Bunney, W. E., Jr., and Sokoloff, L. (1982) Cerebral glucography with positron tomography: use in normal subject and in patients with schizophrenia. *Arch. Gen. Psychiatry* **39**, 251–259.

Campbell, M. and Palij, M. (1985) Behavioral and cognitive measures used in psychopharmacological studies of infantile autism. *Psychopharmacol. Bull.* **21**, 1047–1052.

Christie, J. E. and Crow, T. J. (1971) Turning behaviour as an index of the action of amphetamines and ephedrines on central dopamine-containing neurones. *Br. J. Pharmacol.* **43**, 658–667.

Collins, R. L. (1985) On the inheritance of direction and degree of asymmetry, in *Cerebral Lateralization in Nonhuman Species* (Glick, S. D., ed.), Academic, Orlando, FL, pp. 41–71.

Connolly, J. F., Gruzelier, H. H., and Kleinman, D. M. (1979) Lateralized abnormalities in hemisphere-specific tachistoscopic tasks in psychiatric patients and controls, in *Hemisphere Asymmetries of Function in Psychopathology* (Gruzelier, J. and Flor-Henry, P., eds.), Elsevier-North Holland Biomedical, New York, pp. 491–509.

Courchesne, E., Yeung-Courchesne, R., Press, G. A., Hesselink, J. R., and Jernigan, T. L. (1988) Hypoplasia of cerebellar vermal lobules VI and VII in autism. *N. Engl. J. Med.* **318**, 1349–1354.

Crescimanno, G., Sorberam, F., La Gruttam, V., and Amato, G. (1982) Effects of motor cortex removal on circling behaviour. *Arch. Int. Physiol. Biochem.* **90**, 355–359.

Crow, T. J. (1990) Temporal lobe asymmeties as the key to the etiology of schizophrenia. *Schizophr. Bull.* **16**, 433–443.

Crowne, D. P. and Pathria, M. N. (1982) Some attentional effects of unilateral frontal lesions in the rat. *Behav. Brain Res.* **6**, 25–39.

Donaldson, I., Dolphin, A., Jenner, P., Marsden, C. D., and Pycock, C. (1976) The roles of noradrenaline and dopamine in contraversive circling behavior seen after unilateral electrolytic lesions of the locus coeruleus. *Eur. J. Pharmacol.* **39**, 179–191.

Edwards, D. R. and Bristol, M. M. (1991) Autism: early identification and management in family practice. *Am. Fam. Physician* **44**, 1755–1764.

Flor-Henry, P. (1969) Psychosis and temporal lobe epilepsy: a controlled investigation. *Epilepsia* **10**, 363–395.

Garber, H. J. and Ritvo, E. R. (1992) Magnetic resonance imaging of the posterior fossa in autistic adults. *Am. J. Psychiatry* **149(2)**, 245–247.

Glick, S. D. and Cox, R. F. (1978) Nocturnal rotation in normal rats: correlation with amphetamine-induced rotation and effects of nigrostriatal lesions. *Brain Res.* **150**, 149–161.

Glick, S. D. and Ross, D. A. (1981) Lateralization of function in the rat brain mechanisms may be operative in humans. *Trends Neurosci.* **4(8)**, 196–199.

Glick, S. D. and Shapiro, R. M. (1985) Functional and neurochemical mechanisms of cerebral lateralization in rats, in *Cerebral Lateralization in Nonhuman Species* (Glick, S. D., ed.), Academic, Orlando, FL, pp. 157–183.

Glick, S. D., Ross, D. A., and Hough, L. B. (1982) Lateral asymmetry of neurotransmitters in human brain. *Brain Res.* **234**, 53–63.

Gordon, H. W., Busdiecker, E. C., and Bracha, H. S. (1992) The relationship between leftward turning bias and visuospatial ability in humans. *Intern. J. Neurol.* **128**, 29–36.

Gospe, S. M., Jr., Mora, B. J., and Glick, S. D. (1990) Measurement of spontaneous rotational movement (circling) in normal children. *J. Child Neurol.* **5**, 31–34.

Greenstein, S. and Glick, S. D. (1975) Improved automated apparatus for recording rotation (circling behavior) in rats or mice. *Pharmacol. Biochem. Behav.* **3**, 507–510.

Gruzelier, J. (1973) Bilateral asymmetry of skin conductance orienting activity and levels in schizophrenics. *Biol. Psychol.* **1**, 21–41.

Gruzelier, J. and Venables, P. (1974) Bimodality and lateral asymmetry of skin conductance orienting activity in schizophrenics: replication and evidence of lateral asymmetry in patients with depression and disorders of personality. *Biol. Psychol.* **8**, 55–73.

Gur, R. E. (1977) Motoric laterality imbalance in schizophrenia: a possible concomitant of left hemisphere dysfunction. *Arch. Gen. Psychiatry* **34**, 33.

Gur, R. E., Gur, R., Brett, E., Skolnick, E., Caroff, S., Obrist, W. D., Resnick, S., and Reivich, M. (1985) Brain function in psychiatric disorders III: rCBF in unmedicated schizophrenics. *Arch. Gen. Psychol.* **42**, 329–334.

Harvey, S. A., Nelson, E., Haller, J. W., and Early, T. (1993) Lateralized attentional abnormality in schizophrenia is correlated with severity of symptoms. *Biol. Psychiatry* **33**, 93–99.

Hays, W. (1981) *Statistics,* 3rd ed. Holt, Rinehart, and Winston, New York.

Heilman, K. M. and Valenstein, E. (1972) Frontal lobe neglect in man. *Neurology* **22**, 660–664.

Jerussi, T. P. and Glick, S. D. (1976) Drug-induced rotation in normal rats without lesions: behavioral and neurochemical indices of a normal asymmetry in nigro-striatal function. *Psychopharmacology* **47**, 249–260.

Losonczy, M. F., Song, I. S., Mohs, R. C., Small, M. S., Davidson, M., Johns, C. A., and Davis, K. L. (1986) Correlates of lateral ventricular size in chronic schizophrenia. *Am. J. Psychol.* **143**, 976–981.

Lyon, N. and Satz, P. (1991) Left turning (swivel) in medicated chronic schizophrenic patients. *Schizophr. Res.* **4**, 53–58.

Lyon, N., Satz, P., Fleming, K., Green, M. F., and Bracha, S. (1992) Left turning (swivel) in manic patients. *Schizophr. Res.* **7**, 71–76.

McGlone, J. (1980) Sex differences in human brain asymmetry: a critical survey. *Behav. Brain Sci.* **3**, 215–263.

Morihisa, J. M., Duffy, F. H., and Wyatt, R. D. (1983) Brain electrical activity mapping (BEAM) in schizophrenic patients. *Arch. Gen Psychol.* **40**, 719–728.

Pijnenburg, L. J., Honig, W. M. M., and Van Rossum, J. M. (1975) Inhibition of d-amphetamine-induced locomotor activity by injection of haloperidol into the nucleus accumbens of the rat. *Psycho-Pharmacologia* **41**, 87–95.

Poschel, P. and Ninteman, F. (1964) Excitatory (antidepressant) effects of monoamine oxidase inhibitors on the reward system of the brain. *Life Sci.* **3**, 903–910.

Posner, M., Early, T. S., Reiman, E., Pardo, P., and Dhawan, M. (1988) Asymmetries in hemispheric control of attention in schizophrenia. *Arch. Gen. Psychiatry* **45**, 814–821.

Pycock, C. J. (1980) Turning behavior in animals [commentary]. *Neuroscience* **5**, 515–528.

Pycock, C. J. and Marsden, C. D. (1978) The rotating rodent: a two component system? *Eur. J. Pharmacology* **47**, 167–175.

Quesney, L. F. (1986) Clinical and EEG features of complex partial seizures of temporal lobe origin. *Epilepsia* **27(Suppl. 2)**, 27–45.

Reynolds, G. P. (1983) Decreased concentrations and lateral asymmetry of amygdala dopamine in schizophrenia. *Nature* **305**, 524–529.

Reynolds, G. P., Czudek, C., Bzowej, N., and Seeman, P. (1987) Dopamine receptor asymmetry in schizophrenia. *Lancet.*

Richmond, G., Gulasekaram, B., Costa, J., Swati, R., Bracha, H. S., and Potkin, S. G. (1992) Circling behavior following treatment with placebo, haloperidol and clozapine. *Society of Biological Psychiatry Annual Meeting.*

Schneider, R. C., Calhoun, H. D., and Crosby, E. C. (1968) Vertigo and rotational movement in cortical and subcortical lesions. *J. Neurol. Sci.* **6**, 493–516.

Schopler, E., Reichler, R. J., and Renner, B. R. (1985) *The Childhood Autism Rating Scale—CARS.* Irvington Publication, New York.

Seeman, P. and Guan, H.-C. Dopamine D2 receptors are elevated mostly on the right-side in schizophrenic brain, compatible with leftward turning of schizophrenic patients. *Neuropsychopharmacology,* in press.

Serafentinides, E. A. (1972) Laterality and voltage in the EEG of psychiatric patients. *Dis. Nervous Syst.* **33,** 622,623.

Sorosky, A. D., Ornitz, E. M., Brown, M. B., and Ritvo, E. R. (1968) Systematic observations of autistic behavior. *Arch. Gen. Psychiatry* **8,** 439–449.

Springer, S. P. and Deutsch, G. (1981) *Left Brain, Right Brain.* W. H. Freeman, San Francisco.

Stewart, R. J., Morency, M. A., and Beninger, R. J. (1985) Differential effects of interfrontocortical microinjections of sulpiride into the medial prefrontal cortex on circling behavior of rats. *Prog. Neuropsychopharmacol. Biol. Psychiatry* **9,** 735–738.

Ungerstedt, U. and Arbuthnott, G. W. (1970) Quantitative recording of behavior in rats after 6-hydroxydopamine lesions of the nigrostriatal dopamine system. *Brain Res.* **24,** 485–493.

Volkmar, F. R. and Cohen, D. J. (1991) Comorbid association of autism and schizophrenia. *Am. J. Psychiatry* **148,** 1705–1707.

Weintraub, S. and Mesulam, M. M. (1989) Neglect: hemispheric specialization, behavioral components and anatomical correlates, in *Handbook of Neuropsychology,* vol. 2 (Boller, F. and Grafman, J., eds.), Elsevier, B.V., pp. 357–374.

Zimmerman, Z., Glick, S. D., and Jerussi, T. P. (1974) Neurochemical correlete of a spatial preference in rats. *Science* **185,** 623–625.

CHAPTER 13

Clinical Assessment of Tourette's Syndrome

R. D. Shytle, A. A. Silver, and Paul R. Sanberg

Introduction

Gilles de la Tourette syndrome (TS) is a neuropsychiatric and behavioral disorder with childhood onset that is characterized largely by the expression of both motoric and vocal tics that can range from relatively mild to very severe over the course of a patient's lifetime (Robertson, 1989). Although the exact pathogenesis is still not known, current hypotheses implicate disinhibition of cortical-striatal-thalamocortical minicircuits (Leckman et al., 1990). In addition, there is now strong evidence that TS is a hereditary motor disorder with the most accepted genetic model being an autosomal dominant mode of inheritance with incomplete penetrance and a variable expression (Price et al., 1988; van de Wetering and Heutink, 1993). Like many other hereditary psychiatric disorders, a major obstacle in linking a particular DNA sequence with the disorder in all affected family members is developing a clear and rigorous definition of the phenotype. This has been especially difficult with TS, since a growing body of evidence now indicates that, in addition to the expression of both motoric and vocal tics, many of these patients also exhibit other motoric abnormalities, including obsessive-compulsive disorder (OCD), attention-

From: *Motor Activity and Movement Disorders*
P. R. Sanberg, K. P. Ossenkopp, M. Kavaliers, Eds. Humana Press Inc., Totowa, NJ

deficit-hyperactivity disorder (ADHD), and associated visual-motor deficits (Singer and Rosenberg, 1989; Silver and Hagin, 1990; Bornstein and Yang, 1991; Cohen and Leckman, 1994). Since tics can vary in intensity and can be voluntarily suppressed during a clinical examination (Koller and Biary, 1989), they can sometimes go unnoticed when other associated symptoms like hyperactivity, impulsiveness, or obsessions and compulsions dominate the clinical picture (van de Wetering and Heutink, 1993). Therefore, it is important to have precise criteria for consistent diagnosis among researchers as well as assessment tools that can characterize the overall clinical picture of these patients for the ultimate delineation of the gene(s) and environmental factors involved with the expression of the TS phenotype.

Furthermore, the need for accurate assessment of each of these motoric dimensions in the same patient is also important for evaluating the results of biochemical investigations and therapeutic interventions. For example, Leckman et al. (1994a) recently found significant reductions in cerebrospinal fluid (CSF) oxytocin levels in patients with OCD, but not in patients with OCD associated with a personal or family history of TS. Moreover, the reductions in CSF oxytocin for these OCD-pure patients were significantly correlated with the severity of their OCD symptoms. Thus, the ability to classify patients into various groups and concurrently monitor their symptom severity led to an important biochemical discovery.

Likewise, it is often found that successful pharmacological treatments for one motoric abnormality can often affect others differently. For example, methylphenidate or dextroamphetamine treatment for symptoms of hyperactivity have been reported to exacerbate tic and OCD symptom frequency and severity in patients diagnosed with ADHD (Sverd et al., 1989; Borcherding et al., 1990; Gadow et al., 1992; Robertson and Eapen, 1992). Also, although the use of serotonin uptake inhibitors like fluoxetine can reduce OCD symptoms (Riddle et al., 1992), they have been reported to precipitate tic symptoms in patients comorbid with TS (Gatto et al., 1994). Furthermore, preliminary findings suggest that nicotine/haloperidol cotherapy can reduce tic symptoms (Sanberg et al., 1988, 1989; McConville, 1991, 1992; Silver et al., 1993, 1995; Dursan et al.,

1994; Reveley et al., 1994), and problems with inattention (Sanberg et al., 1988), but not OCD symptoms in TS patients (Silver et al., submitted). Therefore, it is important to con-sider a wide range of diagnostic and assessment tools when conducting psychopharmacological research in these patients.

In light of the growing need for more rigorous assessment of TS patients, the chapter presented here was designed to review the diagnostic and assessment tools available for characterizing motoric abnormalities common to patients diagnosed with TS.

Definitions and Characteristics of Tics

Tics are usually defined as sudden, rapid and brief, recurrent, nonrhythmic, stereotyped motor movements (motor tics) or sounds (vocal tics) that are experienced as irresistible, but can be suppressed for varying lengths of time (Koller and Biary, 1989; Lang, 1992; Tourette Syndrome Classification Study Group, 1993; American Psychiatric Association, 1994).

Both motor and vocal tics can be classified as either simple or complex in appearance (Table 1), although the boundary is not well defined. Simple tics are classified as "meaningless," abrupt, brief, and often repetitive movements occurring in a single and isolated fashion, whereas complex tics are classified as "purposeful," distinct, coordinated patterns of sequential movements (requiring multiple muscle groups) that may appear as if performing a common stereotypical motor act (Jankovic and Fahn, 1986; Tourette Syndrome Classification Study Group, 1993; American Psychiatric Association, 1994).

Because of the wide range of simple and complex motor tics present in patients with tic disorders, they are sometimes difficult to distinguish from symptoms found in other motor disorders, like myoclonic or choreic jerks found in Huntington's chorea, choreoform movements of Sydenham's chorea, dystonic movements of torsion dystonia, or the rapid eye blinking of idiopathic blepharospasm. However, distinctions can be made by ruling out the other diagnoses (Tourette Syndrome Classification Study Group, 1993), for example, by observing a relationship between motor and vocal tics or by determining the

Table 1
Classification of Simple and Complex Tics

Motor examples		Vocal examples
Simple tics: rapid, "meaningless" movements or sounds		
Eyeblinking	Shoulder shrugs	Coughing
Eye movements	Arm movements	Throat clearing
Nose movements	Hand movements	Sniffing
Mouth movements	Abdominal jerks	Grunting
Facial movements	Leg, foot, or toe movements	Whistling
Head jerks	Tensing body parts	Animal sounds
Complex tics: slower, "purposeful" movements or sounds		
Eye gestures	Writing tics	Syllables
Mouth gestures	Dystonic postures	Words
Facial gestures	Bending or gyrating	Phrases
Head gestures	Rotating	Coprolalia (obscene words)
Shoulder gestures	Touching, tapping, grooming	Echolalia (repeating sounds heard from others)
Arm/hand gestures	Sexual movements	

age of onset. Moreover, unlike many other movement disorders, tics are not always present (except in extremely severe cases), but rather periodically arise out of a background of normal activity over periods ranging from days, weeks, or months.

All forms of tics may be exacerbated by stressful situations (Surwillo et al., 1978; Lombroso et al., 1991; Leckman et al., 1993), but are often attenuated during absorbing activities that require sustained attention, such as reading or playing video games. Often tics are elicited in response to internal and sometimes external stimuli (Eapen et al., 1993). By age 10 or 11, many children with TS and as many as 80% of older TS patients recognize and report "premonitory" experiences, often somatic sensations like an itch, stretch, pressure, tingling, tightness, or other, which build in intensity in those parts of the body where the tic occurs (Bliss, 1980; Silver, 1988; Kurlan et al., 1989; Lang, 1991; Cohen and Leckman, 1992; Leckman et al., 1993). Emitting a tic response provides relief as the intensity of the sensation is rapidly, but only briefly, diminished. These tics have been given a variety of different names, such as "sensory," "premonitory," or "attentional" tics (Shapiro et al., 1987b; Kurlan et al., 1989; Kane, 1994). Less prevalent are those "reflex tics" elicited by external stimuli like hearing other people speaking, coughing, or sniffing. Whether the experiences come from internal or external stimuli, these patients are remarkably sensitive to changes occurring in their sensory world (Cohen and Leckman, 1992) and often report repeating behaviors (tics and/or compulsions) until things "feel" or "look" "just right" (Leckman et al., 1994).

Classification of Tic Disorders

TS has been widely reported among diverse racial and ethnic groups, occurring approx 1 case/1000 for males and 1/10,000 for females (Burd et al., 1986; Comings et al., 1990; Apter et al., 1993). The age of onset can occur between 2 and 18 yr, with the average age of onset occurring around 7 yr (Carter et al., 1994). The lifetime prevalence of less defined, but more common tic disorders, like transient tic disorder, has been reported to be between 12 and 25% of the general population (Robertson, 1989; Weiner and Lang, 1989).

The DSM-IV (American Psychiatric Association, 1994) outlines diagnostic criteria for four separate tic disorder categories. Tourette's disorder (Table 2), chronic motor or vocal tic disorder, transient tic disorder, or tic disorder not otherwise specified are distinguished from one another depending on the duration, age of onset, and the variety of tics present. For example, transient tic disorder includes motor and/or vocal tics occurring for at least 4 mo but for no longer than 12 consecutive months, whereas both Tourette's disorder and chronic motor or vocal tic disorder each have a duration of longer than 12 consecutive months, but are distinguished by whether or not both motor and vocal tics have been present during the illness and by the unvarying intensity of tics in chronic tic disorder. Tic disorder not otherwise specified would be appropriate for those cases not meeting criteria for the others.

Recently, the Tourette Syndrome Association (TSA), recognizing the need for developing more precise criteria to define

Table 2
DSM-IV Diagnostic Criteria for Tourette's Disorder[a]

A. Both multiple motor and one or more vocal tics have been present at some time during the illness, although not necessarily concurently
B. The tics occur many times a day (usually in bouts) nearly every day or intermittently throughout a period of more than 1 yr, and during this period, there was never a tic-free period of more than 3 consecutive months
C. The disturbance causes marked distress or significant impairment in social, occupational, or other important areas of functioning
D. The onset is before age 18 yr
E. The disturbance is not caused by the direct physiological effects of a substance (e.g., stimulants) or a general medical condition (e.g., Huntington's disease or postviral encephalitis)

[a]American Psychiatric Association, 1994.

the syndrome and related tic disorders, organized a task force (Tourette Syndrome Classification Study Group) to develop more specific definitions. The group decided to keep the basic definitions for the categories outlined in the DSM-IV (with the exception of increasing the inclusive age from 18 to 21 yr), but divided each of those into two new categories: "definite," in which tic symptoms have been observed by a reliable witness, and "historical," in which the putative symptoms have not been reliably observed. In addition, they developed definitions for other categories that are not covered by the existing DSM-IV criteria. These additional tic disorder categories, chronic single tic disorder, definite tic disorder–diagnosis deferred, and probable Tourette syndrome, are defined in detail in the study group's recently published manuscript (Tourette Syndrome Classification Study Group, 1993).

Obsessive-Compulsive Symptoms in TS

Even the first clinical descriptions suggested that obsessive-compulsive symptoms (OCS) were associated with TS (Gilles de la Tourette, 1885; Meige and Feindel, 1907), an observation that led to more detailed and exact questions about the occurrence of OCD in TS family-genetic and epidemiological studies (Pauls et al., 1984), as well as in neurobiological reports (Insel, 1992).

Although OCD occurs in roughly 1 case/100 in the general population (Swedo et al., 1989, 1992; Cross National Collaborative Group, 1994), it has now been estimated that roughly 30–60% of TS patients report obsessive thoughts and compulsive rituals that emerge several years after the onset of motor tics, during the preadolescent or early adolescent years (Fernando, 1967; Morphew and Sim, 1969; Nee et al., 1980; Yaryura-Tobias et al., 1981; Jagger et al., 1982; Montgomery et al., 1982; Frankel et al., 1986; Comings and Comings, 1987a,b; Grad et al., 1987; Cath et al., 1992; Leonard et al., 1992; Robertson et al., 1993). More recently, there is evidence supporting the hypothesis that these disorders are various genotypical expressions of a TS spectrum disorder that may be under the control of a single major gene (Frankel et al., 1986; Pauls and Leckman, 1986; Pauls et al., 1986a,b, 1991; Pitman et al., 1987; Walkup et al., 1988; Eapen et al., 1993).

In addition, it appears that sex may determine whether the expression of the disorder is primarily TS or OCD (Santangelo et al., 1994). For example, in family-genetic studies, although the male to female ratio for TS has been reported to range from 3:1 to 10:1, with females being more likely to have OCD without tics, when the definition of being affected was expanded to include any of the spectrum disorders (TS, chronic tic disorder [CT], or OCD), the male-to-female ratio among family members dropped to around 1.6:1 (Pauls and Leckman, 1986; Pauls et al., 1991), suggesting that the sex difference is more a difference in prevalence of the expressed spectrum disorder as opposed to the prevalence of the disorder *per se* (Santangelo et al., 1994).

Interestingly, the lack of evidence for a relationship between tic severity and the severity of OC argues for the relative independence of these phenomena (Leckman et al., 1993, 1994a,b), as well as the necessity to monitor each of these phenomena concurrently in future longitudinal, family-genetic, and pharmacological treatment studies.

Attentional Problems and Hyperkinesis in TS

Although the occurrence of ADHD in variable degrees for the general population ranges between 1 and 20%, it has been reported in several studies that somewhere between 20 and 90%

of TS patients may also have attention deficit and hyperactivity symptoms consistent with the diagnosis of ADHD (Cohen et al., 1979, 1984; Jagger et al., 1982). Since, in many children, these symptoms occur prior to the onset of tics, proposals have been made suggesting that these ADHD problems are prodromal or the earliest manifestation of the biological vulnerability to the eventual tic disorder (Comings and Comings, 1984; Comings, 1987a,b). Since ADHD symptoms include excessive hyperactivity, impulsiveness, temper tantrums, and poor self-esteem, these symptoms are often more problematic than the tics themselves.

However, there have been controversies in the literature about the exact relationship between these two disorders. For example, some have argued that the high rates of ADHD among TS patients that have been reported may reflect a referral bias (Berkson, 1946); that is, those TS children with severe behavioral problems are more likely to be seen by a physician than those who are not (Cohen and Leckman, 1994). In fact, referral bias has been demonstrated for this population of TS/ADHD patients (Caine et al., 1988).

In their initial family-genetic studies, Pauls et al. (1986c), unlike the relationship between TS and OCD, did not find a tight association between TS and ADHD. More recently, however, analyses of a larger cohort of TS patients revealed, in contrast to their earlier study, that TS and ADHD were associated at a greater than chance expectancy within the families (Pauls et al., 1993). Moreover, Knell and Comings (1993), using methods to avoid ascertainment bias, examined 338 first-degree relatives of 131 TS probands and found that in relatives with TS, 61% had attention-deficit disorder (ADD) and 36% had ADHD, whereas in those with chronic tics, 41 and 26% had ADD and ADHD, respectively. Therefore, the overall evidence does suggest a genetic relationship between TS and ADHD symptoms.

However, one issue that has not been adequately resolved involves the relationship between ADHD symptoms and tic severity. For example, many of our patients have reported that their attentional problems stem from being distracted by their premonitory urges and subsequent tic responses (Silver, Shytle, and Sanberg, unpublished observations). This interpretation is consistent with the neuropsychological findings of Randolph et al. (1993), who found that in monozygotic twins with TS, objective measures of attention and distractibility were the strongest in distinguishing between twins with differing tic severity. Therefore, the relationship between TS and ADHD in some cases could be more the result of secondary tic severity disrupting attentional processes than a common gene for both disorders.

With respect to the motor abnormalities, little research has been published regarding the relationship between TS and generalized hyperkinesis. To our knowledge, very little has been reported regarding the percentage of TS patients who are simply hyperactive. Most, if not all, of the above-mentioned family-genetic studies used the DSM diagnostic criteria for ADD and ADHD, of which the latter does have questions regarding hyperactivity. However, it is possible that some patients may have hyperactivity without attentional problems and, thus, do not meet criteria for either ADD or ADHD. Therefore, more research needs to be done to investigate the symptoms of generalized hyperkinesis and any relationships between tics and/or OCS in patients with TS.

Limitations in Assessment of TS Patients

As pointed out by Leckman et al. (1988), there are a number of difficulties with the objective quantification of tic severity and other comorbid symptoms found in patients with TS. For example, the expression of tics is often complex, and it is sometimes difficult to classify tics as either simple or complex and whether or not a behavioral manifestation should be counted as one or many tics. Another considerable problem is the variability of symptom expression over temporal parameters of weeks and months (Fahn, 1982). Furthermore, unlike many other hyperkinetic motor disorders, the ability to surpress tic expression voluntarily (Koller and Biary, 1989) can pose problems to efforts of quantitative assessment in the clinical setting of a clinician's office. The variability of symptoms among patients is also important to consider. For example, asking a parent to rate on a five-point scale their child's tic severity over the past week may mean different things to different parents regarding their childrens symptoms. All of these issues, including the clinician's experience with tic disorders, provide potential limitations to the standardization and quantification of assessment tools used in this area of research.

Current Methods of Assessment

Self-Report Instruments for TS Patients

As pointed out by Halvorsen (1990), self-report instruments are valuable to both clinicians and researchers for a number of reasons: They are easy to administer, they have the potential to summarize and characterize a large amount of historical information in a short period of time from large diverse samples, scoring can be automated, and they can often be easier to validate than observational methods.

A number of historical self-report devices have been constructed for research in family-genetic studies, epidemiological studies, and longitudinal studies of TS (Leckman et al., 1988). The Movement Disorder Questionnaire (Shapiro et al., 1989), which can be completed by either parent or patient, is a comprehensive questionnaire emphasizing family movement disorders; genetic, medical, and psychiatric histories; and patient perinatal, maturational, medical, psychiatric, psychological, educational, tic, and treatment histories. The Tourette's Syndrome Questionnaire (Jagger et al., 1982), originally developed for an epidemiological survey, is also well suited for systematically obtaining relevant historical information as well as the impact of TS on the individual's life (Leckman et al., 1988). Another well-known parent/patient self-report device was created for the landmark needs assessment survey conducted for the TSA of Ohio (Stefl, 1983; Bornstein et al., 1990).

The Tourette Syndrome Symptom List (TSSL), a clinical instrument developed at Yale by Cohen et al. (1984) to assist parents and/or patients in making daily or weekly ratings of tic behaviors, has the advantage of classifying tics on the basis of complexity and whether they are motor or vocal in nature. Although this self-report instrument is particularly useful in pharmacological studies, as pointed out by Leckman et al. (1988), it is important for the guardian or parent who is rating a child not to keep him or her under constant "tic surveillance," since this may create an aversive situation and perhaps may influence the expression of tic behavior by the child.

Although there are a number of self-report global impairment rating scales used in psychiatry (e.g., Endicott et al., 1976; Asberg et al., 1977; Shaffer et al., 1983), to our knowledge, none are currently available that concurrently characterize the multidimensional motor abnormalities (e.g., tics, OCD, and ADHD) found in TS patients. Although lengthy, the Child Behavior Checklist (Achenbach and Edelbrock, 1983), a standardized symptom-rating questionnaire including a wide range of childhood emotional/behavioral problems, some of which are relevant to TS patients, has been used successfully in longitudinal studies of TS (Carter et al., 1994). In addition, ancillary scales, such as the Conner's Parent and Teacher Questionnaires (Goyette et al., 1978), are particularly useful for documenting impulsivity/hyperactivity symptoms in open-trial drug studies (e.g., Sanberg et al., 1989).

According to Leckman et al. (1988), although there are a number of obsessive-compulsive (OC) self-report devices available, many are either unpublished or not validated, and few are used routinely in studies with TS patients. However, there are at least three OC symptom inventories designed specifically for OCD patients that may show promise in the assessment of OCS in patients with TS. The Maudsley Obsessional Compulsive Inventory (Hodgson and Rachman, 1977), a 30-item true/false questionnaire inquiring about a variety of OC symptoms was developed for OCD studies. However, several methodological flaws were found with this inventory, most importantly, an item selection bias (Goodman et al., 1989; Baer, 1994). Another potentially useful instrument is the Leyton Obsessional Inventory (Berg et al., 1986) developed at NIMH for diagnostic assessment of childhood OCD. This instrument contains 44 OC items, which are presented to the child on separate cards, with the child responding by dropping the card into either one of two slots marked yes or no. Perhaps the most widely used inventory for OCD is the Yale-Brown Obsessive-Compulsive Scale (Y-BOCS) Symptom Checklist (Goodman et al., 1989), which has been used in a number of TS studies to characterize comorbid OCS (e.g., Leckman et al., 1994a,b). This inventory corrects for the item selection bias of the Maudsley Obsessional Compulsive Inventory and has recently been found by factor analysis to distinguish between those patients with pure OCD and those with OCD/tic disorder comorbidity (Baer, 1994).

The recently introduced Motor tic, Obsessions and compulsions, Vocal tic Evaluation Survey (MOVES) may become particularly useful in future longitudinal and pharmacological treatment studies (Gaffney et al., 1994). This self-report instrument includes 20 questions that can be filled out by the parent and/or patient in approx 5 min and generates scores on five subscales: motor tics, vocal tics, obsessions, compulsions, and associated symptoms. These subscale scores can be combined to form a tic or an OC subscale, both of which were found to correlate well with other widely used clinician rating scales for TS and OCD symptoms. Unfortunately, it does not, however, contain questions regarding attentional deficits or symptoms of hyperactivity. Although a scale exists for self-report rating of "just right" phenomena (Leckman et al., 1993, 1994b), it has never been published and is apparently still under development.

Video-Tape Monitoring

An objective method for rating tic symptoms has been with the use of video recording of patients. Video taping has proven a valuable adjunct to clinical rating for drug trials (Shapiro et al., 1989; Leckman et al, 1991; McConville et al., 1991; Reveley et al., 1994; Silver et al., 1995) and challenge studies (Lombroso et al. 1991; Chappell et al., 1992). Although such a technique can be novel to a patient like a child and cause tic suppression, not to mention being labor-intensive on the part of the clinician, high levels of reliability can be achieved, and meaningful changes in response to treatment can be successfully documented (McConville et al., 1991; Silver et al., 1995). In fact, in a recent extensive report, Chappell and colleagues (1994) found that valid estimates of both motor and vocal tic frequency could be obtained from video-tape counts of at least 5 min in duration. Furthermore, their video tic count data were highly reliable and correlated well with established clinical rating scales.

Clinician Rating Scales

Although developed for patient and/or parent self-report during clinical trials with pimizide, the Tourette Syndrome Severity Scale (TSSS) (Shapiro et al., 1983; Shapiro and Shapiro,

1984) can also be used by the clinician for documenting both simple and complex motor and vocal tics. However, without the aid of video recording, it may be difficult to score during a brief interview. Furthermore, the five ordinal scales vary in both their range and weightings (Leckman et al., 1989).

Another scale, the Tourette Syndrome Global Scale (TSGS) developed at Yale by Harcherick et al. (1984), was designed to make clinical judgments about the frequency and severity of tics. This multidimensional scale has two domains, tics and social functioning, both of which contribute equally to the global score (0 [no impairment] to 100 [severe impairment]). Although this scale was found to be both valid and reliable (Harcherick et al., 1984; Leckman et al., 1988), the distinction between simple and complex motor tics proved difficult in many cases. In addition, this scale has some undesirable psychometric properties, including the exaggeration of the total score owing to small changes in tic severity (Harcherick et al., 1984; Leckman et al., 1988).

The more recent Yale Global Tic Severity Scale (YGTSS) is a radically revised version of the TSGS based on the clinical experience of Leckman and colleagues (1989). This scale includes a tic symptom inventory that is filled out based on the patient's personal recall of tics occurring over the previous week. Using this inventory as a guide, the clinician then rates the severity of both motor and vocal tics on five separate dimensions: number, frequency, intensity, complexity, and interference. In addition, there is also a separate rating of global impairment that characterizes the impact of the disorder on the patient's social functioning, self esteem, and so forth, over the previous week. Data from 105 patients with tic disorders, aged 5–51 yr, were used to demonstrate the construct, convergent, and discriminate validity of this instrument (Leckman et al., 1989). Although others have been validated and deemed reliable (e.g., Goetz et al., 1987) and newer scales are in development (Walkup, et al., 1992; Tourette Syndrome Classification Study Group, 1993), the YGTSS is perhaps the most widely used clinical assessment rating scale currently available for assessing tic symptoms.

As with self-report devices, there are few clinical rating scales that also take into account tics, ADHD, and OCD concurrently. However, the Clinical Global Impression Scales for

TS, OCD, and ADD (Leckman et al., 1988) does show promise. To our knowledge, it has not been validated or deemed reliable, and we are not aware of any studies using it extensively. Most studies are now using multiple scales to assess these other motoric dimensions. For example, there are a number of clinical rating scales for OCD severity, including the Assessor's OC Scale (Philpott, 1975) and the Beaumont four-point scale for OCD (Beaumont, 1975). However, the well validated and reliable Y-BOCS (Goodman et al., 1989) is used in more recent studies to characterize the OCS often found in TS patients (e.g., Leckman et al., 1993, 1994a). However, few studies incorporate clinical rating scales to assess hyperactivity and attentional problems in TS patients (Leckman et al., 1988).

Summary and Future Directions

TS and related tic disorders are complex, and often afflicted patients exhibit a variety of the motoric abnormalities. Currently, there are a number of diagnostic and assessment tools available for characterizing motor and vocal tics and the obsessions and compulsions, but less so for the other problems of hyperactivity and visual motor deficits. In light of the increased use of video taping as an objective measure of tic severity, it may be useful to construct an adjunctive rating scale for hyperactivity, since we have found that many patients are restless in their seats during video recording. In addition, it would be interesting to determine whether or not TS patients exhibit biased rotational turning behavior as has been observed in other psychiatric populations (*see* Chapter 9), especially since many of these TS patients have been reported to have asymmetrical disturbances in the basal ganglia and related regions of the brain (Braun et al., 1993; Peterson et al., 1993; Singer et al., 1993). Interestingly, many children with TS have problems with handwriting (Silver and Hagin, 1990) and patients with Huntington's chorea (Phillips et al., 1994) exhibit similar problems. Objective methods of assessment of handwriting have been developed for these latter patients (Phillips et al., 1994) and could be extended to studies involving TS patients.

In summary, in addition to the motor and vocal tics exhibited by patients with tic disorders, it is now clear that other motoric abnormalities often coexist. Therefore, the need for more extensive assessment tools is warranted for research with this neuropsychiatric population, in particular, to investigate the relationship between these abnormalities for future family-genetic, longitudinal, and pharmacological studies.

Acknowledgments

This review was supported in part by grants from the Smokeless Tobacco Research Council and The National Institute of Neurological Disease and Stroke (RO1 NS. 32067-01A1). Also, thanks to Jeff Jordan for assistance with the development of this manuscript.

References

Achenbach, T. M. and Edelbrock, C. (1983) *Manual for the Child Behavior Checklist and Revised Child Behavior Profile*. University of Vermont Department of Psychology, Burlington, VT.

American Psychiatric Association (1994) *Diagnostic and Statistical Manual of Mental Disorders*, 4th ed., revised. American Psychiatric Association, Washington, DC.

Apter, A., Pauls, D. L., Bleich, A., Zohar, A. H., Kron, S., Ratzoni, G., Dycian, A., Kotler, M., Weizman, A., Gadot, N., and Cohen, D. J. (1993) An epidemiological study of Gilles de la Tourette's syndrome in Israel. *Arch. Gen. Psychiatry* **50**, 734–738.

Asberg, M., Montgomery, S. A., and Perris, C. (1977) A comprehensive psychopathological rating scale. *Acta Psychiatr. Scand.* **1271**, 5–27.

Baer, L. (1994) Factor analysis of symptom subtypes of obsessive compulsive disorder and their relation to personality and tic disorders. *J. Clin. Psychiatry* **55(Suppl.)**, 18–23.

Beaumont, G. (1975) Obsessional and phobic disorders: a new rating scale for obsessional and phobic states. *Scott. Med. J.* **20**, 25–32.

Berg, C., Rapoport, J. L., and Flament, M. (1984) The Leyton obsessional inventory—child version. *J. Am. Acad. Adolesc. Child Psychiatry* **25**, 84–91.

Berkson, J. (1946) Limitations of the application of fourfold table analysis to hospital data. *Biometrics* **2**, 47–51.

Bliss, J. (1980) Sensory experiences of Gilles de la Tourette syndrome. *Arch. Gen. Psychiatry* **37**, 1343–1347.

Borcherding, B. G., Keysor, C. S., Rapoport, J. L., Elia, J., and Amass, J. (1990) Motor/vocal tics and compulsive behaviors on stimulant drugs: is there a common vulnerability? *Psychiatry Res.* **33**, 83–94.

Bornstein, R. A. and Yang, V. (1991) Neuropsychological performance in medicated and unmedicated patients with Tourette's disorder. *Am. J. Psychiatry* **149**, 468–471.

Bornstein, R. A., Stefl, M., and Hammond, L. (1990) A survey of Tourette syndrome patients and their families: the 1987 Ohio Tourette Survey. *J. Neuropsychiatry Clin. Neurosci.* **2**, 275–281.

Braun, A. R., Stoetter, B., Randolph, C., Hsiao, J. K., Vladar, K., Gernert, J., Carson, R. E., Herscovitch, P., and Chase, T. N. (1993) The functional neuroanatomy of Tourette's syndrome: an FDG-PET study. I. Regional changes in cerebral glucose metabolism differentiating patients and controls. *Neuropsychopharmacology* **9**, 277–291.

Burd, L., Kerbeshian, L., Wikenheiser, M., and Fisher, W. (1986) Prevalence of Gilles de la Tourette's syndrome in North Dakota adults. *Am. J. Psychiatry* **143**, 787.

Caine, E. D., McBride, M. C., Chiverton, M. S., Bamford, K. A., Rediess, S., and Shiao, J. (1988) Tourette's syndrome in Monroe county school children. *Neurology* **38**, 472–475.

Carter, A. S., Pauls, D. L., Leckman, J. F., and Cohen, D. J. (1994) A prospective longitudinal study of Gilles de la Tourette's syndrome. *J. Am. Acad. Child Adolesc. Psychiatry* **33**, 377–385.

Cath, D. C., Hoogduin, C. A. L., van de Wetering, B. J. M., van Woerkom, C. A. M., Roos, R. A. C., and Rooymans, H. G. M. (1992) Tourette syndrome and obsessive-complusive disorder: an analysis of associated phenomena. *Adv. Neurology* **58**, 33–41.

Chappell, P. B., Leckman, J. F., Riddle, M. A., et al. (1992) Neuroendocrine and behavioral effects of naloxone in Tourette's syndrome. *Adv. Neurology* **58**, 253–262.

Chappell, P. B., McSwiggan-Hardin, M., Walker, D., Cohen, D. J., and Leckman, J. F. (1994) Videotape tic counts in the assessment of Tourette's syndrome, stability, reliability, and validity. *J. Am. Acad. Child Adolesc. Psychiatry* **33**, 386–393.

Cohen, D. J. and Leckman, J. F. (1992) Sensory phenomena associated with Gille de la Tourette syndrome. *J. Clin. Psychiatry.* **53**, 319–323.

Cohen, D. J. and Leckman, J. F. (1994) Developmental psychopathology and neurobiology of Tourette's syndrome. *J. Am. Acad. Child Adolesc. Psychiatry* **33**, 2–15.

Cohen, D. J., Shaywitz, B. A., Young, J. G., Carbonari, C. M., Nathanor, J. A., Lieberman, D., and Bowers, M. B. (1979) Central biogenic amine metabolism in children with the syndrome of chronic mulitple tics of Gilles de la Tourette: norepinephrine, serotonin and dopamine. *J. Am. Acad. Adolesc. Child Psychiatry* **18**, 320–341.

Cohen, D. J., Leckman, F. J., and Shaywitz, B. A. (1984) The Tourette's syndrome and other tics, in *Diagnosis and Treatment in Pediatric Psychiatry* (Shaffer, D., Ehrhanlt, A. A., Greenhill, I., eds.), MacMillian Free Press, New York, pp. 3–28.

Comings, D. E. (1987a) A controlled study of Tourette syndrome. I–VI. *Am. J. Hum. Genet.* **41**, 701–838.

Comings, D. E. (1987b) A controlled study of Tourette syndrome, VII: summary: a common genetic disorder causing disinhibition of the limbic system. *Am. J. Hum. Genet.* **41**, 839–866.

Comings, D. E. and Comings, B. G. (1984) Tourette syndrome and attention deficit disorder with hyperactivity—are they genetically related? *J. Am. Acad. Adolesc. Child Psychiatry* **23**, 138–144.

Comings, D. E. and Comings, B. G. (1987) Hereditary agoraphobia and obsessive-compulsive behavior in relatives of patients with Gilles de la Tourette's syndrome. *Br. J. Psychiatry* **151**, 195–199.

Comings, D. E., Himes, J. A., and Comings, B. G. (1990) An epidemiological study of Tourette's syndrome in a single school district. *J. Clin. Psychiatry* **51**, 463–469.

Cross National Collaborative Group (1994) The cross national epidemiology of obsessive compulsive disorder. *J. Clin. Psychiatry* **55(Suppl. 3)**, 5–10.

Dursan, S. M., Reveley, M. A., Bord, R., and Stirton, F. (1994) Long lasting improvement of Tourette's syndrome with transdermal nicotine. *Lancet* **344**, 157.

Eapen, N., Pauls, D. L., and Robertson, M. M. (1993) Evidence for autosomal dominant transmission in Gilles de la Tourette syndrome: United Kingdom Cohort Study. *Br. J. Psychiatry* **162**, 593–596.

Eapen, V., Moriarty, J., and Robertson, M. M. (1994) Stimulus induced behaviours in Tourette's syndrome. *J. Neurology Neurosurg. Psychiatry* **57**, 853–855.

Endicott, J., Spitzer, R. L., Fleiss, J. L., et al. (1976) The Clinical Global Scale: a procedure for measuring overall severity of psychiatric disturbance. *Arch. Gen. Psychiatry* **33**, 766–771.

Fahn, S. (1982) The clinical spectrum of tics, in *Gilles de la Tourette Syndrome* (Friedhoff, A. J. and Chase, T. N., eds.), Raven, New York, pp. 341–344.

Fernando, S. J. M. (1967) Gilles de la Tourette's syndrome. *Br. J. Psychiatry* **113**, 607–617.

Frankel, M., Cummings, J. L., Robertson, M. M., Trimble, M. R., Hill, M. A., and Benson, F. (1986) Obsessions and compulsions in Gilles de la Tourettes's syndrome. *Neurology* **36**, 378–382.

Gadow, K. D., Nolan, E. E., and Sverd, J. (1992) Methylphenidate in hyperactive boys with comormid tic disorder, II: short-term behavioral effects in school settings. *J. Am. Acad. Child Adolesc. Psychiatry* **31**, 462–471.

Gaffney, G. R., Sieg, K., and Hellings, J. (1994) The MOVES: a self-rating scale for Tourette's syndrome. *J. Child Adolesc. Psychopharm.* **4**, 269–280.

Gatto, E., Pikielny, R., and Micheli, F. (1994) Fluoxetine in Tourette's syndrome. *Am. J. Psychiatry* **151**, 946–947.

Gilles de la Tourette, G. (1885) Etude sur une affection nerveuse caracterisee par de l'incoordination motrice accompagnee d'echolalie et de coprolalie. *Archives de Neurologie* **9**, 19–42, 158–200.

Goetz, C. G., Tanner, C. M., Wilson, R. S., Shannon, K. M. (1987) A rating scale for Gilles de la Tourette's syndrome: description, reliability, and validity of data. *Neurology* **37**, 1542–1544.

Goodman, W. K., Price, L. H., Rasmussen, S. A., Mazure, C., Fleischman, R. L., Hill, C. L., Heninger, G. R., and Charney, D. S. (1989) Yale-Brown Obsessive Compulsive Scale. *Arch. Gen. Psychiatry* **46**, 1006–1011.

Goyette, C. H., Connors, C. K., and Ulrich, R. F. (1978) Normative data on revised Connor's parent and teacher rating scales. *J. Abnormal Child Psych.* **6**, 221–236.

Grad, L. R., Pelcovitz, D., Olson, M., Matthews, M., and Grad, G. J. (1987) Obsessive complusive symptomatology in children with Tourette's syndrome. *J. Am. Acad. Child Adolesc. Psychiatry* **26**, 69–73.

Halvorsen, J. G. (1990) Designing self-report instruments for family assessment. *Fam. Med.* **22**, 478–484.

Harcherick, D. F., Leckman, J. F., Detlor, J., and Cohen, D. J. (1984) A new instrument for clinical studies of Tourette's syndrome. *J. Am. Acad. Adolesc. Child Psychiatry* **23**, 153–160.

Hodgson, R. J. and Rachman, S. (1977) Obsessional-compulsive complaints. *Behav. Res. Therap.* **15**, 389–395.

Insel, T. R. (1992) Neurobiology of obsessive compulsive disorder: a review. *Int. Clin. Psychopharmacol.* **7(Suppl. 1)**, 31–33.

Jagger, J., Prusoff, B. A., Cohen, D. J., Kidd, K. K., Carbonari, C. M., and John, K. (1982) The epidemiology of Tourette's syndrome: a pilot study. *Schizophr. Bull.* **8**, 267–278.

Jankovic, J. and Fahn, S. (1986) The phenomenology of tics. *Movement Disord.* **1**, 17–26.

Kane, M. J. (1994) Premonitory urges as "attentional tics" in Tourette's syndrome. *J. Am. Acad. Child Adolesc. Psychiatry* **33**, 805–808.

Knell, E. R. and Comings, D. E. (1993) Tourette's syndrome and attention-deficit hyperactivity disorder: evidence for a genetic relationship. *J. Clin. Psychiatry* **54**, 331–337.

Koller, W. C. and Biary, N. M. (1989) Volitional control of involuntary movements. *Movement Disord.* **4**, 153–156.

Kurlan, R., Lichter, D., and Hewitt, D. (1989) Sensory tics in Tourette's syndrome. *Neurology* **39**, 731–734.

Lang, A. (1991) Patient perception of tics and other movement disorders. *Neurology* **41**, 223–228.

Lang, A. E. (1992) Clinical phenomenology of tic disorders. *Adv. Neurology* **58**, 25–32.

Leckman, J. F., Towbin, K. E., Ort, S. I., and Cohen, D. J. (1988) Clinical assessment of tic disorder severity, in *Tourette's Syndrome and Tic Disorders: Clinical Understanding and Treatment* (Cohen, D. J., Brunn, R. D., and Leckman, J. F., eds.), Wiley, New York, pp. 56–78.

Leckman, J. F., Riddle, M. A., Hardin, M. T., Ort, S. I., Swartz, K. L., Stevenson, J., and Cohen, D. J. (1989) The Yale Global Tic Severity Scale. (YGTSS): initial testing of a clinician-rated scale of tic severity. *J. Am. Acad. Child Adolesc. Psychiatry* **28**, 566–573.

Leckman, J. F., Knorr, A., Rasmusson, A., and Cohen, D. J. (1990) Another frontier for basal ganglia research: Tourette's syndrome and related disorders. *Trends Neurosci.* **3**, 207–211.

Leckman, J. F., Hardin, M. T., Riddle, M. R., Stevenson, J., Ort, S., and Cohen, D. J. (1991) Clonidine treatment of Gilles de la Tourette's syndrome. *Arch. Gen. Psychiatry* **48**, 324–328.

Leckman, J. F., Walker, D. E., and Cohen, D. J. (1993) Premonitory urges in Tourette syndrome. *Am. J. Psychiatry* **150**, 98–102.

Leckman, J. F., Goodman, W. K., North, W. G., Chappell, P. B., Price, L. H., Pauls, D. L., Anderson, G. M., Riddle, M. A., McSwiggan-Hardin, M., McDougle, C. J., Barr, L. C., and Cohen, D. J. (1994a) Elevated cerebrospinal fluid levels of oxytocin in obsessive-compulsive disorder: comparison with Tourette's syndrome and healthy controls. *Arch. Gen. Psychiatry* **51**, 782–792.

Leckman, J. F., Walker, D. E., Goodman, W. K., Pauls, D. L., and Cohen, D. J. (1994b) "Just right" perceptions associated with compulsive behavior in Tourette's syndrome. *Am. J. Psychiatry* **151**, 675–680.

Leonard, H. L., Swedo, S. E., Rapoport, J. L., Rickler, K. C., Topol, D., Lee, S., and Rettew, D. (1992) Tourette syndrome and obsessive-compulsive disorder. *Adv. Neurology* **58**, 83–93.

Lombroso, P. J., Mack, G., Scahill, L., King, L., King, R., and Leckman, J. F. (1991) Exacerbation of Tourette's syndrome associated with thermal stress: a family study. *Neurology* **41**, 1984–1987.

McConville, B. J., Fogelson, M. H., Norman, A. B., et al. (1991) Nicotine potentiation of haloperidol in reducing tic frequency in Tourette's disorder. *Am. J. Psychiatry* **148**, 793–794.

McConville, B. J., Sanberg, P. R., Fogelson, M. H., et al. (1992) The effects of nicotine plus haloperidol compared to nicotine only and placebo nicotine only in reducing tic severity and frequency in Tourette's disorder. *Biol. Psychiatry* **31**, 832–840.

Meige, H. and Feindel, F. (1907) *Tics and Their Treatment.* Sidney Appleton, London.

Montgomery, M. A., Clayton, P. J., and Friedhoff, A. J. (1982) Psychiatric illness in Tourette syndrome patients and first-degree relatives, in *Gilles de la Tourette Syndrome, Advances in Neurology,* vol. 35 (Friedhoff, A. J., Chase, T. N., eds.), Raven, New York, pp. 335–339.

Morphew, J. A. and Sim, M. (1969) Gilles de la Tourette's syndrome. A clinical and psychopathological study. *Br. J. Med. Psychol.* **42**, 293–301.

Nee, L. E., Canine, E. D., Polinsky, R. J., Eldridge, R., and Ebert, M. H. (1980) Gilles de la Tourette syndrome: a clinical and family study of 50 cases. *Ann. Neurology* **7**, 41–49.

Pauls, D. L. and Leckman, J. F. (1986) The inheritance of Gilles de la Tourette syndrome and associate behaviors. Evidence for autosomal dominant transmission. *N. Engl. J. Med.* **315**, 993–997.

Pauls, D. L., Kruger, S. D., Leckman, J. F., Cohen, D. J., and Kidd, K. K. (1984) The risk of Tourette's syndrome and chronic multiple tics among relatives of Tourette's syndrome patients obtained by direct interview. *J. Am. Acad. Child Psychiatry* **23**, 134–137.

Pauls, D. L., Leckman, J. F., Towbin, K. E., Zahner, G. E. P., and Cohen, D. J. (1986a) A possible genetic relationship between Tourette's syndrome and obsessive-complusive disorder. *Psychopharm. Bull.* **22**, 730–733.

Pauls, D. L., Towbin, K. E., Leckman, J. F., Zahner, G. E. P., and Cohen, D. J. (1986b) Gilles de la Tourette syndrome and obsessive compulsive disorder: evidence supporting an etiological relationship. *Arch. Gen. Psychiatry* **43**, 1180–1182.

Pauls, D. L., Hurst, C. R., Kidd, K. K., Kruger, S. D., Leckman, J. F., and Cohen, D. J. (1986c) Tourette sydrome and attention deficit disorder: evidence against a genetic relationship. *Arch. Gen. Psychiatry* **43**, 1177–1179.

Pauls, D. L., Raymond, C. I., Leckman, J. F., and Stevenson, J. M. (1991) A family study of Tourette's syndrome. *Am. J. Hum. Genet.* **48**, 154–163.

Pauls, D. L., Leckman, J. F., and Cohen, D. J. (1993) Familial relationship between Gilles de la Tourette's syndrome, attention deficit disorder, learning disabilities, speech disorders, and stuttering. *J. Am. Acad. Child Adolesc. Psychiatry* **32**, 1044–1050.

Peterson, B., Riddle, M. A., Cohen, D. J., Katz, L. D., Smith, J. C., Hardin, M. T., and Leckman, J. F. (1993) Reduced basal ganglia volumes in Tourette's syndrome using three-dimensional reconstruction techniques from magnetic resonance images. *Neurology* **43**, 941–949.

Phillips, J. G., Bradshaw, J. L., Chiu, E., and Bradshaw, J. A. (1994) Characteristics of handwriting of patients with Huntington's disease. *Movement Disord.* **9**, 521–530.

Philpott, R. (1975) Recent advances in the behavioral measurement of obsessional illness: difficulties common to these and other measures. *Scott. Med. J.* **20**, 33–40.

Pitman, R. K., Green, R. C., Jenlike, M. A., and Mesulam, M. M. (1987) Clinical comparison of Tourette's disorder and obsessive compulsive disorder. *Am. J. Psychiatry* **144**, 1166–1171.

Price, R. A., Pauls, D. L., Kruger, S. D., and Caine, E. D. (1988) Family data support a dominant major gene for Tourette syndrome. *Psychiatry Res.* **24**, 251–261.

Randolph, C., Hyde., T., M., Gold, J. M., Goldberg, T. E., and Weinberger, D. R. (1993) Tourette's syndrome severity to neuropsychological function. *Arch. Neurology* **50**, 725–728.

Reveley, M. A., Bird, R., Stirton, R. F., and Dursun, S. M. (1994) Microstructural analysis of the symptoms of Tourette's syndrome and the effects of a trial use of transdermal nicotine patch. *J. Psychopharmacol. Suppl.* **A30**, 117.

Riddle, M. A., Scohill, L., King, R. A., Anderson, G. M., Ort, S. I., Smith, J. C., Leckman, J. F., and Cohen, D. J. (1992) Double-blind, crossover trial of fluoxetine and placebo in children and adolescents with obsessive-compulsive disorder. *J. Am. Acad. Child Adolesc. Psychiatry* **31**, 1062–1069.

Robertson, M. M. (1989) The Gilles de la Tourette syndrome: the current status. *Br. J. Psychiatry* **154**, 147–169.

Robertson, M. M. and Eapen, V. (1992) Pharmacologic controversy of CNS stimulants in Gilles de la Tourette's syndrome. *Clin. Neuropharmacol.* **15**, 408–425.

Robertson, M. M., Cannon, S., Baker, J., and Flynn, D. (1993) The psychopathology of the Gilles de la Tourette syndrome: a controlled study. *Br. J. Psychiatry* **162**, 114–117.

Sanberg, P. R., Fogelson, H. M., Manderscheid, P. Z., Parker, K. W., Norman, A. B., and McConville, B. J. (1988) Nicotine gum and haloperidol in Tourette's syndrome. *Lancet* **1**, 592.

Sanberg, P. R., McConville, B. J., Fogelson, H. M., Manderscheid, P. Z., Parker, K. W., Blythe, M. M., Klykylo, W., and Norman, A. B. (1989) Nicotine potentiates the effects of haloperidol in animals and patients with Tourette's syndrome. *Biomed. Pharmacother.* **43**, 19–23.

Santangelo, S. L., Pauls, D. L., Goldstein, J. M., Faraone, S. V., Tsuang, M. T., and Leckman, J. F. (1994) Tourette's syndrome: What are the influences of gender and comorbid obsessive-compulsive disorder? *J. Am. Acad. Child Adolesc. Psychiatry* **33**, 795–804.

Shaffer, D., Gould, M. S., Brasic, J., Ambrosini, P., Fisher, P., Bird, H., and Aluwahalia, S. (1983) A children's global assessment scale. *Arch. Gen. Psychiatry* **40**, 1228–1231.

Shapiro, A. K. and Shapiro, E. (1984) Controlled study of pimozide vs. placebo in Tourette's syndrome. *J. Am. Acad. Adolesc. Child Psychiatry* **23**, 161–173.

Shapiro, A. K., Shapiro, E., and Eisenkraft, G. J. (1983) Treatment of Gilles de la Tourette syndrome with pimozide. *Am. J. Psychiatry* **140,** 1183–1186.

Shapiro, A. K., Shapiro, E. S., Young, J. G., and Feinberg, T. E. (1988) Sensory tics, in *Gilles de la Tourette Syndrome,* 2nd ed. (Shapiro, A. K., Shapiro, E. S., Young, J. G., and Feinberg T. E., eds.), Raven, New York, pp. 356–360.

Shapiro, E. S., Shapiro, A. K., Fulop, G., Hubbard, M., Mandeli, J., Nordie, J., and Phillips, R. A. (1989) Controlled study of haloperidol, pimozide, and placebo for the treatment of Gilles de la Tourette's syndrome. *Arch. Gen. Psychiatry* **46,** 722–730.

Silver, A. A. (1988) Intrapsychic processes and adjustment in Tourette's syndrome, in *Tourette's Syndrome and Tic Disorders, Clinical Understanding and Treatment* (Cohen, D. J., Brunn, R. D., and Leckman, J. F., eds.), Wiley, New York, pp. 198–206.

Silver, A. A. and Hagin, R. A. (1990) Gilles de la Tourette's Syndrome, in *Disorders of Learning Childhood* (Noshpitz, J. D., ed.), Wiley, New York, pp. 469–508.

Silver, A. A. and Sanberg, P. R. (1993) Transdermal nicotine patch and potentiation of haloperidol in Tourette's Syndrome. *Lancet* **342,** 182.

Silver, A. A., Shytle, R. D., Philipp, M. K., and, Sanberg, P. R. (1995) Transdermal nicotine in Tourette's syndrome, in *The Effects of Nicotine on Biological Systems. Advances in Pharmacological Sciences* (Clarke, P. B. S., Quik, M., and Thurau, K., eds.), Birkhauser, Boston, MA.

Silver, A. A., Shytle, R. D., Philipp, M. K., and and Sanberg, P. R. (submitted) Transdermal nicotine in Tourette's syndrome: a video tape report of long-term effect. *Movement Disord.*

Singer, H. S. and Rosenberg, L. A. (1989) The development of behavioral and emotional problems in Tourette syndrome. *Pediatr. Neurology* **5,** 41–44.

Singer, H. S., Reiss, A. L., Brown, J. E., Aylward, E. H., Shih, B., Chee, E., Harris, E. L., Reader, M. J., Chase, G. A., Bryan, R. N., and Cenckla, M. B. (1993) Volumetric MRI changes in basal ganglia of children with Tourette's syndrome. *Neurology* **43,** 950–956.

Stefl, M. E. (1983) The Ohio Tourette Study. School of Planning, University of Cincinnati.

Surwillo, W. W., Shafti, M., and Barrett, C. L. (1978) Gilles de la Tourette: a 20 month study of the effects of stressful life events and haloperidol on symptom frequency. *J. Nerv. Ment. Dis.* **166,** 812–816.

Sverd, J., Gadow, K., and Paolicelli, L. M. (1989) Methylphenidate treatment of attention-deficit hyperactivity disorder in boys with Tourette's syndrome. *J. Am. Acad. Child Adolesc. Psychiatry* **4,** 574–579.

Swedo, S. E., Rapoport, J. L., Leonard, H., Lenane, M., and Cheslow, D. (1989) Obsessive-compulsive disorder in children and adolescents. Clinical phenomenology of 70 consecutive cases. *Arch. Gen. Psychiatry* **46,** 335–341.

Swedo, S. E., Leonard, H. L., and Rapoport, J. L. (1992) Childhood-onset obsessive compulsive disorder. *Psychiatr. Clin. N. Am.* **15,** 767–775.

Tourette Syndrome Classification Study Group. (1993) Definitions and classification of tic disorders. *Arch. Neurology* **50,** 1013–1016.

Walkup, J. T., Leckman, J. F., Price, A. R., Hardin, M. T., Ort, S. I., and Cohen, D. J. (1988) The relationship between obsessive compulsive disorder and Tourette's syndrome: a twin study. *Psychopharmacol. Bull.* **24,** 375–379.

Walkup, J. T., Rosenberg, L. A., Brown, J., and Singer, H. S. (1992) The validity of instruments measuring tic severity in tourette's syndrome. *J. Am. Acad. Child Adolesc. Psychiatry* **31(3),** 472–477.

Weiner, W. D. and Lang, A. E. (1989) Gilles de la Tourette syndrome, in *Movement Disorders: A Comprehensive Survey* (Friedhoff, A. J. and Chase, T. N., eds.), Futura, Mount Kisco, NY, pp. 531–568.

van de Wetering, B. J. M. and Heutink, P. (1993) The genetics of the Gilles de la Tourette syndrome: a review. *J. Lab. Clin. Med.* **21(5),** 638–645.

Yaryura-Tobias, J. A., Neziroglu, F., Howard, S., and Fuller, B. (1981) Clinical aspects of Gilles de la Tourette syndrome. *Orthomol. Psychiatry* **10,** 263–268.

Index